TOPICS IN

Discrete Mathematics

TOPICS IN
DISCRETE MATHEMATICS

[For B.A. / B.Sc. III Course of Kurukshetra University, Kurukshetra and MDU, Rohtak and other Computer Science Courses of Various Universities]

By

Dr. SATINDER BAL GUPTA
B.Tech. (CSE), MCA, UGC-NET, Ph.D. (CS)
Professor, Deptt. of Computer Science & Applications
Vaish College of Engineering
Rohtak (Haryana)

PARMANAND GUPTA
B.Sc. (Hons.), M.Sc. (Delhi)
M.Phil (KU), Pre. Ph.D. (IIT Delhi)
Associate Professor & Head, Deptt. of Mathematics
Indira Gandhi National College, Ladwa
and
Member, Faculty of Sciences
KURUKSHETRA UNIVERSITY
Haryana

UNIVERSITY SCIENCE PRESS
(An Imprint of Laxmi Publications Pvt. Ltd.)
An ISO 9001:2008 Company
BENGALURU • CHENNAI • COCHIN • GUWAHATI • HYDERABAD
JALANDHAR • KOLKATA • LUCKNOW • MUMBAI • RANCHI • NEW DELHI
BOSTON (USA) • ACCRA (GHANA) • NAIROBI (KENYA)

TOPICS IN DISCRETE MATHEMATICS

Printed and bound in India
Typeset at : Goswami Printers, Delhi
New Edition
ISBN : 81-7008-952-2

Published in India by
UNIVERSITY SCIENCE PRESS
(An Imprint of Laxmi Publications Pvt. Ltd.)
An ISO 9001:2008 Company
113, GOLDEN HOUSE, DARYAGANJ, NEW DELHI-110002, INDIA
Telephone : 91-11-4353 2500, 4353 2501
Fax : 91-11-2325 2572, 4353 2528
www.laxmipublications.com info@laxmipublications.com

Branches

Bengaluru	080-26 75 69 30
Chennai	044-24 34 47 26
Cochin	0484-237 70 04, 405 13 03
Guwahati	0361-254 36 69, 251 38 81
Hyderabad	040-27 55 53 83, 27 55 53 93
Jalandhar	0181-222 12 72
Kolkata	033-22 27 43 84
Lucknow	0522-220 99 16
Mumbai	022-24 91 54 15, 24 92 78 69
Ranchi	0651-220 44 64

C—11781/016/04

Printed at : Ajit Printing Press, Delhi.

PREFACE

It gives us a great pleasure in bringing out the first edition of the book 'Topics in Discrete Mathematics'. The need for Discrete Mathematics arises due to selection of certain applications from various areas of the field. The subject 'Discrete Mathematics' is newly introduced in various degree colleges. Students face lot of problems in this subject as no book covers the whole syllabi and also due to lack of solved problems in various books. This book is based on the experience gained in teaching a course on the subject. Our intention is to cover all the topics in the syllabi alongwith a number of solved and unsolved problems and solutions of questions from the previous question papers of different universities. We are sure that this book will help a student to grip each topic without any difficulty.

Salient Features :

- To facilitate the readers with the subject matter in a simple, clear and systematic manner.
- Neat and well-labelled diagrams.
- Every mathematical step with a large clarity.
- A lot of solved problems at the end of the chapters and various exercises, wherever necessary.

We would seek the indulgence of our readers in informing us about the errors and shortcomings noticed by them in the book.

—AUTHORS

SYLLABUS

B.A./B.Sc. Part III

B.M. 304: DISCRETE MATHEMATICS

Max. Marks
For B.Sc. : 45
For B.A. : 30
Time : 3 Hours

Note. The paper-setter is requested to set ten questions in all selecting questions (sectionwise) as indicated in the syllabus. The candidate is required to attempt five questions, selecting atleast one question from each section.

SECTION I
(Three Questions)

SETS

Countability of sets : Cardinality of sets (Priniciples of inclusion and exclusion), Partially ordered (p.o.) sets, dual p.o. sets, Totally ordered sets, Chains, antichains, Isomorphem of ordered sets, Hasse diagram.

(Pigeonhole Principle and its applications).

Discrete probability : Mutually exclusive and exhaustive system of events, axiom of calculus of probability, Law of addition of probability, conditional probability, Multiplication law of probability, independed events, Baye's theorem, probability distribution, binomial distribution.

SECTION II
(Two Questions)

Recurrence relations : Explicit formula for a sequence, Fibonacci sequence, solution of recurrence relations using back tracking, Linear recurrence relations with constant coefficients, homogeneous solutions, Particular solution and total solution of Linear recurrence relations, discrete numeric functions, generating functions, Convolution of numeric functions, solution of recurrence relations by the method of generating functions.

Propositions : Definition and Types, Basic logical operations, Truth tables, De Morgan's and commutative laws.

SECTION III

(Three Questions)

Lattices : Definition, examples and simple properties of lattices, lattices as algebraic system, equivalence of two definitions, definition and examples of distributive and complemented lattices.

Boolean Algebra : Definition, examples and basic properties of Boolean algebras, Switching algebras, Boolean expressions, Fundamental product, sum-of-products forms, Logic gates and circuits—OR gate, AND gate, NOT gate, simple logic circuits, Boolean Functions and their truth tables.

SECTION IV

(Two Questions)

Graphs : Definition and examples, multigraphs, Euler's first theorem of graphs theory, complete graphs, regular graph, bipartite graph, subgraph, Paths and circuits, Eulerian paths, and circuits, weighted graphs, Shortest paths, Travelling Salesman Problem, definition and examples of planar graphs.

Trees : Definition and simple examples of trees.

Finite State Machines : Simple examples of finite states machines, Equivalence of finite state machines.

CONTENTS

SECTION I

Chapter 1

Sets

DEFINITION

A set is defined as a collection of distinct objects of same type or class of objects.

The objects of a set are called elements or members of the set. Objects can be numbers, alphabets, names etc.

e.g., $A = \{1, 2, 3, 4, 5\}$

A is a set of numbers containing elements 1, 2, 3, 4 and 5.

* A set is usually denoted by capital letters A, B, C, D, P, Q, R, S, T etc.

* Elements of the set are defined by p, q, r, t or p_1, q_1, r_1, t_1 etc.

SET FORMATION

The set can be formed by two ways :

(*i*) Tabular form of a set

(*ii*) Builder form of a set.

(*i*) **Tabular Form of a Set.** If a set is defined by actually listing its members, *e.g.,* if set P contains elements a, b, c, d then it is expressed as $P = \{a, b, c, d\}$.

This is called tabular form of a set.

(*ii*) **Builder Form of a Set.** If a set is defined by the properties which its elements must satisfy *e.g.,*

$P = \{x : x \in N, x \text{ is a multiple of } 3\}$.

$R = \{x : x > 1 \text{ and } x < 10 \text{ and } x \text{ is an odd integer}\}$.

$T = \{x : x \text{ is even less than } 9\}$.

STANDARD NOTATIONS

$x \in A$	x belongs to A or x is an element of set A.
$x \notin A$	x does not belong to set A.
ϕ	Empty set.
U	Universal set.
N	The set of all natural numbers.
I	The set of all integers.
I_0	The set of all non-zero integers.

I_+	The set of all + ve integers.
C, C_0	The sets of all complex, non-zero complex numbers respectively.
Q, Q_0, Q_+	The sets of rational, non-zero rational, + ve rational numbers respectively.
R, R_0, R_+	The sets of real, non-zero real, +ve real number respectively.

FINITE SET

If a set consists of specific number of different elements then that set is called finite set.

e.g.,

$$P = \{x : x \in N, 3 < x < 9\}.$$
$$Q = \{2, 4, 6, 8\}.$$
R = {months of year}.

INFINITE SET

If a set consists of infinite number of different elements or if the counting of different elements of the set does not come to an end, the set is called infinite set. *e.g.*,

I = {The set of all integers}

$E = \{x : x \in N, x$ is a multiple of $2\}$.

COUNTABLY INFINITE OR DENUMERABLE SET

A set is said to be countably infinite if there is a one-to-one correspondence between the elements in the set and the elements in N.

e.g., — The set of all non-negative odd integers {1, 3, 5, 7, ...} is a countably infinite set since there is one-to-one correspondence between all +ve odd integers and all natural numbers.

— The set of all positive multiples of 5 {0, 5, 10, 15, ...} is a countably infinite set.

UNCOUNTABLE INFINITE OR NON-DENUMERABLE SET

A set is said to be uncountable infinite if it is infinite and there is no one-to-one correspondence between the elements in the set and the elements in N.

e.g., — The set of real numbers is an uncountably infinite

— The set of real numbers between 0 and 1 is uncountably infinite.

COUNTABLE SET

A set is called countable set if it is finite or denumerable set.

e.g., 1. Every finite set is a countable set.

2. Every Denumerable set is countable set.

UNCOUNTABLE SET

A set is called uncountable set if it is infinite and is not cardinally equivalent to N or is a non-denumerable set.

e.g. 1. The set of complex numbers is an uncountable set.

2. The set of real numbers between two integers is an uncountable set.

Theorem I. *Every subset of a countable set is countable.*

Proof. Let A is a countable set. Then A is either finite or denumerable set.

First consider that A is finite and we know that every subset of a finite set is finite and hence it is countable.

Secondly, let A is countably infinite set then A can be written as

$$A = (a_1, a_2, a_3, \ldots\ldots) \qquad \ldots(i)$$

Let B is a subset of A then if B = ϕ then B is countably but if B $\neq \phi$ then B can be expressed as B = $(a_{n1}, a_{n2}, a_{n3}, \ldots\ldots)$, where $a_{ni} \in A$

From (*i*), we can say that B is countable.

Hence proved.

Theorem II. *A countable union of countable sets in countable. If S_i is a countable set or for each integer $i \geq 0$, then set $S = \bigcup_{i=0}^{\infty} S_i$ is countable.*

Proof. Firstly, if all the S_i's are empty, then S is empty and therefore countable.

Secondly, when S_i's are not empty, we describe a way of listing the elements of the Union. Consider the two-dimensional array in which the elements of S_i are listed in the *i*th row. Since S_i may be finite or infinite, these rows are likely to be of unequal length as shown in figure.

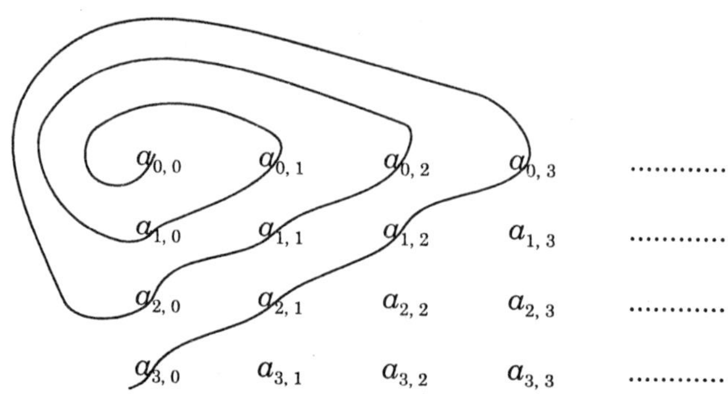

Now consider the path shown in figure, this path will hit each element of S at least once. It will hit an element more than once if that element belongs to S_i for more than one *i*.

Suppose we define the function *f* by

$$f(n) = \begin{cases} a_{0,0} & \text{if } n = 0 \\ \text{The first element of } S - [f(0), \ldots, f(n-1)] \\ \text{on the path if } n > 0 \end{cases}$$

Then *f* is one-to-one and on to (bijection) either from {0, 1, 2, 3, ..., *m*} to S or from N to S, depending on whether the set S is finite or infinite. Therefore, S is a countable set.

Example 1. *Show that the set of real numbers in [0, 1] is uncountable set.*

Sol. Let us assume that the set of real numbers in [0, 1] is countable. Then all the real number in $0 \le n \le 1$ can be listed in some order, say, $x_1, x_2, x_3, \ldots\ldots$ The representation of these real numbers is as follows.

$$x_1 = 0.\, a_{11}\, a_{12}\, a_{13}\, a_{14} \ldots\ldots$$
$$x_2 = 0.\, a_{21}\, a_{22}\, a_{23}\, a_{24} \ldots\ldots$$
$$x_3 = 0.\, a_{31}\, a_{32}\, a_{33}\, a_{34} \ldots\ldots$$
$$\ldots \quad \ldots \quad \ldots \quad \ldots$$
$$\ldots \quad \ldots \quad \ldots \quad \ldots$$

where each a_{ij} is one of the numbers of the set {0, 1, 2, 3, 4, 5, 6, 7, 8, 9}. Now form a new real number y by the decimal.

$$y = 0.\, b_1\, b_2\, b_3\, b_4 \ldots\ldots$$

where $\quad b_1 = 1 \quad$ if $a_{ii} = 9$

$\quad = 9 - a_{ii}$ if $a_{ii} = 0, 1, 2, 3, 4, 5, 6, 7, 8,$ for all i.

For those numbers which can be expressed in two different decimal expansions *e.g.*, 1/2 = 0.50000000 ... = 0.49999999..., we choose the expansion which ends with nines. This ensures a unique representation of all numbers.

The number $0.\, b_1\, b_2\, b_3\, b_4\, b_5 \ldots$ is a real number between 0 and 1 that does not have trailing 0's. Thus, the real number y is not equal to any of $x_1, x_2, x_3, \ldots$, since it differs from the first number in the first digit, the second number in the second digit, the ith number in the ith digit, and so on. Since there is a real number y between 0 and 1 that is not in the list which contradicts the assumption that this set is countably infinite so the set of real numbers between 0 and 1 is uncountable.

Theorem III. *Every infinite set has a denumerable subset.*

Proof. Let A be an infinite set, since A is non empty, there exists an element $a_1 \in A$ and because A is an infinite set (not finite), there exists an element $a_2 \ne a_1$ and $a_2 \in A$. Hence for any natural number n, there are n distinct element $a_1, a_2, a_3, \ldots\ldots, a_n$ of A. Since A is infinite, we can find a_{n+1} in A distinct from these n elements. Thus, we obtain a set $\{a_1, a_2, a_3, \ldots\ldots, a_n, a_{n+1}, a_{n+2}, \ldots\ldots\}$ which is a subset of A and is equivalent to N.

EQUALITY OF SETS

Two sets A and B are said to be equal and written as A = B if both have the same elements. Therefore, every element which belongs to A is also an element of the set B and every element which belongs to the set B is also an element of the set A.

$$A = B \iff (x \in A \iff x \in B).$$

If there is some element in a set A that does not belong to set B or vice-versa then $A \ne B$ *i.e.*, A is not equal to B.

* A set does not change if one or more of its elements are repeated.

* A set does not change if we change the order in which its elements are tabulated. *e.g.*,

(*a*) $\quad A = \{x : x < 10$ and x is even$\}$

$B = \{2, 4, 6, 8\}$

$C = \{x : x > 1$ and $x < 10$ and x is even$\}$

All the three sets are equal.

(*b*)
$$P = \{r, s, t\}$$
$$Q = \{r, r, s, t\}$$
$$R = \{s, r, t\}$$
$$S = \{r, s, t, t\}$$

All the four sets P, Q, R, S are equal.

(*c*)
$$A = \{a, b, c\}$$
$$B = \{b, a, c\}$$
$$C = \{b, a\}$$

The sets A and B are equal but C is not equal to either A or B.

DISJOINT SETS

Two sets A and B are said to be disjoint if no element of A is in B and no element of B is in A. *e.g.,*
$$R = \{a, b, c\},\ S = \{k, l, m\}$$
R and S are disjoint sets.

FAMILY OF SETS

If a set A contains elements which are itself sets then it is called family of sets or a set of sets. *e.g.,*
$$A = \{\{1, 2\}, \{3, 4\}, \phi\}$$
A is set of sets.

SUBSET OF A SET

If every element of a set A is also an element of a set B then A is called subset of B and is written as $A \subseteq B$. B is called superset of A.
$$A \subseteq B = \{x : x \in A \quad \Rightarrow \quad x \in B\}$$

(*a*) **Proper Subset.** If A is subset of B and $A \neq B$ then A is said to be proper subset of B. If A is a proper subset of B then B is not subset of A *i.e.,* there is atleast one element in B which is not in A. *e.g.,*

(*i*) Let
$$A = \{2, 3, 4\}$$
$$B = \{2, 3, 4, 5\}$$
A is a proper subset of B.

(*ii*) The null set ϕ is a proper subset of every set.

(*b*) **Improper Subset.** If A is subset of B and A = B, then A is said to be an improper subset of B. *e.g.,*

(*i*) $A = \{2, 3, 4\}$, $B = \{2, 3, 4\}$

A is improper subset of B.

(*ii*) Every set is improper subset of itself.

NULL SET OR EMPTY SET

The set that contains no element is called the null set or the empty set and is denoted by ϕ. *e.g.,*

(a) $P = \{x : x^2 = 4, x \text{ is odd}\}$.

(b) $Q = \{x : x^2 = 9, x \text{ is even}\}$.

(c) $R = \{x : x^2 = 9, 2x = 4\}$.

The set $\phi = \{0\}$ is not a null set because 0 is the element of the set.

The set $A = \{\phi\}$ is not a null set because set ϕ is the element of the set.

POWER SET

The power set of any given set A is the set of all subsets of A and is denoted by P(A). If A has n elements then P(A) has 2^n elements.

If $A = \{r, s\}$ then its subsets are ϕ, $\{r\}$, $\{s\}$, $\{r, s\}$

$P(A) = \{\phi, \{r\}, \{s\}, \{r, s\}\}$ which has $2^2 = 4$ elements.

UNIVERSAL SET

If all the sets under investigation are subsets of a fixed set U, then the set U is called universal set. *e.g.*, in plane geometry, the universal set consists of all the points in the plane.

COMPARABILITY

Two sets A and B are comparable if one of them is subset of other, *i.e.*, $A \subseteq B$ or $B \subseteq A$. If $A \subseteq B$ and $B \subseteq A$, then $A = B$. Set ϕ is comparable to every set. Every set is comparable to the universal set U.

Two sets A and B are said to be incomparable if $A \not\subseteq B$ and also $B \not\subseteq A$ *i.e.*, there is at least one element in A not in B and vice-versa.

Theorem I. *Prove that, the null set ϕ is a subset of every set.*

Proof. Suppose that ϕ is not a subset of A *i.e.*, $\phi \not\subset A$.

Then there exists an element $x \in \phi$ such that $x \notin A$.

But ϕ is the null set and hence for every x, $x \notin \phi$ because null set does not contain any element.

From above, $x \in \phi$ and $x \notin \phi$ which is contradiction. Hence, $\phi \not\subset A$ is wrong supposition.

Therefore, ϕ is a subset of A. Hence proved.

Theorem II. *Prove that, every set is a subset of itself i.e., $A \subseteq A$.*

Proof. Let $x \in A \Rightarrow x \in A \quad \therefore \quad A \subseteq A$

As every element belonging to set A is also an element of set A. Therefore, A is subset of A or itself. Hence proved.

Theorem III. *Prove that, if $A \subseteq B$ and $B \subseteq C$, then $A \subseteq C$.*

Proof. Let x be any element of the set A.

Since, $A \subseteq B \quad \therefore \quad x \in A \Rightarrow x \in B$

$B \subseteq C \quad \therefore \quad x \in B \Rightarrow x \in C \qquad \therefore \quad x \in A \Rightarrow x \in C$

$\therefore \quad A \subseteq C$. Hence proved.

Theorem IV. *Prove that, if $A \subseteq B$, $B \subseteq C$, $C \subseteq A$, then $A = C$.*

Proof. If $A \subseteq B$ and $B \subseteq C$, then $A \subseteq C$

$C \subseteq A$ (Given)

$\therefore$ $A \subseteq C$ and $C \subseteq A$

$\therefore$ $A = C$. Hence proved.

Theorem V. *Prove that, if $A \subset \phi$, then $A = \phi$.*

Proof. $\phi \subset A$ (As ϕ is subset of every set).

But $A \subset \phi$ (Given)

$\therefore$ $A = \phi$. Hence proved.

Example 2. *How many subsets can be formed from a set of n elements ? How many of these will be proper and how many improper ?*

Sol. There are ${}^nC_1$ subsets each consisting of one of the elements of the given set.

There will be ${}^nC_2$ subsets each consisting of any two of the n elements of the given set.

Also, there will be ${}^nC_3$ subsets each consisting of any three of the n elements of given set.

At last, there will be ${}^nC_n$ subset consisting of all the n elements of the given set.

Also, there will be one set ϕ.

Hence, the total number of subsets will be

$$({}^nC_1 + {}^nC_2 + {}^nC_3 + \ldots\ldots + {}^nC_n) + 1 \qquad \text{(For the null set)}$$

or $${}^nC_0 + {}^nC_1 + {}^nC_2 + \ldots\ldots + {}^nC_n = 2^n \qquad \because \; {}^nC_0 = 1.$$

Therefore, 2^n subsets formed from n elements of a set. Out of 2^n subsets $2^n - 1$ subsets will be proper and 1 (one) subset improper *i.e.,* the set itself.

OPERATIONS ON SETS

The basic set operations are :

1. Union of Sets. Union of the sets A and B is defined to be the set of all those elements which belong to A or B or both and is denoted by $A \cup B$.

$$A \cup B = \{x : x \in A \quad \text{or} \quad x \in B\}$$

e.g., Let $A = \{1, 2, 3\}$, $B = \{3, 4, 5, 6\}$

$$A \cup B = \{1, 2, 3, 4, 5, 6\}.$$

2. Intersection of Sets. Intersection of two sets A and B is the set of all those elements which belong to both A and B and is denoted by $A \cap B$.

$$A \cap B = \{x : x \in A \text{ and } x \in B\}$$

e.g., Let $A = \{a, b, c, d\}$, $B = \{a, b, l, m\}$

$$A \cap B = \{a, b\}.$$

3. Difference of Sets. The difference of two sets A and B is a set of all those elements which belong to A but do not belong to B and is denoted by $A - B$.

$$A - B = \{x : x \in A \text{ and } x \notin B\}$$

e.g., Let $A = \{a, b, c, d\}$, $B = \{d, l, m, n\}$

$$A - B = \{a, b, c\}.$$

4. Complement of a Set w.r.t. a Universal Set. The complement of a set A is a set of all those elements of the universal set which do not belong to A and is denoted by A^c.

$$A^c = U - A = \{x : x \in U \text{ and } x \notin A\} = \{x : x \notin A\}$$

e.g., Let U is the set of all natural numbers.

$$A = \{1, 2, 3\}$$

$$A^c = \{\text{all natural numbers except } 1, 2, 3\}.$$

5. Symmetric Difference of Sets. The symmetric difference of two sets A and B is the set containing all the elements that are in A or in B but not in both and is denoted by A ⊕ B *i.e.*,

$$A \oplus B = (A \cup B) - (A \cap B)$$

e.g., (*i*) Let $A = \{a, b, c, d\}$

$B = \{a, b, l, m\}$

$A \oplus B = \{c, d, l, m\}$

(*ii*) Let $A = \{1, 2, 3, 4\}$

$A \oplus \phi = \{1, 2, 3, 4\}$

(*iii*) Let $A = \{l, m, n, m\}$

$A \oplus A = \phi.$

EXERCISES

1. Describe each of the following sets in tabular form
   (*i*) $A = \{x : x \text{ is a prime number}, x < 25\}$ (*ii*) $B = \{x : x \text{ is an integer}, -5 < x < 5\}$
   (*iii*) $C = \{x : x \text{ is a perfect square}, x < 100\}$ (*iv*) $D = \{x : x \text{ is divisible by } 5, x < 100\}$
2. Describe each of the following sets in builder form
   (*i*) $A = \{1, 3, 5, 7, 9\}$ (*ii*) $B = \{3, 6, 9, 12, 15, 18, ...\}$
   (*iii*) $C = \{1, 1/2, 1/3, 1/4, 1/5, ...\}$ (*iv*) $D = \{1, 2, 3, 5, 7, 11, 13, 17, ...\}$
3. Determine whether {5} is an element of the set, for each of the following sets ?
   (*i*) {{5}, {{5}}} (*ii*) {{{5}}}
   (*iii*) {5, {5}} (*iv*) {{5}, {5, {5}}}
4. Show that the set of integers that are multiples of 2 are countable.
5. Show that A – B is an uncountable set if A is uncountable and B is a countable set.
6. Determine the power set of $A = \{(a, b), c\}$.
7. Find two sets A and B such that $A \in B$ and $A \subseteq B$.
8. Determine all the subsets of the set $A = \{1, 2, 3\}\}$.
9. Let $A = \{a, b, c, d, e, f\}$, $B = \{c, f, g, h, i, j\}$ and $C = \{b, c, e, h, j\}$, then find
   (*i*) $A \cup B$ (*ii*) $A \cap B$ (*iii*) $A - B$ (*iv*) $B - A$ (*v*) $(A \cap B) \cup C$
   (*vi*) $A \cap (B \cap C)$ (*vii*) $A \oplus B$ (*viii*) $(A \oplus B) \oplus B$ (*ix*) $B - B$ (*x*) $(A \cup B) \oplus A$

ALGEBRA OF SETS

A set is obtained from a set formula by replacing the variables by definite sets. When the set variables appearing in two set formulas are replaced by any sets and both the set formulas are equal as sets, then we call that both the set formula are equal. The equality of set formulas are called set identities. Some of the identities describe certain properties of the operations involved. These properties describe an algebra called algebra of sets.

Table I

(*i*) **Idempotent Laws** (*a*) $A \cup A = A$ (*b*) $A \cap A = A$ (*ii*) **Associative Laws** (*a*) $(A \cup B) \cup C = A \cup (B \cup C)$ (*b*) $(A \cap B) \cap C = A \cap (B \cap C)$ (*iii*) **Commutative Laws** (*a*) $A \cup B = B \cup A$ (*b*) $A \cap B = B \cap A$ (*iv*) **Distributive Laws** (*a*) $A \cup (B \cap C) = (A \cup B) \cap (A \cup C)$ (*b*) $A \cap (B \cup C) = (A \cap B) \cup (A \cap C)$ (*v*) **De Morgan's Laws** (*a*) $(A \cup B)^c = A^c \cap B^c$ (*b*) $(A \cap B)^c = A^c \cup B^c$	(*vi*) **Identity Laws** (*a*) $A \cup \phi = A$ (*b*) $A \cap U = A$ (*c*) $A \cup U = U$ (*d*) $A \cap \phi = \phi$ (*vii*) **Complement Laws** (*a*) $A \cup A^c = U$ (*b*) $A \cap A^c = \phi$ (*c*) $U^c = \phi$ (*d*) $\phi^c = U$ (*viii*) **Involution Law** (*a*) $(A^c)^c = A$

The Table I shows the laws of algebra of sets.

Example 3. *Prove (a)* $A \cup A = A$ *(b)* $A \cap A = A$.

Sol. (*a*) To prove $A \cup A = A$

Since, $B \subset A \cup B$, therefore $A \subset A \cup A$

Let $x \in A \cup A \Rightarrow x \in A$ or $x \in A \Rightarrow x \in A$

$\therefore \quad A \cup A \subset A$

As $A \cup A \subset A$ and $A \subset A \cup A \Rightarrow A = A \cup A$. Hence proved.

(*b*) To prove $A \cap A = A$

Since, $A \cap B \subset B$, therefore $A \cap A \subset A$

Let $x \in A \Rightarrow x \in A$ and $x \in A$

$\Rightarrow \quad x \in A \cap A \qquad \therefore \quad A \subset A \cap A$

As $A \cap A \subset A$ and $A \subset A \cap A \Rightarrow A = A \cap A$. Hence proved.

Example 4. *Prove (a)* $(A \cup B) \cup C = A \cup (B \cup C)$

(b) $(A \cap B) \cap C = A \cap (B \cap C)$.

Sol. (*a*) To prove $(A \cup B) \cup C = A \cup (B \cup C)$

Let some $x \in (A \cup B) \cup C$

$\Rightarrow \quad (x \in A$ or $x \in B)$ or $x \in C$

$\Rightarrow \quad x \in A$ or $x \in B$ or $x \in C$

$\Rightarrow \quad x \in A$ or $(x \in B$ or $x \in C)$

$\Rightarrow \quad x \in A$ or $x \in B \cup C$

$\Rightarrow \quad x \in A \cup (B \cup C)$.

Similarly, if some $x \in A \cup (B \cup C)$, then $x \in (A \cup B) \cup C$.

Thus, any $x \in A \cup (B \cup C) \Leftrightarrow x \in (A \cup B) \cup C$. Hence proved.

(*b*) To prove $(A \cap B) \cap C = A \cap (B \cap C)$

Let some $x \in A \cap (B \cap C) \Rightarrow x \in A$ and $x \in B \cap C$

$\Rightarrow$ $x \in A$ and ($x \in B$ and $x \in C$) $\Rightarrow$ $x \in A$ and $x \in B$ and $x \in C$

$\Rightarrow$ ($x \in A$ and $x \in B$) and $x \in C$ $\Rightarrow$ $x \in A \cap B$ and $x \in C$

$\Rightarrow$ $x \in (A \cap B) \cap C$.

Similarly, if some $x \in A \cap (B \cap C)$, then $x \in (A \cap B) \cap C$

Thus, any $x \in (A \cap B) \cap C \Leftrightarrow x \in A \cap (B \cap C)$. Hence proved.

Example 5. *Prove (a) $A \cup B = B \cup A$ (b) $A \cap B = B \cap A$.*

Sol. (*a*) To prove $A \cup B = B \cup A$

$A \cup B = \{x : x \in A \text{ or } x \in B\}$

$= \{x : x \in B \text{ or } x \in A\}$ ($\because$ Order is not preserved in case of sets)

$= B \cup A$. Hence proved.

(*b*) To prove $A \cap B = B \cap A$

$A \cap B = \{x : x \in A \text{ and } x \in B\}$

$= \{x : x \in B \text{ and } x \in A\}$

($\because$ Order is not preserved in case of sets)

$= B \cap A$. Hence proved.

Example 6. *Prove (a) Intersection of sets is distributive w.r.t. union of sets i.e.,*

$$A \cap (B \cup C) = (A \cap B) \cup (A \cap C)$$

(*b*) *Union of sets is distributive w.r.t. intersection of sets i.e.,*

$$A \cup (B \cap C) = (A \cup B) \cap (A \cup C).$$

Sol. (*a*) To prove $A \cap (B \cup C) = (A \cap B) \cup (A \cap C)$

Let $x \in A \cap (B \cup C) \Rightarrow x \in A$ and $x \in B \cup C$

$\Rightarrow$ ($x \in A$ and $x \in A$) and ($x \in B$ or $x \in C$)

$\Rightarrow$ ($x \in A$ and $x \in B$) or ($x \in A$ and $x \in C$)

$\Rightarrow$ $x \in A \cap B$ or $x \in A \cap C$

$\Rightarrow$ $x \in (A \cap B) \cup (A \cap C)$

Therefore, $A \cap (B \cup C) \subset (A \cap B) \cup (A \cap C)$ ...(*i*)

Again, let $y \in (A \cap B) \cup (A \cap C) \Rightarrow y \in A \cap B$ or $y \in A \cap C$

$\Rightarrow$ ($y \in A$ and $y \in B$) or ($y \in A$ and $y \in C$)

$\Rightarrow$ ($y \in A$ or $y \in A$) and ($y \in B$ or $y \in C$)

$\Rightarrow$ $y \in A$ and $y \in B \cup C$

$\Rightarrow$ $y \in A \cap (B \cup C)$

Therefore, $(A \cap B) \cup (A \cap C) \subset A \cap (B \cup C)$ ...(*ii*)

Combining (*i*) and (*ii*), we get $A \cap (B \cup C) = (A \cap B) \cup (A \cap C)$. Hence proved.

(*b*) To prove $A \cup (B \cap C) = (A \cup B) \cap (A \cup C)$

Let $x \in A \cup (B \cap C) \Rightarrow x \in A$ or $x \in B \cap C$

$\Rightarrow$ ($x \in A$ or $x \in A$) or ($x \in B$ and $x \in C$)

$\Rightarrow$ ($x \in A$ or $x \in B$) and ($x \in A$ or $x \in C$)

$\Rightarrow$ $x \in A \cup B$ and $x \in A \cup C$

$\Rightarrow \quad x \in (A \cup B) \cap (A \cup C)$

Therefore, $A \cup (B \cap C) \subset (A \cup B) \cap (A \cup C)$ ...(*i*)

Again, let $y \in (A \cup B) \cap (A \cup C) \Rightarrow y \in A \cup B$ and $y \in A \cup C$

$\Rightarrow \quad (y \in A$ or $y \in B)$ and $(y \in A$ or $y \in C)$

$\Rightarrow \quad (y \in A$ and $y \in A)$ or $(y \in B$ and $y \in C)$

$\Rightarrow \quad y \in A$ or $y \in B \cap C$

$\Rightarrow \quad y \in A \cup (B \cap C)$

Therefore, $(A \cup B) \cap (A \cup C) \subset A \cup (B \cap C)$ ...(*ii*)

Combining (*i*) and (*ii*), we get $A \cup (B \cap C) = (A \cup B) \cap (A \cup C)$. Hence proved.

Example 7. *Prove De Morgan's Laws*

(*a*) $(A \cup B)^c = A^c \cap B^c$ (*b*) $(A \cap B)^c = A^c \cup B^c$.

Sol. (*a*) To prove $(A \cup B)^c = A^c \cap B^c$

Let $x \in (A \cup B)^c \Rightarrow x \notin A \cup B$ $(\because a \in A \Leftrightarrow a \notin A^c)$

$\Rightarrow x \notin A$ and $x \notin B$

$\Rightarrow x \in A^c$ and $x \in B^c$

$\Rightarrow x \in A^c \cap B^c$

Therefore, $(A \cup B)^c \subset A^c \cap B^c$ ...(*i*)

Again, let $x \in A^c \cap B^c \Rightarrow x \in A^c$ and $x \in B^c$

$\Rightarrow x \notin A$ and $x \notin B$

$\Rightarrow x \notin A \cup B$

$\Rightarrow x \in (A \cup B)^c$

Therefore, $A^c \cap B^c \subset (A \cup B)^c$ ...(*ii*)

Combining (*i*) and (*ii*), we get $A^c \cap B^c = (A \cup B)^c$. Hence proved.

(*b*) To prove $(A \cap B)^c = A^c \cup B^c$

Let $x \in (A \cap B)^c \Rightarrow x \notin A \cap B$ $(\because a \in A \Leftrightarrow a \notin A^c)$

$\Rightarrow x \notin A$ or $x \notin B$

$\Rightarrow x \in A^c$ or $x \in B^c$

$\Rightarrow x \in A^c \cup B^c$

$\therefore \quad (A \cap B)^c \subset A^c \cup B^c$ ...(*i*)

Again, let $x \in A^c \cup B^c \Rightarrow x \in A^c$ or $x \in B^c$

$\Rightarrow x \notin A$ or $x \notin B$

$\Rightarrow x \notin A \cap B$

$\Rightarrow x \in (A \cap B)^c$

$\therefore \quad A^c \cup B^c \subset (A \cap B)^c$ ....(*ii*)

Combining (*i*) and (*ii*), we get $(A \cap B)^c = A^c \cup B^c$. Hence proved.

Example 8. *Prove*

(*a*) $A \cup \phi = A$ (*b*) $A \cap \phi = \phi$ (*c*) $A \cup U = U$ (*d*) $A \cap U = A$.

Sol. (*a*) To prove $A \cup \phi = A$

Let $x \in A \cup \phi \Rightarrow x \in A$ or $x \in \phi$

$\Rightarrow x \in A$ $(\because x \notin \phi$, as ϕ is the null set)

Therefore, $x \in A \cup \phi \Rightarrow x \in A$

Hence $A \cup \phi \subset A$.

We know that $A \subset A \cup B$ for any set B.

But for $B = \phi$, we have $A \subset A \cup \phi$

From above, $A \subset A \cup \phi, A \cup \phi \subset A \Rightarrow A = A \cup \phi$. Hence proved.

(*b*) To prove $A \cap \phi = \phi$

If $x \in A$, then $x \notin \phi$ ($\because$ ϕ is null set)

Therefore, $x \in A, x \notin \phi \Rightarrow A \cap \phi = \phi$. Hence proved.

(*c*) To prove $A \cup U = U$

Every set is a subset of universal set

$\therefore$ $A \cup U \subseteq U$

Also, $U \subseteq A \cup U$

Therefore, $A \cup U = U$. Hence proved.

(*d*) To prove $A \cap U = A$

We know $A \cap U \subset A$ ...(*i*)

So we have to show that $A \subset A \cap U$

Let $x \in A \Rightarrow x \in A$ and $x \in U$ ($\because$ $A \subset U$ so $x \in A \Rightarrow x \in U$)

$\therefore$ $x \in A \Rightarrow x \in A \cap U$

$\therefore$ $A \subset A \cap U$ ...(*ii*)

From (*i*) and (*ii*), we get $A \cap U = A$. Hence proved.

Example 9. *Prove* (*a*) $A \cup A^c = U$ (*b*) $A \cap A^c = \phi$

(*c*) $U^c = \phi$ (*d*) $\phi^c = U$.

Sol. (*a*) To prove $A \cup A^c = U$

Every set is a subset of U

$\therefore$ $A \cup A^c \subset U$ ...(*i*)

We have to show that $U \subseteq A \cup A^c$

Let $x \in U \Rightarrow x \in A$ or $x \notin A$

$\Rightarrow x \in A$ or $x \in A^c \Rightarrow x \in A \cup A^c$

$\therefore$ $U \subseteq A \cup A^c$ ...(*ii*)

From (*i*) and (*ii*), we get $A \cup A^c = U$. Hence proved.

(*b*) To prove $A \cap A^c = \phi$

As ϕ is a subset of every set

$\therefore$ $\phi \subseteq A \cap A^c$ ...(*i*)

We have to show that $A \cap A^c \subseteq \phi$

Let $x \in A \cap A^c \Rightarrow x \in A$ and $x \in A^c$

$\Rightarrow x \in A$ and $x \notin A$ (A contradictory statement)

$\Rightarrow x \in \phi$

$\therefore$ $A \cap A^c \subset \phi$ ...(*ii*)

From (*i*) and (*ii*), we get $A \cap A^c = \phi$. Hence proved.

(*c*) To prove $U^c = \phi$

Let $x \in U^c \Leftrightarrow x \notin U \Leftrightarrow x \in \phi$

$\therefore$ $U^c = \phi$. Hence proved. (As U is a universal set) (*d*) To prove $\phi^c = U$

Let $x \in \phi^c \Leftrightarrow x \notin \phi \Leftrightarrow x \in U$ (As ϕ is an empty set)

$\therefore$ $\phi^c = U$. Hence proved.

Example 10. *Prove* $(A^c)^c = A$.

Sol. Let $x \in (A^c)^c \Leftrightarrow x \notin A^c \Leftrightarrow x \in A$

$\therefore$ $(A^c)^c = A$. Hence proved.

CARDINALITY OF A SET

The total numbers of unique elements in the set is called the cardinality of the set. The cardinality of the countably infinite set is countably infinite *e.g.*,

1. Let $P = \{k, l, m, n\}$

The cardinality of the set P is 4.

2. Let A is the set of all non-negative even integers *i.e.*,

$$A = \{0, 2, 4, 6, 8, 10, \ldots\ldots\ldots\}$$

As A is countably infinite set hence the cardinality.

VENN DIAGRAMS

The Venn diagram represents sets as regions in a plane. These are used to understand the relationships between sets. Results from a Venn diagram usually are not considered to constitute a proof, but the results in the Venn diagram can be converted into a proof.

Venn Diagrams and Set Operations. Venn diagrams are used to represent set operations. In Venn diagrams, shadings are used to indicate which regions represent which sets. The results are obtained by determining which regions in the diagram are shaded in which fashion.

To draw Venn diagrams following points are to be taken into consideration :

(*i*) All the sets are usually represented by circles.

(*ii*) The universal set is represented by a rectangle.

(*iii*) The complement of the set is represented by that portion of the universal set which is not in the set.

Example 11. *Draw Venn diagram to represent the sets A, A^c and U.*

Sol. The shaded portion represents the set A. (Fig. 1).

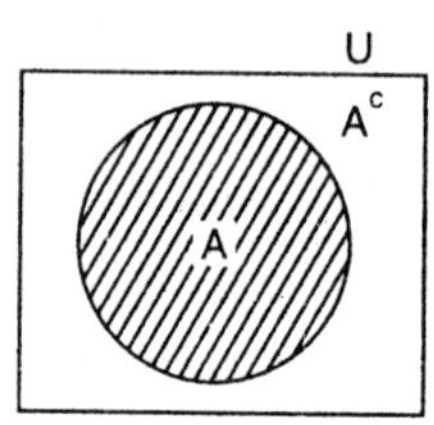

Fig. 1

The unshaded portion represents the set A^c w.r.t. the universal set represented by the rectangle.

Example 12. *Draw the Venn diagrams for the following :*

(i) $A \cup B$ *(ii)* $A \cap B$.

Sol. *(i)* The Venn diagram for $A \cup B$ is shown in Fig. 2.

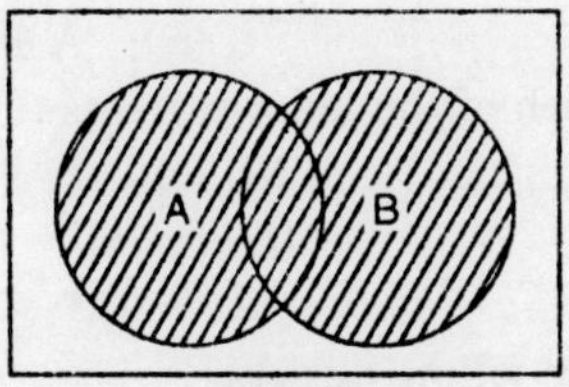

$A \cup B$

Fig. 2

(ii) The Venn diagram for $A \cap B$ is shown in Fig. 3.

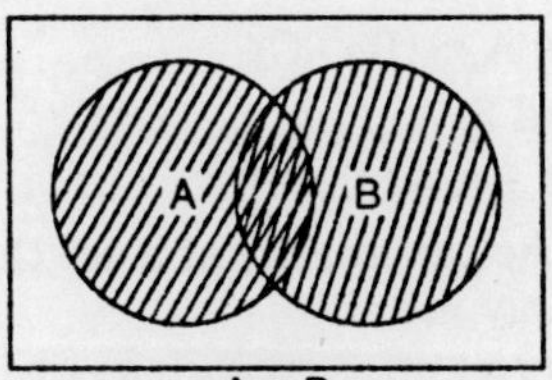

$A \cap B$

Fig. 3

The region shaded in both directions represents $A \cap B$.

Example 13. *Draw the Venn diagram for* $A^c \cap B^c \cap C^c$.

Sol. The Venn diagram for $A^c \cap B^c \cap C^c$ is shown in Fig. 4.

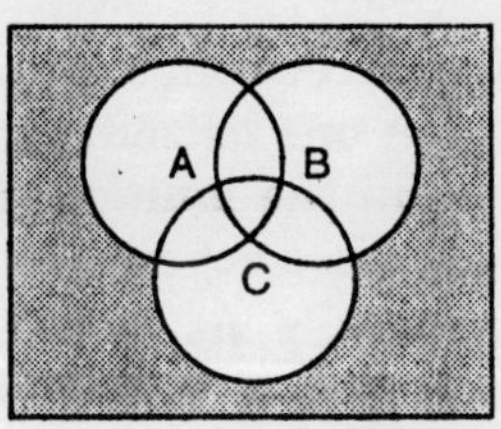

$A^c \cap B^c \cap C^c$

Fig. 4

The shaded portion represents $A^c \cap B^c \cap C^c$.

EXERCISES

1. Prove that

(i) $(A \cap B) - C = (A - C) \cap (B - C)$ *(ii)* $A \cap (B - C) = (A \cap B) - (A \cap C)$

(iii) $A - B = B' - A'$

2. Show that if A and B are sets, then $A - B = A \cap \overline{B}$
3. Show that if A and B are sets, then $(A \cap B) \cup (A \cap \overline{B}) = A$
4. Show that if A, B and C are sets, then $(A - B) - C = (A - C) - (B - C)$
5. Prove that
   (*i*) $(A - B) \cap (B - A) = \phi$ (*ii*) $(A \cap B) \cap (A - B) = \phi$
6. Prove that
   (*i*) $A \oplus \phi = A$ (*ii*) $A \oplus A = \phi$.
7. Determine the cardinality of each of the following sets :
   (*i*) ϕ (*ii*) $\{\phi\}$
   (*iii*) $\{\phi, \{\phi\}, \{\phi, \{\phi\}\}\}$ (*iv*) $\{\phi, \{\phi\}\}$.
8. Find a set whose cardinality is 0.
9. Determine the cardinality of the following sets :
   (*i*) $A = \{1, 3, 5, 7, 9, 11, 13, 15, ...\}$ (*ii*) $B = \{a, e, i, o, u\}$
   (*iii*) $C = \{2, 4, 6, 8, 10\}$
10. Draw a Venn diagram for the symmetric difference of the sets A and B.
11. Draw a Venn diagram to show $(A \cup B)' \cap (A \cup C)'$
12. Draw the Venn diagram to show $A \cap (B \cup C)$.
13. Show that for any two sets A and B
$$A - B = A - (A \cap B).$$
14. Give an example to show that the cardinality of a set that is the intersection of two countably infinite sets may also be countably infinite.
15. Let A, B and C be sets such that $(A \cap B \cap C) = \phi$, $(A \cap B) \neq \phi$, $(A \cap C) \neq \phi$, and $(B \cap C) \neq \phi$. Draw the corresponding venn diagram.
16. Draw a Venn diagram to represent the following facts for the sets P, Q, R and S :
   (*i*) $P \cap Q \neq \phi$ (*ii*) $S \subseteq Q \subseteq R$ and (*iii*) $P \cap S = \phi$.
   (*MCA, IGNOU, June, 2002*)

MULTISETS

Definition. A multiset is an unordered collection of elements, in which the multiplicity of an element may be one or more than one or zero. The multiplicity of an element is the number of times the element repeated in the multiset. In other words, we can say that an element can appear any number of times in a set *e.g.*,

$$A = \{l, l, m, m, n, n, n, n\}$$
$$B = \{a, a, a, a, a, c\}.$$

OPERATIONS ON MULTISETS

1. Union of Multisets. The union of two multisets A and B, is a multiset such that the multiplicity of an element is equal to the maximum of the multiplicity of an element in A and B and is denoted by $A \cup B$, *e.g.*,

Let $A = \{l, l, m, m, n, n, n, n\}$

$B = \{l, m, m, m, n\}$, then $A \cup B = \{l, l, m, m, m, n, n, n, n\}$.

2. Intersection of Multisets. The intersection of two multisets A and B, is a multiset such that the multiplicity of an element is equal to the minimum of the multiplicity of the element in A and B, and is denoted by $A \cap B$, *e.g.*,

Let $A = \{l, l, m, n, p, q, q, r\}$

$B = \{l, m, m, p, q, r, r, r, r\}$, then $A \cap B = \{l, m, p, q, r\}$.

3. Difference of Multisets. The difference of two multisets A and B, is a multiset such that the multiplicity of an element is equal to the multiplicity of the element in A minus the multiplicity of the element in B if the difference is +ve, and is equal to 0 if the difference is 0 or negative *e.g.,*

Let $A = \{l, m, m, m, n, n, n, p, p, p\}$

$B = \{l, m, m, m, m, n, r, r, r\}$, then $A - B = \{n, n, p, p, p\}$.

4. Sum of Multisets. The sum of two multisets A and B, is a multiset such that the multiplicity of an element is equal to the sum of the multiplicities of the element in A and B, *e.g.,*

Let $A = \{l, m, n, p, r\}$

$B = \{l, l, m, n, n, n, p, r, r\}$

then $A + B = \{l, l, l, m, m, n, n, n, n, p, p, r, r, r\}$

5. Cardinality of a Multiset. The cardinality of a multiset is the number of distinct elements in a multiset without considering the multiplicity of an element *e.g.,*

Let $A = \{l, l, m, m, n, n, n, p, p, p, p, q, q, q\}$

The cardinality of the multiset A is 5.

ORDERED PAIRS

An ordered pair consists of two elements such that one of them is designated as first member and other as second member. If p is the first element and q is the second element, then the ordered pair is written as (p, q).

e.g., Let $A = \{1, 2\}$, then possible ordered pairs of $A \times A$

$A \times A = \{(1, 1), (1, 2), (2, 1), (2, 2)\}$

CARTESIAN PRODUCT OF TWO SETS

The cartesian product of two sets P and Q in that order is the set of all ordered pairs whose first member belongs to the set P and second member belongs to set Q and is denoted by $P \times Q$ *i.e.,*

$$P \times Q = \{(x, y) : x \in P, y \in Q\}.$$

Example 14. *Let $P = \{a, b, c\}$ and $Q = \{k, l, m, n\}$. Determine the cartesian product of P and Q.*

Sol. The cartesian product of P and Q is

$$P \times Q = \begin{Bmatrix} (a, k), (a, l), (a, m), (a, n) \\ (b, k), (b, l), (b, m), (b, n) \\ (c, k), (c, l), (c, m), (c, n) \end{Bmatrix}.$$

Example 15. *Let R = {1, 2, 3} and S = {4, 5, 6}. Determine the cartesian product.*

Sol. The cartesian product of R and S is

$$R \times S = \begin{Bmatrix} (1,4),(1,5),(1,6) \\ (2,4),(2,5),(2,6) \\ (3,4),(3,5),(3,6) \end{Bmatrix}.$$

Example 16. *Prove that (A × B) ∩ (P × Q) = (A ∩ P) × (B ∩ Q).*

Sol. Let $(x, y) \in (A \times B) \cap (P \times Q)$

$\Rightarrow$ $(x, y) \in (A \times B)$ and $(x, y) \in (P \times Q)$

$\Rightarrow$ $(x \in A$ and $y \in B)$ and $(x \in P$ and $y \in Q)$

$\Rightarrow$ $(x \in A$ and $x \in P)$ and $(y \in B$ and $y \in Q)$

$\Rightarrow$ $x \in (A \cap P)$ and $y \in (B \cap Q)$

$\Rightarrow$ $(x, y) \in (A \cap P) \times (B \cap Q)$

Therefore, $(A \times B) \cap (P \times Q) \subset (A \cap P) \times (B \cap Q)$ ...(*i*)

Now, conversely let $(x, y) \in (A \cap P) \times (B \cap Q)$

$\Rightarrow$ $x \in (A \cap P)$ and $y \in (B \cap Q)$

$\Rightarrow$ $(x \in A$ and $x \in P)$ and $(y \in B$ and $y \in Q)$

$\Rightarrow$ $(x \in A$ and $y \in B)$ and $(x \in P$ and $y \in Q)$

$\Rightarrow$ $(x, y) \in (A \times B)$ and $(x, y) \in (P \times Q)$

$\Rightarrow$ $(x, y) \in (A \times B) \cap (P \times Q)$

Therefore, $(A \cap P) \times (B \cap Q) \subset (A \times B) \cap (P \times Q)$ ...(*ii*)

From (*i*) and (*ii*), we have

$(A \times B) \cap (P \times Q) = (A \cap P) \times (B \cap Q)$. Hence proved.

Example 17. *Prove that A × (B ∩ C) = (A × B) ∩ (A × C).*

Sol. Let $(x, y) \in A \times (B \cap C)$ $\Rightarrow$ $x \in A$ and $y \in B \cap C$

$\Rightarrow$ $x \in A$ and $(y \in B$ and $y \in C)$

$\Rightarrow$ $x \in A$ and $y \in B$ and $x \in A$ and $y \in C$

$\Rightarrow$ $(x, y) \in A \times B$ and $(x, y) \in A \times C$

$\Rightarrow$ $(x, y) \in (A \times B) \cap (A \times C)$

Therefore, $A \times (B \cap C) \subset (A \times B) \cap (A \times C)$ ...(*i*)

Now, conversely let $(x, y) \in (A \times B) \cap (A \times C)$

$\Rightarrow$ $(x, y) \in A \times B$ and $(x, y) \in A \times C$

$\Rightarrow$ $(x \in A$ and $y \in B)$ and $(x \in A$ and $y \in C)$

$\Rightarrow$ $x \in A$ and $y \in B$ and $y \in C$

$\Rightarrow$ $x \in A$ and $y \in B \cap C$

$\Rightarrow$ $(x, y) \in A \times (B \cap C)$

Therefore, $(A \times B) \cap (A \times C) \subset A \times (B \cap C)$ ...(*ii*)

From (*i*) and (*ii*), we have $A \times (B \cap C) = (A \times B) \cap (A \times C)$. Hence proved.

Example 18. *Prove that A × (B ∪ C) = (A × B) ∪ (A × C).*

Sol. Let $(x, y) \in A \times (B \cup C)$ $\Rightarrow$ $x \in A$ and $y \in B \cup C$

$\Rightarrow$ $x \in A$ and $(y \in B$ or $y \in C)$ $\Rightarrow$ $(x \in A$ and $y \in B)$ or $(x \in A$ and $y \in C)$

$\Rightarrow$ $(x, y) \in A \times B$ or $(x, y) \in (A \times C)$ $\Rightarrow$ $(x, y) \in (A \times B) \cup (A \times C)$

Therefore, $A \times (B \cup C) \subset (A \times B) \cup (A \times C)$ ...(*i*)

Now, conversely let $(x, y) \in (A \times B) \cup (A \times C)$

$\Rightarrow$ $(x, y) \in (A \times B)$ or $(x, y) \in (A \times C)$ $\Rightarrow$ $(x \in A$ and $y \in B)$ or $(x \in A$ and $y \in C)$

$\Rightarrow$ $x \in A$ and $y \in B$ or $y \in C$ $\Rightarrow$ $x \in A$ and $y \in (B \cup C)$

$\Rightarrow$ $(x, y) \in A \times (B \cup C)$

Therefore, $(A \times B) \cup (A \times C) \subset A \times (B \cup C)$ ...(*ii*)

From (*i*) and (*ii*), we have $A \times (B \cup C) = (A \times B) \cup (A \times C)$. Hence proved.

Example 19. *Prove that if $A \subset B$, then $A \times C \subset B \times C$.*

Sol. Let $(x, y) \in A \times C$ $\Rightarrow$ $x \in A$ and $y \in C$

$\Rightarrow$ $x \in B$ and $y \in C$ $(\because\ A \subset B)$

$\Rightarrow$ $(x, y) \in B \times C$

Therefore, $A \times C \subset B \times C$. Hence proved.

Example 20. *If S and T have n elements in common. Show that $S \times T$ and $T \times S$ have n^2 elements in common.*

Sol. Suppose, a set R, consisting of n common elements of S and T.

Therefore, $R \subset S$ and $R \subset T$

Let $(x, y) \in (R \times R)$ $\Leftrightarrow x \in R$ and $y \in R$

$\Leftrightarrow (x \in R$ and $y \in R)$ and $(x \in R$ and $y \in R)$

$\Leftrightarrow (x \in S$ and $y \in T)$ and $(x \in T$ and $y \in S)$ $(\because\ R \subset S ; R \subset T)$

$\Leftrightarrow (x, y) \in (S \times T)$ and $(x, y) \in (T \times S)$

$\Leftrightarrow (x, y) \in (S \times T) \cap (T \times S)$

Therefore, $(R \times R) = (S \times T) \cap (T \times S)$.

The right hand side contains ordered pairs common to both $S \times T$ and $T \times S$.

The left hand side $R \times R$ has n^2 elements $(\because$ R has n elements$)$

Since, the two sets are equal, both have the same number of elements.

Hence, $S \times T$ and $T \times S$ have n^2 common elements.

EXERCISES

1. If $A = \{2, 3\}$, $B = \{3, 4\}$ and $C = \{4, 5\}$, then determine

   (*i*) $(A \times B) \cup (A \times C)$ (*ii*) $(A \times B) \cap (A \times C)$

   (*iii*) $A \times (B \cup C)$ (*iv*) $A \times (B \cap C)$.

2. Show that $(A \times B)' = \overline{A} \times \overline{B}$ for all sets A and B.
3. Show that $A \times (B - C) = (A \times B) - (A \times C)$ for all sets A, B and C.
4. Show that $A \times \phi = \phi$ for any set A.
5. If A and B are non empty sets and $A \times B = B \times A$, what can we conclude about A and B ?

SOLVED PROBLEMS

Problem 1. *Write the following sets in builder form :*

(a) $A = \{1, 3, 5, 7, 9, ...\}$ (b) $B = \{1, 8, 27, 64, ...\}$

(c) $R = \{a, e, i, o, u\}$ (d) $S = \{2, 3, 5, 7, 11, 13, 17, 19, ...\}$

(e) T = *{JAWAHAR LAL NEHRU, INDIRA GANDHI, ..., ATAL BEHARI VAJPAI}.*

Sol. (a) $\{x : x$ is an odd +ve integer$\}$.

(b) $\{x : x$ is a +ve integer, x is a perfect cube$\}$.

(c) $\{x : x$ is a letter of alphabet and x is a vowel$\}$.

(d) $\{x : x$ is a +ve integer and x is a prime number$\}$.

(e) $\{x : x$ was prime minister of India$\}$.

Problem 2. *Write the following sets in builder from :*

(a) $A = \{2, 4, 6, 8, 10, 12, 14\}$ (b) $K = \{3, 6, 9, 12, 15, 18, ...\}$

(c) L = *{PUNJAB, HARYANA, DELHI, UP, ..., BIHAR}.*

Sol. (a) $\{x : x$ is a +ve integer divisible by 2 and less than 15$\}$.

(b) $\{x : x$ is +ve integer and is multiple of 3$\}$.

(c) $\{x : x$ is the stage of India$\}$.

Problem 3. *Write the following sets in tabular form :*

(a) $A = \{x : x^2 = 9\}$ (b) $B = \{x : x$ *is a multiple of 3 and* $0 < x < 20\}$

(c) $C = \{x : x$ *is a +ve even integer*$\}$ (d) $D = \{x : x$ *is a multiple of 5*$\}$.

Sol. (a) A = {3} (b) B = {3, 6, 9, 12, 15, 18}

(c) C = {2, 4, 6, 8, 10, ...} (d) D = {5, 10, 15, 20, 25, ...}

Problem 4. *Write the following sets in tabular form :*

(a) $A = \{x : x$ *is a +ve integer and a perfect square*$\}$

(b) $B = \{x : x$ *is a divisor of 24*$\}$

(c) $C = \{x : x$ *is a multiple of 3 and 5*$\}$

(d) $D = \{x : x$ *is a multiple of 3 or 5*$\}$

(e) $T = \{x : x$ *is a letter in the alphabet*$\}$.

Sol. (a) A = {1, 4, 9, 16, 25, 36, ...} (b) B = {1, 2, 3, 4, 6, 8, 12, 24}

(c) C = {15, 30, 45, 60, ...} (d) D = {3, 5, 6, 10, 15, ...}

(e) T = $\{a, b, c, d, e, ...\}$.

Problem 5. *Determine finite or infinite sets among the sets in Problem 4.*

Sol. (a) Infinite. There are infinite integers that are perfect square.

(b) Finite. Divisor of 24 are finite.

(c) Infinite. There are infinite numbers that are multiple of 3 and 5.

(d) Infinite. There are infinite numbers that are multiple of 3 or 5.

(e) Finite. There are 26 letters in the alphabet.

Problem 6. *Determine finite or infinite sets among the following :*

(*i*) *R = {month in a year}* (*ii*) *S = {lines through the origin}*
(*iii*) *T = {positive odd integer}* (*iv*) *K = {cities in India}*
(*v*) *L = {members of the Parliament}* (*vi*) *M = {positive integer between 1 and – 1}.*

Sol. (*i*) Finite. Twelve months in a year.

(*ii*) Infinite. There are infinite number of lines passing through origin.

(*iii*) Infinite. There are infinite number of +ve odd integers.

(*iv*) Finite. There are finite number of cities in India.

(*v*) Finite. The members of the parliament are finite.

(*vi*) Finite. The set is empty. There is no +ve integer between 1 and – 1.

Problem 7. *Which of the following sets are equal :*

(*i*) *A = {a, b, c, d}* (*ii*) *B = {a, a, b, b, c, d, d, d}*
(*iii*) *K = {d, a, c, b}* (*iv*) *L = {x : x is the letter of alphabet before e}*
(*v*) *M = {First four letters of the alphabet}.*

Sol. All the sets are equal. In set B multiple number of same elements does not affect the set. In set K, the order does not change the set.

Problem 8. *Which of the following sets are equal :*

(*a*) $R = \{x : x^2 = 7\}$ (*b*) $S = \{x : x + 4 = 4\}$
(*c*) $T = \{x : x^2 + 2 = 8\}$.

Sol. The sets R and T are equal as they are empty sets. The set S is not, because element 0 belongs to set S.

Problem 9. *Determine the power set P(A) of the set A = {1, 2, 3}.*

Sol. P(A) = {{1}, {2}, {3}, {1, 2}, {1, 3}, {2, 3}, {1, 2, 3}, ϕ}.

Problem 10. *Determine the power sets of the following sets :*

(*a*) *{a}* (*b*) *{{a}}* (*c*) *{ϕ, {ϕ}}.*

Sol. (*a*) {{*a*}, ϕ} (*b*) {{{*a*}}, ϕ} (*c*) {{ϕ}, {{ϕ}}, {ϕ, {ϕ}}, ϕ}.

Problem 11. *Let A = {ϕ, {ϕ}}. Determine whether the following statements are true or false.*

(*a*) $\phi \in P(A)$ (*b*) $\phi \subseteq P(A)$ (*c*) $\{\phi\} \subseteq P(A)$
(*d*) $\{\phi\} \subseteq A$ (*e*) $\{\phi\} \in P(A)$ (*f*) $\{\phi\} \in A$
(*g*) $\{\{\phi\}\} \subseteq P(A)$ (*h*) $\{\{\phi\}\} \subseteq A$ (*i*) $\{\{\phi\}\} \in P(A)$
(*j*) $\{\{\phi\}\} \in A$.

Sol. The power P(A) = {{ϕ}, {{ϕ}}, {ϕ, {ϕ}}, ϕ}

(*a*) True (*b*) True (*c*) True (*d*) True
(*e*) True (*f*) True (*g*) True (*h*) True
(*i*) True (*j*) False.

Problem 12. *Let A = {a, {a}}. Determine whether each of the following is true or false :*

(*a*) $\phi \in P(A)$ (*b*) $\phi \subseteq P(A)$ (*c*) $\{a\} \in P(A)$ (*d*) $\{a\} \subseteq P(A)$
(*e*) $\{\{a\}\} \in P(A)$ (*f*) $\{\{a\}\} \subseteq P(A)$ (*g*) $\{a, \{a\}\} \in P(A)$ (*h*) $\{a, \{a\}\} \subseteq P(A)$
(*i*) $\{\{\{a\}\}\} \in P(A)$ (*j*) $\{\{\{a\}\}\} \subseteq P(A)$.

Sol. The power set P(A) = $\{\{a\}, \{\{a\}\}, \{a, \{a\}\}, \phi\}$

(*a*) True (*b*) True (*c*) True (*d*) False

(*e*) True (*f*) True (*g*) True (*h*) False

(*i*) False (*j*) True.

Problem 13. *Let A = {ϕ}. Let B = P(P(A)).*

(*a*) *Is $\phi \in B$? $\phi \subseteq B$?* (*b*) *Is $\{\phi\} \in B$? $\{\phi\} \subseteq B$?*

(*c*) *Is $\{\{\phi\}\} \in B$? $\{\{\phi\}\} \subseteq B$?*

Sol. The power set of A *i.e.,* P(A) = $\{\{\phi\}, \phi\}$

$$B = P(P(A)) = \{\{\phi\}, \{\{\phi\}\}, \{\{\phi\}, \phi\}, \phi\}$$

(*a*) Yes, $\phi \in$ B and $\phi \subseteq$ B as ϕ is a subset of every set.

(*b*) Yes, $\{\phi\} \in$ B. Also, $\{\phi\} \subseteq$ B as ϕ is an element in B.

(*c*) Yes, $\{\{\phi\}\} \in$ B and also $\{\{\phi\}\} \subseteq$ B since $\{\phi\}$ is an element of set B.

Problem 14. *Suppose*

U = {1, 2, 3, 4, 5, 6, 7, 8, 9}

A = {1, 4, 9}

P = {x : x ∈ U and x is a perfect square}

R = {1, 2, 3, 5, 7, 9}

D = {2, 3, 5, 7}

N = {x : x ∈ U and x is a prime number}

ϕ

(*a*) *Determine which sets are subsets of others.*

(*b*) *Determine which sets are proper subsets of others.*

(*c*) *Determine pair of sets which are disjoint.*

(*d*) *Determine pair of sets which are comparable.*

(*e*) *Determine pair of sets which are incomparable.*

Sol. (*a*) All the sets are subsets of U since the elements of every set belong to U. Set ϕ is subset of all other sets. Set A is subset of P. Set D is subset of R and N.

(*b*) All the sets are proper subsets of U since they are not equal to U. Set D is proper subset of R. Set N is proper subset of R. Set ϕ is proper subset of all the other sets.

(*c*) The pairs A and D, A and N, P and D, and P and N are disjoint sets.

(*d*) All the sets are comparable with the set U.

Set D is comparable with R as D $\subseteq$ R

Set N is comparable with R as N $\subseteq$ R

All sets comparable with ϕ as $\phi \subseteq$ Every set.

(*e*) The pairs A and R, A and D, A and N, P and R, P and D, P and N, are incomparable.

Problem 15. *Determine the cardinalities of the sets :*

(*a*) *P = {n^7 : n is a positive integer}* (*b*) *Q = {n^{109} : n is a positive integer}*

(*c*) $P \cup Q$ (*d*) $P \cap Q$.

Sol. (*a*) The cardinality of the set P is infinite as the number of +ve integers are infinite.

(*b*) The cardinality of the set Q is infinite as the number of positive integers are infinite.

(*c*) Since sets P and Q are both infinite, their union is also infinite and hence cardinality.

(*d*) The cardinality of P ∩ Q is one because for $n = 1$

$$1^7 = 1^{109}$$

But for $n = 2$; $2^7 \neq 2^{109}$ and so on.

Hence, P ∩ Q contains only one element.

Problem 16. *Let N is the set of all natural numbers. Let P denotes all finite subsets of N. What is the cardinality of set P ? Give reason.*

Sol. The number of subsets of any set is given by 2^n, where n is number of elements in the set. As the number of subsets of set N is 2^n, hence there are 2^n number of subsets in set P. Therefore, the cardinality of set S is 2^n.

Problem 17. *Let U = {1, 2, 3, 4, 5, 6, 7, 8, 9} and suppose that*

A = {1, 2, 3, 4, 5}, B = {4, 5, 6, 7}, C = {5, 6, 7, 8, 9}, D = {1, 3, 5, 7, 9}.

Find the following subsets of U :

(*a*) $A \cup B$ (*b*) $A \cap B$ (*c*) A'

(*d*) $C - D$ (*e*) $(C \cup D)'$.

Sol. (*a*) A ∪ B = {1, 2, 3, 4, 5, 6, 7} (*b*) A ∩ B = {4, 5}

(*c*) A′ = {6, 7, 8, 9} (*d*) C – D = {6, 8}

(*e*) (C ∪ D)′ = {2, 4}.

Problem 18. *Determine the following sets :*

(*a*) $\phi \cup \{\phi\}$ (*b*) $\phi \cap \{\phi\}$ (*c*) $\{\phi\} \cup \{a, \phi, \{\phi\}\}$ (*d*) $\{\phi\} \cap \{a, \phi, \{\phi\}\}$

(*e*) $\phi \oplus \{a, \phi, \{\phi\}\}$ (*f*) $\{\phi\} \oplus \{a, \phi, \{\phi\}\}$.

Sol. (*a*) {ϕ} (*b*) ϕ

(*c*) {*a*, ϕ, {ϕ}} (*d*) {ϕ}

(*e*) {*a*, ϕ, {ϕ}} (*f*) {*a*, {ϕ}}.

Problem 19. *Determine whether each of the following statements is true or false. Explain your answer.*

(*a*) $A \cup P(A) = P(A)$ (*b*) $A \cap P(A) = A$

(*c*) $\{A\} \cup P(A) = P(A)$ (*d*) $\{A\} \cap P(A) = A$

(*e*) $A - P(A) = A$ (*f*) $P(A) - \{A\}$.

Sol. (*a*) A ∪ P(A) = P(A).

False. P(A) contains all subsets of A but does not contain all elements of A. Therefore, A ∪ P(A) ≠ P(A).

(*b*) A ∩ P(A) = A.

False. Since power set of A contains all subsets of A but no elements of A, hence there is no element common to both the sets.

(*c*) {A} ∪ P(A) = P(A)

True. P(A) contains all subsets of A and {A} is also an element of P(A).

(*d*) {A} ∩ P(A) = A.

False. {A} is common to both {A} and P(A) because P(A) contains all subsets of A but their intersection is not A but {A}.

(*e*) A – P(A) = A **False.**

(*f*) P(A) – {A} = P(A).

False. {A} is also an element of P(A). Therefore, P(A) – {A} ≠ P(A).

Problem 20. *Let* $A = \{\phi, a\}$. *Construct the following sets :*

(*a*) $A - \phi$ (*b*) $\{\phi\} - A$ (*c*) $A \cup P\{A\}$ (*d*) $A \cap P\{A\}$.

Sol. (*a*) $\{\phi, a\}$ (*b*) ϕ

(*c*) $\{\phi, a\} \cup \{\{\phi\}, \{a\}, \{\phi, a\}, \phi\} = \{\phi, a, \{\phi\}, \{a\}, \{\phi, a\}\}$

(*d*) $\{\phi, a\} \cap \{\{\phi\}, \{a\}, \{\phi, a\}, \phi\} = \{\phi\}$.

Problem 21. *Let A, B, C be sets. Under what conditions is each of the following true ?*

(*a*) $(A - B) \cup (A - C) = \phi$ (*b*) $(A - B) \cup (A - C) = \phi$

(*c*) $(A - B) \cap (A - C) = \phi$ (*d*) $(A - B) \oplus (A - C) = \phi$.

Sol. (*a*) This is true if B and C are null sets or A and B are disjoint sets and A and C are disjoint sets.

(*b*) This is true if A = B = C or A, B and C are null sets.

(*c*) This is true if A = B = C or all three sets are null sets.

(*d*) This is true if $A \subseteq B$ and $A \subseteq C$.

It is also true if $A \subseteq B$ and $A \not\subseteq C$ but B = C.

Problem 22. *What can you say about P and Q if*

(*a*) $P \cap Q = P$ (*b*) $P \cup Q = P$ (*c*) $P \oplus Q = P$ (*d*) $P \cap Q = P \cap Q$.

Sol. (*a*) This tells that $P \subseteq Q$. (*b*) This tells that P = Q or $Q \subseteq P$.

(*c*) This tells that Q is a null set. (*d*) This tells that P = Q.

Problem 23. *Let A, B, C be arbitrary sets :*

(*a*) *Show that* $(A - B) - C = A - (B \cup C)$.

(*b*) *Show that* $(A - B) - C = (A - C) - B$.

(*c*) *Show that* $(A - B) - C = (A - C) - (B - C)$.

Sol. (*a*) $(A - B) - C = A - (B \cup C)$

Let $x \in (A - B) - C \Rightarrow x \in (A - B), x \notin C$

$\Rightarrow x \in A, x \notin B, x \notin C$

$\Rightarrow x \in A, x \notin B$ or C

$\Rightarrow x \in A, x \notin B \cup C$

$\Rightarrow x \in A - (B \cup C)$.

$$\therefore \quad A - (B \cup C) \subseteq (A - B) - C \qquad ...(i)$$

Conversely, let $x \in A - (B \cup C)$

$\Rightarrow x \in A, x \notin B \cup C$

$\Rightarrow x \in A, x \notin B, x \notin C$

$\Rightarrow x \in A - B, x \notin C$

$\Rightarrow x \in (A - B) - C$

$$\therefore \quad (A - B) - C \subseteq A - (B \cup C) \qquad ...(ii)$$

From (*i*) and (*ii*), we get $(A - B) - C = A - (B \cup C)$.

(*b*) $(A - B) - C = (A - C) - B$

Let $x \in (A - B) - C$

$\Rightarrow x \in A - B, x \notin C \quad \Rightarrow x \in A, x \notin B, x \notin C$

$\Rightarrow x \in A - C, x \notin B \quad \Rightarrow x \in (A - C) - B$

$\therefore \quad (A - C) - B \subseteq (A - B) - C$ ...(*i*)

Conversely, let $x \in (A - C) - B$

$\Rightarrow x \in (A - C), x \notin B \quad \Rightarrow x \in A, x \notin C, x \notin B$

$\Rightarrow x \in A, x \notin B, x \notin C \quad \Rightarrow x \in (A - B), x \notin C$

$\Rightarrow x \in (A - B) - C$

$\therefore \quad (A - B) - C \subseteq (A - C) - B$ ...(*ii*)

From (*i*) and (*ii*), we get $(A - B) - C = (A - C) - B$.

(*c*) $(A - B) - C = (A - C) - (B - C)$

Let $x \in (A - B) - C \quad \Rightarrow x \in (A - B), x \notin C$

$\Rightarrow x \in A, x \notin B, x \notin C \quad \Rightarrow x \in A, x \notin C, x \notin B, x \notin C$

$\Rightarrow x \in A - C, x \notin B - C \quad \Rightarrow x \in (A - C) - (B - C)$

$\therefore \quad (A - C) - (B - C) \subseteq (A - B) - C$ ...(*i*)

Conversely, let $x \in (A - C) - (B - C)$

$\Rightarrow x \in (A - C), x \notin (B - C) \quad \Rightarrow x \in A, x \notin C, x \notin B, x \notin C$

$\Rightarrow x \in A, x \notin B, x \notin C \quad \Rightarrow x \in A - B, x \notin C$

$\Rightarrow x \in (A - B) - C$

$\therefore \quad (A - B) - C \subseteq (A - C) - (B - C)$ ...(*ii*)

From (*i*) and (*ii*), we get $(A - B) - C = (A - C) - (B - C)$.

Problem 24. *Given that $P \cup Q = P \cup R$, is it necessary that $Q = R$? Justify your answer.*

Sol. This is not necessary.

$P \cup Q \Rightarrow$ all elements of P or Q are in this set.

$P \cup R \Rightarrow$ all elements of P or R are in this set.

If $Q \subseteq P$ and also $R \subseteq P$. Then $P \cup Q = P \cup R = P$.

Therefore, it is not necessary that $Q = R$.

e.g., Let $P = \{1, 2, 3, 4\}$, $Q = \{1, 2\}$, $R = \{3, 4\}$, then $P \cup Q = P \cup R = P$

But here $Q = R$.

Problem 25. *Given that $P \cap Q = P \cap R$, is it necessary that $Q = R$? Justify your answer.*

Sol. This is not necessary

$P \cap Q$ contains elements common to both P and Q.

$P \cap R$ contains elements common to both P and R.

But it is not necessary that $Q = R$ because set R can have elements of set Q which are elements of set P as well, but it can also have elements other than those in set Q.

e.g., Let $P = \{1, 2, 5, 6\}$, $Q = \{1, 2, 8\}$, $R = \{1, 2, 9\}$

Then $P \cap Q = P \cap R = \{1, 2\}$; but $Q \neq R$.

Problem 26. *Given that $P \oplus Q = P \oplus R$, is it necessary that $Q = R$? Justify your answer.*

Sol. It is necessary that $Q = R$.

Because $P \oplus Q = (P \cup Q) - (P \cap Q)$ *i.e.,* this set contains elements which are in sets P or Q but does not contain elements common to both sets.

Similarly, $P \oplus R = (P \cup R) - (P \cap R)$ *i.e.*, this set contains elements which are in sets P or R but does not contain elements common to both sets.

As $P \oplus Q = P \oplus R$

$\therefore \quad (P \cup Q) - (P \cap Q) = (P \cup R) - (P \cap R)$

$\Rightarrow \quad P \cup Q = P \cup R$ and $P \cap Q = P \cap R$

$\Rightarrow \quad Q = R$

e.g., Let $P = \{1, 2, 3, 4\}, Q = \{1, 5\}, R = \{1, 5\}$

$P \oplus Q = (P \cup Q) - (P \cap Q) = \{1, 2, 3, 4, 5\} - \{1\} = \{2, 3, 4, 5\}$

$P \oplus R = (P \cup R) - (P \cap R) = \{1, 2, 3, 4, 5\} - \{1\} = \{2, 3, 4, 5\}$

$\Rightarrow \quad Q = R.$

Problem 27. *Prove that $(A - B) \cap B = \phi$.*

Sol. $(A - B) \cap B = \phi$

If $x \in B$, then $x \notin A - B$ due to definition of $A - B$

$$x \in B, x \notin A - B \Rightarrow (A - B) \cap B = \phi. \text{ Hence proved.}$$

Problem 28. *Prove that $A \cup B = \phi \Rightarrow A = \phi, B = \phi$.*

Sol. Let $A \cup B = \phi$

Now ϕ is a subset of every set $\Rightarrow \phi \subset A, \phi \subset B$

Since, $A \subset A \cup B, B \subset A \cup B$

Hence, $A \subset \phi, B \subset \phi$

So $A \subset \phi, \phi \subset A \Rightarrow A = \phi$

$B \subset \phi, \phi \subset B \Rightarrow B = \phi$. Hence proved.

Problem 29. *Prove that $A - B \subset B'$.*

Sol. Let $x \in A - B \Rightarrow x \in A$ and $x \notin B$

$\Rightarrow x \in A$ and $x \in B' \Rightarrow x \in A \cap B'$ as $A \cap B' \subset B'$

$\Rightarrow x \in B'$

Therefore, any $x \in A - B \Rightarrow x \in B'$

So $A - B \subset B'$. Hence proved.

Problem 30. *Given that*

$$(A \cap C) \subseteq (B \cap C)$$

$$(A \cap \overline{C}) \subseteq (B \cap \overline{C})$$

Show that $A \subseteq B$.

Sol. For $(A \cap C) \subseteq (B \cap C)$, all elements of set A which are also elements of set C are contained in a set which contains elements common to both B and C. It implies $A \subseteq B$ for elements which are common in A and C, and B and C.

For $(A \cap \overline{C}) \subseteq (B \cap \overline{C})$, elements common to sets A and $\overline{C}$ are contained in a set containing elements common to both B and C. It implies $A \subseteq B$ for elements common in A and $\overline{C}$ and also in B and $\overline{C}$.

Hence, from both conditions it is shown that $A \subseteq B$.

Problem 31. *Show that the set I of all integers is countable.*

Sol. The set I of all integers is an infinite set. If it is countable, it must be countably infinite.

To prove I to be countably infinite, we have to show that there is one-to-one correspondence between N and I. We can define a function from N to I which is one-to-one and onto as shown below.

N =	1	2	3	4	5	6	7	....
I =	0	1	− 1	2	− 2	3	− 3	....

The description of the above function can be described by the explicit formula

$f : \mathrm{I} \to \mathrm{N}$ as

$$f(n) = \begin{cases} \dfrac{n}{2} & \text{if } n \text{ is + ve even integer} \\ \dfrac{(1-n)}{2} & \text{if } n \text{ is + ve odd integer.} \end{cases}$$

Problem 32. *Show that the set of all bit strings is countable.*

Sol. Let the bit strings is of length K. So there are a finite number 2K bit strings of length K. The set of all bit strings is the Union of the bit strings of length K over K = 0, 1, 2, 3, ... Since the union of a countable sets is countable, there are a countable number of bit strings.

Problem 33. *Show that if A is an uncountable set and B is a countable set, then A – B must be uncountable set.*

Sol. Let us assume that A – B is countable. Then, since A = (A – B) ∪ B, hence the elements of A can be listed in a sequence by alternating elements of A – B and elements of B. This contradicts the uncountability of A.

Problem 34. *Show that if X is an uncountable set and X ⊂ Y, then Y is uncountable.*

Sol. Let us assume that Y is countable. Then the elements of X can be listed as x_1, x_2, x_3, ... Since X is a subset of Y, taking the subsequence of $\{x_n\}$ that contains the terms that are in X gives a listing of the elements of X. Since X is uncountable, this is not possible.

Chapter 2

Principle of Inclusion and Exclusion

As we know the cardinality of the set P is the number of unique elements in set P. It is denoted as $|P|$ and read as cardinality of set P.

FIRST PRINCIPLE

If P and Q are disjoint sets, then

$$|P \cup Q| = |P| + |Q|.$$

Theorem I. *Let P and Q be any two non-disjoint sets. Then*

$$|P \cup Q| = |P| + |Q| - |P \cap Q|.$$

Proof. Draw Venn diagram for the above as shown in Fig. 1.

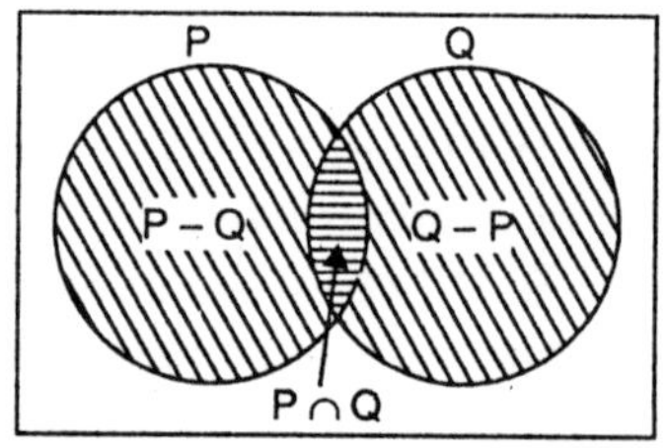

Fig. 1

From figure, we see that $P \cup Q$ can be seen to be the union of three disjoint sets $P - Q$, $Q - P$ and $P \cap Q$.

So $|P \cup Q| = |P - Q| + |Q - P| + |P \cap Q|$ ...(*i*)

Also, $|P| = |P - Q| + |P \cap Q|$ ...(*ii*)

and $|Q| = |Q - P| + |P \cap Q|$ ...(*iii*)

Combining (*ii*) and (*iii*)

$$|P| + |Q| = |P - Q| + |Q - P| + 2|P \cap Q|$$

$$|P| + |Q| = |P \cup Q| + |P \cap Q|$$

(As $|P \cup Q| = |P - Q| + |Q - P| + |P \cap Q|$)

$$|P \cup Q| = |P| + |Q| - |P \cap Q|$$

Hence proved.

Theorem II. *Let P, Q and R are three finite sets. Then*

$$|P \cup Q \cup R| = |P| + |Q| + |R| - |P \cap Q| - |P \cap R| - |Q \cap R| + |P \cap Q \cap R|.$$

Proof. Using theorem I, we have

$$|P \cup (Q \cup R)| = |P| + |Q \cup R| - |P \cap (Q \cup R)|$$

$$= |P| + |Q| + |R| - |Q \cap R| - |P \cap (Q \cup R)| \quad ...(i)$$

As $P \cap (Q \cup R) = (P \cap Q) \cup (P \cap R)$

So $|P \cap (Q \cup R)| = |P \cap Q| + |P \cap R| - |(P \cap Q) \cap (P \cap R)|$

$$= |P \cap Q| + |P \cap R| - |P \cap Q \cap R| \quad ...(ii)$$

Putting (*ii*) in (*i*), we get

$$| P \cup Q \cup R | = | P | + | Q | + | R | - | P \cap Q | - | P \cap R | - | Q \cap R | + | P \cap Q \cap R |$$

Hence proved.

INCLUSION-EXCLUSION PRINCIPLE IN GENERAL

Let $P_1, P_2, \ldots\ldots, P_n$ are finite sets. Then

$$| P_1 \cup P_2 \cup \ldots\ldots \cup P_n | = \sum_{1 \le i \le n} |P_i| - \sum_{1 \le i < j \le n} |P_i \cap P_j| + \sum_{1 \le i < j < k \le n} | P_i \cap P_j \cap P_k | \ldots\ldots + (-1)^{n-1} | P_1 \cap P_2 \cap \ldots\ldots \cap P_n |$$

Example 1. *In a survey of 200 musicians, it was found that 40 wore gloves on the left hand and 39 wore gloves on the right hand. If 160 wore no gloves at all, how many wore a glove on only the right hand ? Only the left hand ? On both hands ?*

Sol. Total number of musicians wore gloves on left, right or both hands *i.e.,*

$$| L \cup R | = 200 - 160 = 40$$

Musicians wore gloves on left hand $| L | = 40$

Musicians wore gloves on right hand $| R | = 39$

Musicians who wore gloves on both hands

$$| L \cap R | = | L | + | R | - | L \cup R | = 40 + 39 - 40 = 39$$

Musicians who wore gloves only on right hand

$$= 40 - 39 = 1$$

Musicians who wore gloves only on left hand

$$= 39 - 39 = 0.$$

Example 2. *Out of 1200 students at a college*

582 took Economics
627 took English
543 took Mathematics
217 took both Economics and English
307 took both Economics and Mathematics
250 took both Mathematics and English
222 took all three courses.

How many took none of the three ?

Sol. Suppose $| A | = 582$, $| B | = 627$, $| C | = 543$, $| A \cap B | = 217$, $| A \cap C | = 307$, $| B \cap C | = 250$, $| A \cap B \cap C | = 222$

The total number of students who took any of three subjects

$$| A \cup B \cup C | = 582 + 627 + 543 - 217 - 307 - 250 + 222 = 1200$$

Students who took none of three

= (total students in the college) – (total students who took any of three subjects)

= 1200 – 1200 = 0.

EXERCISES

1. In a group of 70 persons, 37 like oranges, 52 like mangoes and each person likes at least one of the two fruits. Determine
   (*i*) How many people like both oranges and mangoes.
   (*ii*) How many like oranges but not mangoes.
   (*iii*) How many like mangoes but not oranges.
2. There are 80 students in a class, out of which 60 play cricket, 53 play foot ball and 35 play both of the games. Determine how many students.
   (*i*) play only cricket (*ii*) play only foot ball
   (*iii*) do not play any of the games.
3. Among the positive integers less than or equal to 100, let P be the set of even integers and let Q be the set of integers divisible by 3. Determine how many integers are even or divisible by 3.
4. In a group of students, 60 have a digital camera, 130 have a mobile phone and 40 have both. Determine how many own at least one of these devices ?
5. Among the positive integers less than or equal to 500, let P be the set of odd integers and let Q be the set of integers divisible by 5. Determine how many integers are odd or divisible by 5.
6. In a group of 500 people 300 can speak in English and 80 can speak in Hindi only. Determine
   (*i*) How many can speak both Hindi and English ?
   (*ii*) How many can speak Hindi ?
7. 30 cars were assembled in a factory. The options available were radio, an A/C and white wall tyres. It is known that 15 cars have radios, 8 cars have A/C and 6 cars have white wall tyres. Moreover, 3 cars have all the three options. At least how many of them have no options at all.

(VTU, Karnataka, 2000)

SOLVED PROBLEMS

Problem 1. *40 computer programmers interviewed for a job. 25 knew JAVA, 28 knew ORACLE, and 7 knew neither language. How many knew both languages ?*

Sol. Now
$$| J | = 25$$
$$| O | = 28$$
$$| J \cup O | = 40 - 7 = 33$$

Computer programmers who knew both languages are
$$| J \cap O | = 25 + 28 - 33 = 20.$$

Problem 2. *Among 100 students, 32 study Mathematics, 20 study Physics, 45 study Biology, 15 study Mathematics and Biology, 7 study Mathematics and Physics, 10 study Physics and Biology and 30 do not study any of three subjects.*

(*a*) *Find the number of students studying all three subjects.*

(*b*) *Find the number of students studying exactly one of the three subjects.*

Sol. $| M | = 32 \qquad | P | = 20 \qquad | B | = 45$

$| M \cap B | = 15 \qquad | M \cap P | = 7 \qquad | P \cap B | = 10$

$| M \cup P \cup B | = 100 - 30 = 70.$

(*a*) Number of students studying all three subjects

$| M \cap P \cap B | = 70 - 32 - 20 - 45 + 15 + 7 + 10 = 5.$

(*b*) 5 study all three subjects.

15 – 5 = 10 study Mathematics and Biology but not all the three.

7 – 5 = 2 study Mathematics and Biology but not all the three.

10 – 5 = 5 study Physics and Biology but not all the three.

32 – 10 – 2 – 5 = 15 study only Mathematics.

20 – 2 – 5 – 5 = 8 study only Physics.

45 – 10 – 5 – 5 = 25 study only Biology.

Number of students studying exactly one of three subjects

= 15 + 8 + 25 = 48.

Problem 3. *A survey of 550 television watchers produced the following information :*

285 watch football games

195 watch hockey games

115 watch baseball games

45 watch football and baseball games

70 watch football and hockey games

50 watch hockey and baseball games

100 do not watch any of the three games.

(*a*) *How many people in the survey watch all three games ?*

(*b*) *How many people watch exactly one of the three games ?*

Sol. $| F | = 285$; $| H | = 195$; $| B | = 115$

$| F \cap B | = 45$; $| F \cap H | = 70$; $| H \cap B | = 50$

$| F \cup H \cup B | = 550 - 100 = 450$

(*a*) The number of people watch all three games

$| F \cap H \cap B | = 450 - 285 - 195 - 115 + 45 + 70 + 50 = 20.$

(*b*) 20 watch all three games.

45 – 20 = 25 watch football and baseball but not all three.

70 – 20 = 50 watch football and hockey but not all three.

50 – 20 = 30 watch hockey and baseball but not all three.

285 – 25 – 50 – 20 = 190 watch only football.

195 – 50 – 30 – 20 = 95 watch only hockey.

115 – 25 – 30 – 20 = 40 watch only baseball.

Number of people exactly watch one of the three games

= 190 + 95 + 40 = 325.

Problem 4. *In a survey of 300 students,*

64 had taken a Mathematics course

94 had taken a English course

58 had taken a Computer course

28 had taken both a Mathematics and a Computer course
26 had taken both a English and a Mathematics course
22 had taken both a English and a Computer course
14 had taken all three courses.

(*a*) *How many students were surveyed who had taken non of the three courses ?*

(*b*) *How many had taken only a Computer course ?*

Sol. $| M | = 64 ; | E | = 94 ; | C | = 58$

$| M \cap C | = 28 ; | M \cap E | = 26 ; | E \cap C | = 22$

$| M \cap E \cap C | = 14$

(*a*) $| M \cup E \cup C | = | M | + | E | + | C | - | M \cap C |$
$- | M \cap E | - | E \cap C | + | M \cap E \cap C |$
$= 64 + 94 + 58 - 28 - 26 - 22 + 14 = 154$

Students who had taken none of the courses
$= 300 - 154 = 146.$

(*b*) 14 had taken all three courses.

28 – 14 = 14 had taken both a Mathematics and a Computer but not all three

22 – 14 = 8 had taken both a English and a Computer courses but not all three

58 – 14 – 8 – 14 = 22 had taken only Computer course.

Problem 5. *A survey was conducted among 1000 people. Of these 595 like Metro Channel, 595 like Star Movies and 550 like Zee TV ; 395 of them like Metro Channel and Star Movies, 350 of them like Metro Channel and Zee TV and 400 of them like Star Movies and Zee TV ; 250 of them like Metro Channel, Star Movies and Zee TV.*

(*a*) *How many of them who do not like Metro Channel, do not like Star Movies and do not like Zee TV ?*

(*b*) *How many of them who like Metro Channel, do not like Star Movies and do not like Zee TV ?*

Sol. $| M | = 595 ; | S | = 595 ; | Z | = 550$

$| M \cap S | = 395 ; | M \cap Z | = 350 ; | S \cap Z | = 400$

$| M \cap S \cap Z | = 250$

(*a*) Number of people like at least one of the channels

$| M \cup S \cup Z | = 595 + 595 + 550 - 395 - 350 - 400 + 250 = 845$

The number of people who do not like Metro Channel, do not like Star Movies and do not like Zee TV

= total number of people surveyed – number of people like at least one of the channels
= 1000 – 845 = 155.

(*b*) 250 people like all the three channels

395 – 250 = 145 like Metro and Star Movies but not all the three

350 – 250 = 100 like Metro and Zee TV but not all the three

595 – 145 – 100 – 250 = 100 people like Metro Channel but do not like Star Movies and do not like Zee TV.

Problem 6. *Among the first 500 positive integers :*

(*a*) *Determine the integers which are not divisible by 2, nor by 3, nor by 5.*

(*b*) *Determine the integers which are exactly divisible by one of them.*

Sol. Let A is the number of integers divisible by 2

B is the number of integers divisible by 3

C is the number of integers divisible by 5.

$$|A| = \left[\frac{500}{2}\right] = 250 \,; \qquad |B| = \left[\frac{500}{3}\right] = 166 \,; \qquad |C| = \left[\frac{500}{5}\right] = 100$$

$$|A \cap B| = \left[\frac{500}{2 \times 3}\right] = 83 \,; \qquad |A \cap C| = \left[\frac{500}{2 \times 5}\right] = 50$$

$$|B \cap C| = \left[\frac{500}{3 \times 5}\right] = 33 \,; \qquad |A \cap B \cap C| = \left[\frac{500}{2 \times 3 \times 5}\right] = 16.$$

(*a*) $|A \cup B \cup C| = 250 + 166 + 50 - 83 - 100 - 33 + 16 = 366$

The integers not divisible by 2, 3 and 5 = 500 – 366 = 134.

(*b*) The integers divisible by all the three = 16

83 – 16 = 67 integers are divisible by 2 and 3 but not all the three

50 – 16 = 34 integers are divisible by 2 and 5 but not by all the three

33 – 16 = 17 integers are divisible by 3 and 5 but not by all the three

250 – 67 – 34 – 16 = 133 integers are only divisible by 2

166 – 67 – 17 – 16 = 66 integers are only divisible by 3

100 – 34 – 17 – 16 = 33 integers are only divisible by 5

Total number of integers only divisible by 2, 3 and 5

= 133 + 33 + 66 = 232.

Problem 7. *Among the first 1000 positive integers :*

(*a*) *Determine the integers which are not divisible by 5, nor by 7, nor by 9.*

(*b*) *Determine the integers divisible by 5, but not by 7, not by 9.*

Sol. Let A is the number of integers divisible by 5

B is the number of integers divisible by 7

C is the number of integers divisible by 9.

$$\text{So} \qquad |A| = \left|\frac{1000}{5}\right| = 200 \,; \qquad |B| = \left[\frac{1000}{7}\right] = 142$$

$$|C| = \left[\frac{1000}{9}\right] = 111 \,; \qquad |A \cap B| = \left[\frac{1000}{5 \times 7}\right] = 28$$

$$|A \cap C| = \left[\frac{1000}{5 \times 9}\right] = 22 \,; \qquad |B \cap C| = \left[\frac{1000}{7 \times 9}\right] = 15$$

$$|A \cap B \cap C| = \left[\frac{1000}{5 \times 7 \times 9}\right] = 3.$$

(*a*) The number of integers divisible by 5, 7 and 9

$| A \cup B \cup C | = 200 + 142 + 111 - 28 - 22 - 15 + 3$

$= 391.$

The number of integers not divisible by 5, nor by 7, nor by 9

= Total number of integers – integers divisible by 5, 7 and 9

$= 1000 - 391 = 609.$

(*b*) The integers divisible by all the three integers = 3

$28 - 3 = 25$ integers divisible by 5 and 7 but not by all the three

$22 - 3 = 19$ integers divisible by 5 and 9 but not by all the three

$\therefore$ $200 - 25 - 19 - 3 = 153$ integers divisible by 5 but not by 7, not by 9.

Chapter 3

Partially Ordered Sets

PARTIALLY ORDERED SET

Definition. Consider a relation R on a set S satisfying the following properties :

1. R is reflexive *i.e.*, xRx for every $x \in$ S.
2. R is antisymmetric *i.e.*, if xRy and yRx, then $x = y$.
3. R is transitive *i.e.*, if xRy and yRz, then xRz.

Then R is called a partially order relation and the set S together with partial order is called a partially order set or POSET and is denoted by (S, ≤).

For example :

1. The set N of natural numbers form a poset under the relation '≤' because firstly $x \le x$, secondly, if $x \le y$ and $y \le x$, then we have $x = y$ and lastly if $x \le y$ and $y \le z$, it implies $x \le z$ for all $x, y, z \in$ N.

2. The set N of natural numbers under divisibility *i.e.*, 'x divides y' forms a poset because x/x for every $x \in$ N. Also if x/y and y/x, we have $x = y$. Again if x/y, y/z we have x/z, for every x,y,z $\in$ N.

3. Consider a set S = {1, 2} and power set of S is P(S). The relation of set inclusion ⊆ is a partial order. Since, for any sets A, B, C in P(S), firstly we have A ⊆ A, secondly, if A ⊆ B and B ⊆ A, then we have A = B. Lastly, if A ⊆ B and B ⊆ C, then A ⊆ C. Hence, (P(S), ⊆) is a poset.

Example 1. *Consider a set S = {a, b, c}. Is the relation of set inclusion '⊆' is a partial order on P(S) where P(S) is a power set of S ?*

Sol. The power set of S is

$$P(S) = \{\{a\}, \{b\}, \{c\}, \{a, b\}, \{a, c\}, \{b, c\}, \{a, b, c\}, \phi\}$$

Now consider any sets A, B and C in P(S).

1. Since every A ⊆ A, hence it is reflexive.
2. If A ⊆ B and B ⊆ A, we have A = B. Hence it is antisymmetric.
3. If A ⊆ B, B ⊆ C, we have A ⊆ C. Hence it is transitive.

∴ (P(S), ⊆) is a poset.

Example 2. *Consider a set A = {4, 9, 16, 36}. Is the relation 'divides' a partial order.*

Sol. The relation 'divides' is a partial order if it satisfies the property of reflexivity, antisymmetry and transitivity.

1. Since for every $a \in$ A, we have a/a. Hence, 'divides' is reflexive.

2. If a/b and b/a, we have $a = b$ for any a, $b \in$ A. Hence, 'divides' is antisymmetric.

3. If a/b and b/c, we have a/c for any a, b, $c \in$ A. Hence, the relation 'divides' is a partial order and (A, /) is a poset.

COMPARABLE ELEMENTS

Consider an ordered set A. Two elements a and b of set A are called comparable if

$$a \underset{R}{\leq} b \qquad \text{or} \qquad b \underset{R}{\leq} a.$$

NON-COMPARABLE ELEMENTS

Consider an ordered set A. Two elements a and b of set A are called non-comparable if neither $a \leq b$ nor $b \leq a$.

Example 3. *Consider A = {1, 2, 3, 5, 6, 10, 15, 30} is ordered by divisibility. Determine all the comparable and non-comparable pairs of elements of A.*

Sol. The comparable pairs of elements of A are :

{1, 2}, {1, 3}, {1, 5}, {1, 6}, {1, 10}, {1, 15}, {1, 30}
{2, 6}, {2, 10}, {2, 30}
{3, 6}, {3, 15}, {3, 30}
{5, 10}, {5, 15}, {5, 30}
{6, 30}, {10, 30}, {15, 30}.

The non-comparable pair of elements of A are :

{2, 3}, {2, 5}, {2, 15}
{3, 5}, {3, 10}, {5, 6}, {6, 10}, {6, 15}, {10, 15}.

LINEARLY ORDERED SET

Consider an ordered set A. The set A is called linearly ordered set or totally ordered set, if every pair of elements in A are comparable.

For Example : The set of positive integers I_+ with the usual order $\leq$ is a linearly ordered set.

Example 4. *Consider the set I = {1, 2, 3,......} is ordered by divisibility. Determine whether each of the following subsets of I are linearly ordered or not.*

(*i*) *{2, 4, 8}* (*ii*) *{3, 6, 9, 11}*

(*iii*) *{1}* (*iv*) *{2, 4, 6, 8, 10, ...}.*

Sol. (*i*) The subset is linearly ordered, since every pair of elements is comparable *i.e.*, $2 \mid 4 \mid 8$.

(*ii*) The subset is not linearly ordered, since the pair (3, 11) is not comparable.

(*iii*) The subset is linearly ordered, since the set containing one element is always linearly ordered.

(*iv*) The subset is not linearly ordered since every pair of elements is not comparable *i.e.*, neither 4/6 nor 6/4.

CHAIN

Let us assume that (A, ≤) be a poset. A subset of A is said to be chain if and only if every two elements in the subset are related. The number of elements in a chain is called the length of the chain. If A is chain then it is also called total order relation.

ANTICHAIN

Let us assume that (A, ≤) be a poset. A subset of A is called antichain if no two distinct elements in the subset are related.

Example 5. *Consider the following poset*

$$\{(a, b), (a, c), (a, d), (b, c), (b, d), (a, a), (b, b), (c, c), (d, d)\}$$

Determine all chains and antichains of the poset.

Sol. The following are the chains in the poset with more than one element.

$$\{a, b, c\} \quad \{a, b, d\}$$

$$\{a, b\}, \quad \{b, c\}, \quad \{b, d\}.$$

The following are the antichains in the poset with more than one element.

$$\{c, d\}$$

Note. All the single elements in the poset are chains as well as antichains.

EXERCISES

1. Let (A, ≤) be a poset, let $\leq_R$ be a binary relation on A such that for a and b in A, $a \leq_R b$ iff $b \leq a$. Show that $\leq_R$ is a partial ordering relation.
2. Let (P, ≤) denote a partially ordered set, and n be the length of the longest chain in P. Then the elements in P can be partitioned into n disjoint antichains.
   [**Hint.** Prove it by induction on n.]
3. Determine whether the relation R is a linear order on the set A.
   $$A = R \text{ and } a\, Rb \text{ iff } a \geq b.$$
4. Let R be a linear order on the set A. Show that R^{-1} is also a linear order on A.
5. Let (A, ≤) and (B, ≤) are posets, then show that (A × B, ≤) is a poset, with partial order ≤ defined by $(a, b) \leq (a', b')$ if $a \leq a'$ in A and $b \leq b'$ in B.
6. Determine two incomparable elements in the poset (P({0, 1, 2}), ≤).
7. Which of the following pairs of elements are comparable in the poset (I_+, /) ?
   (*i*) 4, 16 (*ii*) 7, 49 (*iii*) 8, 19
   (*iv*) 3, 3 (*v*) 3, 8 (*vi*) 6, 9.
8. Let (S, R) be a poset. Show that (S_1R^{-1}) is also a poset. Where R^{-1} is the inverse of R.

HASSE DIAGRAMS

It is a useful tool, which completely describes the associated partial order. Therefore, it is also called an ordering diagram. It is very easy to convert a directed graph of a relation on a set A to an equivalent Hasse diagram. Therefore, while drawing a Hasse diagram following points must be remembered.

1. The vertices in Hasse diagram are denoted by points rather than by circles.

2. Since a partial order is reflexive, hence each vertex of A must be related to itself, so the edges from a vertex to itself are deleted in Hasse diagram.

3. Since a partial order is transitive, hence whenever aRb, bRc, we have aRc. Eliminate all edges that are implied by the transitive property in Hasse diagram *i.e.,* Delete edge from a to c but retain the other two edges.

4. If a vertex 'a' is connected to vertex 'b' by an edge *i.e.,* aRb, then vertex 'b' appears above vertex 'a'. Therefore, the arrows may be ommitted from the edges in Hasse diagram.

The Hasse diagram is much simpler than the directed graph of the partial order.

Example 6. *Consider the set A = {4, 5, 6, 7}. Let R be the relation ≤ on A. Draw the directed graph and the Hasse diagram of R.*

Sol. The relation ≤ on the set A is given by

R = {{4, 5}, {4, 6}, {4, 7}, {5, 6}, {5, 7}, {6, 7}, {4, 4}, {5, 5}, {6, 6}, {7, 7}}

The directed graph of the relation R is as shown in Fig. 1.

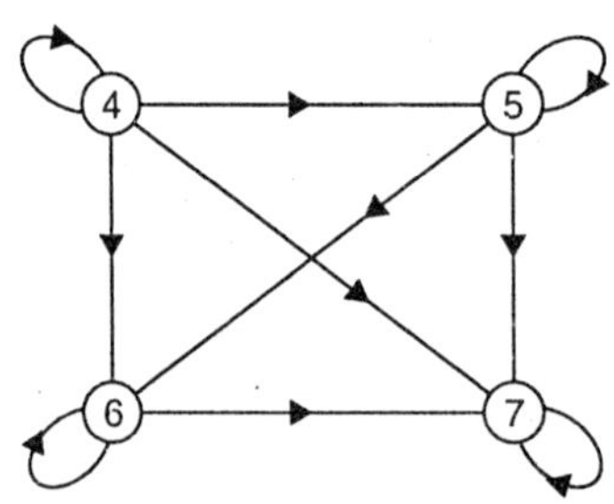

Fig. 1

To draw the Hasse diagram of partial order, apply the following points :

1. Delete all edges implied by reflexive property *i.e.,*
(4, 4), (5, 5), (6, 6), (7, 7).
2. Delete all edges implied by transitive property *i.e.,*
(4, 7), (5, 7) and (4, 6).
3. Replace the circles representing the vertices by dots.
4. Omit the arrows.

The Hasse diagram is as shown in Fig. 2.

Fig. 2

Example 7. *Draw the directed graph of relation determined by the Hasse diagram on the set A = {1, 4, 6, 8} as shown in Fig. 3.*

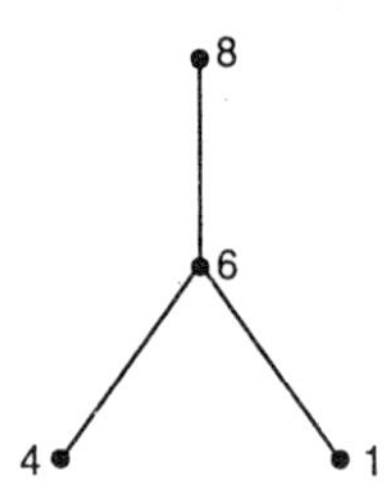

Fig. 3

Sol. The directed graph is shown in Fig. 4.

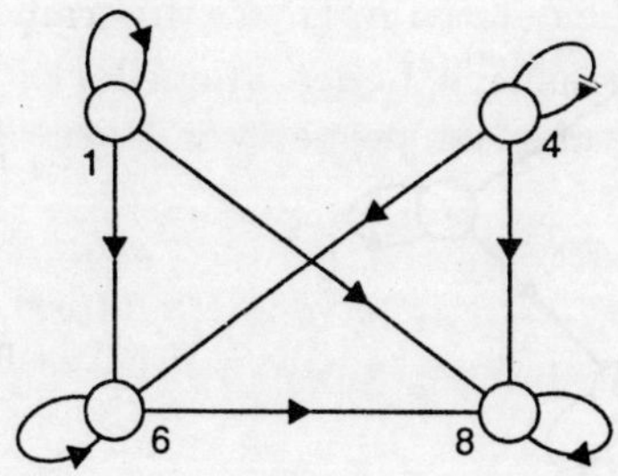

Fig. 4

Example 8. *Determine the Hasse diagram of the partial order having the directed graph as shown in Fig. 5.*

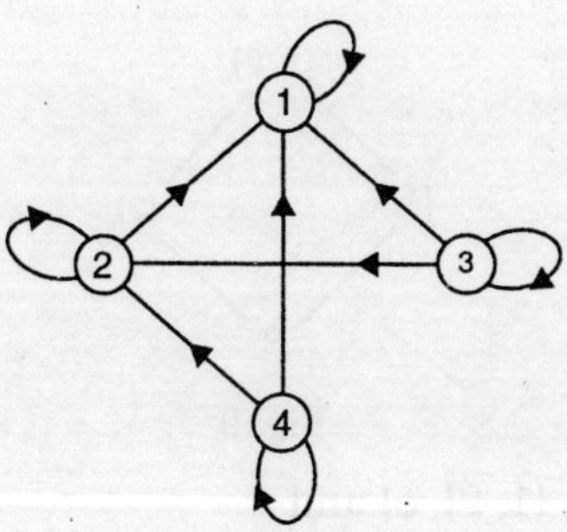

Fig. 5

Sol. The Hasse diagram of the given partial order determine by the directed graph is as shown in Fig. 6.

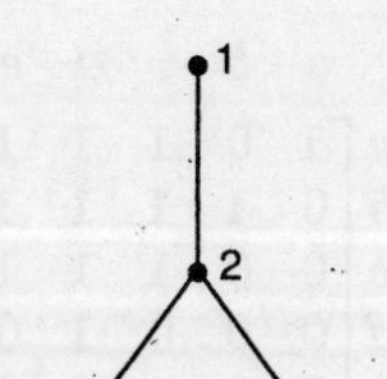

Fig. 6

Example 9. *Consider the set A = {k, l, m, n, p} and the corresponding relation R = {(k, k), (l, l), (m, m), (n, n), (p, p), (k, m), (k, l), (k, n), (k, p), (m, n), (m, p), (n, p), (l, p)}.*

Construct the directed graph and the corresponding Hasse diagram of this partial order.

Sol. The directed graph of the partial order is as shown in Fig. 7.

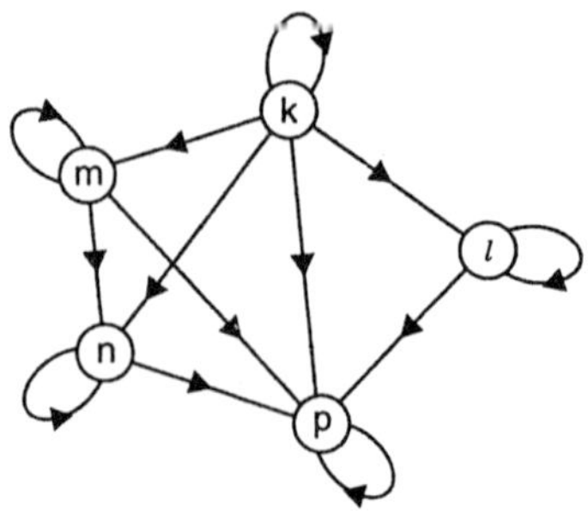

Fig. 7

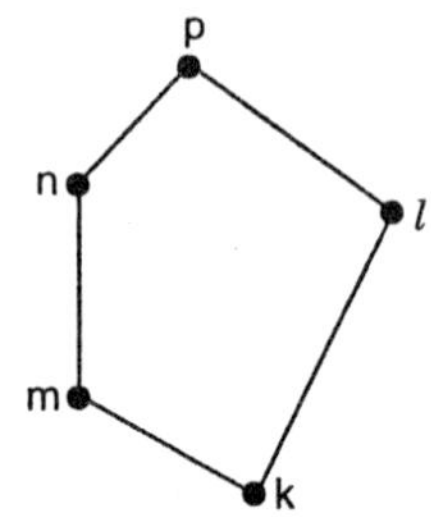

Fig. 8

The Hasse diagram of the partial order is as shown in Fig. 8.

Example 10. *Consider the Hasse diagram as shown in Fig. 9. Determine the value of set A and also determine the set R.*

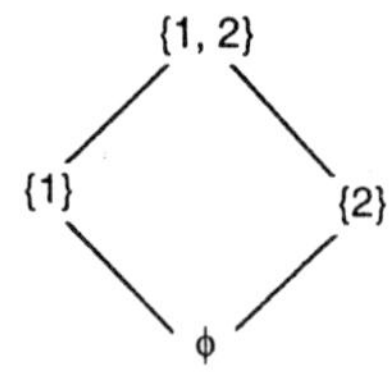

Fig. 9

Sol. The set $A = \{\{1\}, \{2\}, \{1, 2\}, \phi\}$ and

$$R = \{(\{1\}, \{1\}), (\{2\}, \{2\}), (\{1, 2\}, \{1, 2\}), \{\phi, \phi\}, (\{1\}, \{1, 2\}), (\{2\}, \{1, 2\}), (\phi, \{1\}), (\phi, \{2\}), (\phi, \{1, 2\})\}.$$

Here the relation R is (set inclusion) $\subseteq$.

Example 11. *Determine the Hasse diagram of the relation on A = {a, b, c, d, e} whose matrix is shown in Fig. 10.*

$$\begin{array}{c} \\ a \\ b \\ c \\ d \\ e \end{array} \begin{array}{c} \begin{array}{ccccc} a & b & c & d & e \end{array} \\ \begin{bmatrix} 1 & 0 & 1 & 1 & 1 \\ 0 & 1 & 1 & 1 & 1 \\ 0 & 0 & 1 & 1 & 1 \\ 0 & 0 & 0 & 1 & 0 \\ 0 & 0 & 0 & 0 & 1 \end{bmatrix} \end{array}$$

Fig. 10

Sol. The Hasse diagram of the relation on A = {*a*, *b*, *c*, *d*, *e*} is shown in Fig. 11.

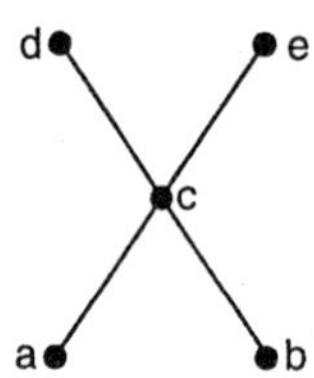

Fig. 11

Example 12. *Let A = {1, 2, 3, 4, 5} be ordered by Hasse diagram.*

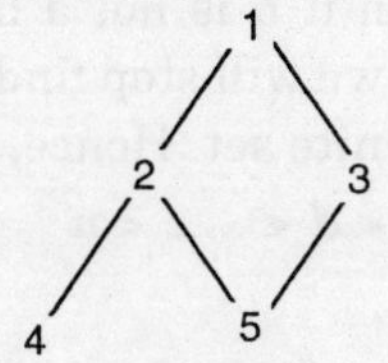

Fig. 12

Insert correct symbol <, >, or ∥ (not comparable) between :

(i) 1 – 5 *(ii) 2 – 3* *(iii) 4 – 1* *(iv) 3 – 4.*

Sol. *(i)* As there is a path from 1 to 5, hence 1 > 5.

(ii) As there is no path from 2 to 3 or vice-versa, hence 2 ∥ 3.

(iii) As there is a path from 4 to 1, hence 4 < 1.

(iv) As there is no path from 3 to 4, hence 3 ∥ 4.

ELEMENTS OF POSET

Maximal element. An element $a \in$ A is called a maximal element of A if there is no element c in A such that $a \le c$.

Minimal element. An element $b \in$ A is called a minimal element of A if there is no element c in A such that $c \le b$.

Note. There can be more than one maximal or more than one minimal element.

Example 13. *Determine all the maximal and minimal elements of the poset whose Hasse diagram is shown in Fig. 13.*

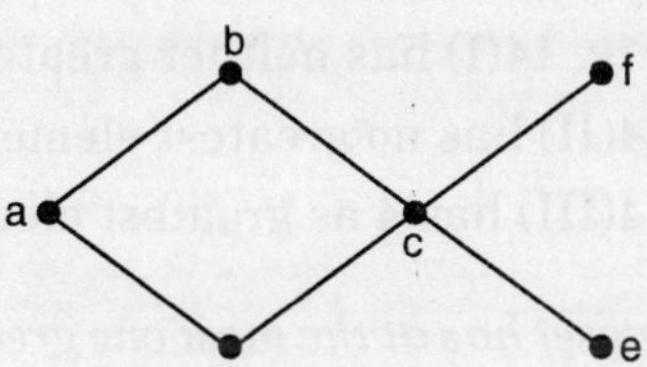

Fig. 13

Sol. The maximal elements are b and f.

The minimal elements are d and e.

Example 14. *Let A = {2, 3, 4, 6, 8, 24, 48} with partial ordering of divisibility. Determine all the maximal and minimal elements of A.*

Sol. The maximal element is 48.

The minimal elements are 2 and 3.

Theorem I. *Prove that the finite non empty poset (A, ≤) has at least one maximal and one minimal element in A.*

Proof. Consider an element $a \in A$. Now if a is not maximal element, we will find another element $b \in A$ such that $a < b$. Again if b is not a maximal element, we will find another element $c \in A$ such that $b < c$. At last, we will stop finding the element, as we cannot continue this process indefinitely since A is a finite set. Hence, we obtain a finite sequence

$$a < b < c < d < \ldots\ldots < n$$

which cannot be extended further.

Therefore, we cannot have $n < x$ for any $x \in A$. So, we conclude that n is a maximal element of (A, ≤). By giving similar arguments we can prove that the finite non empty poset (A, ≤) has at least one minimal element in A.

Greatest element. An element $x \in A$ is called a greatest element of A if for all $a \in A$, $a \leq x$.

Least element. An element $y \in A$ is called a least element of A if, for all $a \in A$, $y \leq a$.

* The greatest element of a poset is denoted by 1 and is called the unit element and the least element of a poset is denoted by 0 and is called the zero element.

Example 15. *Determine the greatest and least elements of the poset whose Hasse diagrams are shown in Fig. 14, if they exist.*

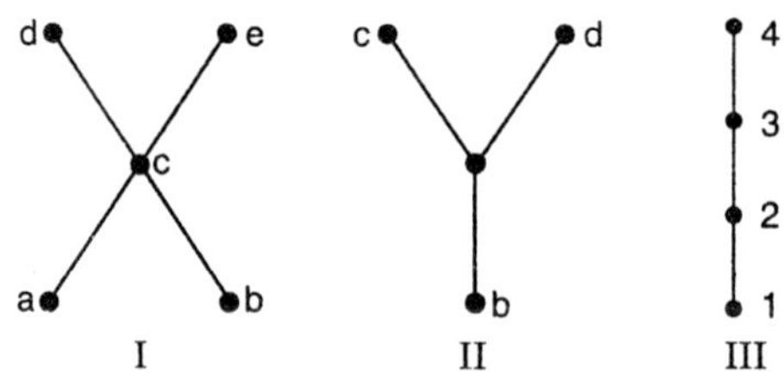

Fig. 14

Sol. The poset shown in Fig. 14(I) has neither greatest nor least element.

The poset shown in Fig. 14(II) has no greatest element but b is the least element.

The poset shown in Fig. 14(III) has 4 as greatest element and 1 as least element.

Theorem II. *Prove that a poset has at the most one greatest element and one least element.*

Proof. Let x and y are two greatest elements of a poset A. Now, since y is a greatest element, so we have $x \leq y$. Also, as x is a greatest element, so we have $y \leq x$ which implies that $x = y$ (*antisymmetric property*). Hence, we can say that if the poset has a greatest element, there exists only one such element.

By similar arguments, we can prove that if the poset has a least element, there exists only one such element.

UPPER BOUND

Consider B be a subset of a partially ordered set A. An element $x \in A$ is called an upper bound of B if $y \leq x$ for every $y \in B$.

LOWER BOUND

Again, consider B be a subset of a partially order set A. An element $z \in A$ is called a lower bound of B if $z \leq x$ for every $x \in B$.

Example 16. *Consider the poset A = {a, b, c, d, e, f, g} be ordered as shown in Fig. 15. Also let B = {c, d, e}. Determine the upper and lower bounds of B.*

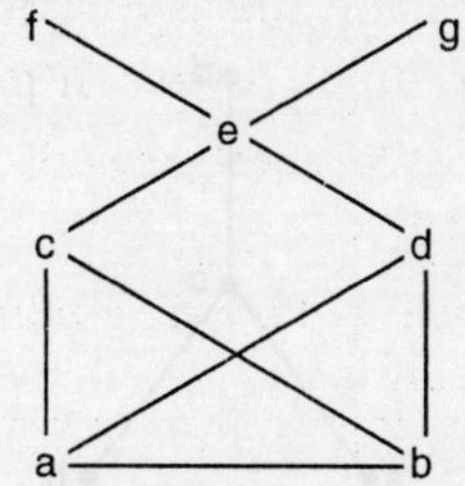

Fig. 15

Sol. The upper bounds of B are *e*, *f* and *g* because every element of B is '≤' *e*, *f* and *g*.

The lower bounds of B are *a* and *b* because *a* and *b* are '≤' every element of B.

Example 17. *Consider the poset A = {1, 2, 3, 4, 5, 6} be ordered as shown in Fig. 16. Also let B = {3, 4}. Determine the upper and lower bounds of B.*

Fig. 16

Sol. The upper bound of B is 4, 5 and 6 because every element of B is '≤' 4, 5 and 6.

The lower bound of B is 3, 2 and 1 because 3, 2 and 1 are '≤' every element of B.

Example 18. *Consider the poset A = {1, 2, 3, 4, 5, 6, 7, 8} be ordered as shown in Fig. 17. Also let B = {3, 4, 5}. Determine the upper and lower bounds of B.*

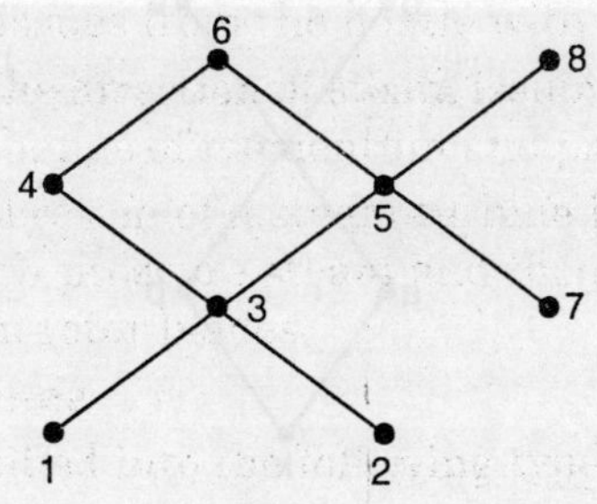

Fig. 17

Sol. The upper bound of B is 6 because every element of B is '≤' 6. The 8 is not an upper bound of B since $4 \nleq 8$.

The lower bound of B is 1, 2 and 3 because 1 and 2 are '≤' every element of B. The 7 is not a lower bound of B since $7 \nleq 3$ and also $7 \nleq 4$.

Example 19. *Consider the poset A = {a, b, c, d} as shown in Fig. 18 and let B = {c}. Determine the upper and lower bounds of B.*

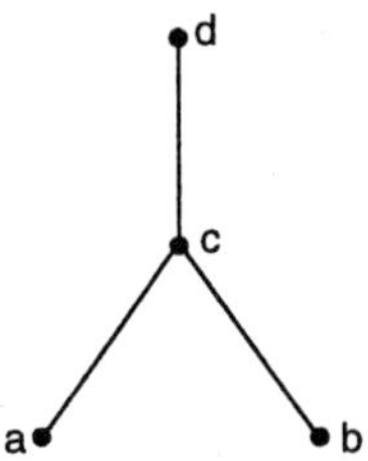

Fig. 18

Sol. The upper bound of B is *c, d*.

The lower bound of B is *a, b, c*.

Note 1. A subset B of a poset A may or may not contain upper bounds or lower bounds in poset A.

2. An upper bound or lower bound of B may or may not belong to the subset B itself.

LEAST UPPER BOUND (SUPREMUM)

Consider B be a subset of a partially ordered set A. An element $x \in$ A is called a least upper bound of B or supremum written as L U B (B) or S U P(B) if x is an upper bound of B and we have $x \leq x'$ for every x' which is an upper bound of B.

GREATEST LOWER BOUND (INFIMUM)

Consider B be a subset of a partially ordered set A. An element $y \in$ A is called a greatest lower bound or infimum of B, written as GLB(B) or INF(B) if y is a lower bound of B and we have $y' \leq y$, for every y' which is a lower bound of B.

Example 20. *Determine the least upper bound and greatest lower bound of B = {a, b, c} if they exist, of the poset whose Hasse diagram is shown in Fig. 19.*

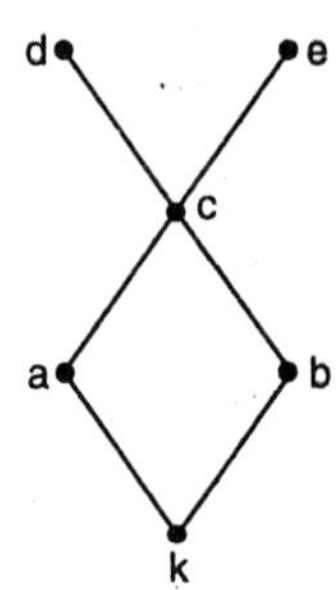

Fig. 19

Sol. The least upper bound is c.

The greatest lower bound is k.

Example 21. *Let D_{100} = {1, 2, 4, 5, 10, 20, 25, 50, 100} and let the relation ≤ be the relation / (divides) be a partial ordering on D_{100}.*

(a) Determine the GLB of B, where B = {10, 20}.

(b) Determine the LUB of B, where B = {10, 20}.

(c) Determine the GLB of B, where B = {5, 10, 20, 25}.

(d) Determine the LUB of B, where B = {5, 10, 20, 25}.

Sol. The Hasse diagram of the poset D_{100} is as shown in Fig. 20.

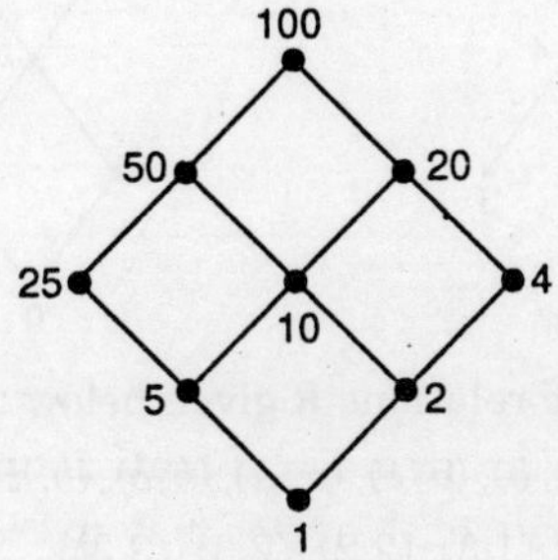

Fig. 20

(*a*) The GLB (B) is 10. (*b*) The LUB (B) is 20.

(*c*) The GLB (B) is 5. (*d*) The LUB (B) is 100.

Note. D_n denote the set of all positive integers which are divisors of n, where n is a + ve integer *e.g.*,

$$D_{10} = \{1, 2, 5, 10\} \text{ as all are divisors of 10.}$$

Example 22. *Determine the least upper bound and greatest lower bound of each B if they exist, of the poset whose Hasse diagram is shown in Fig. 21.*

Fig. 21

(*i*) B = {2, 3} (*ii*) B = {2, 3, 4}

(*iii*) B = {1, 2, 3, 4} (*iv*) B = {1, 2, 3, 4, 5}.

Sol. (*i*) The least upper bound of B is 3. The greatest lower bound of B is 2.

(*ii*) The least upper bound of B is 4. The greatest lower bound of B is 2.

(*iii*) The least upper bound of B is 4. The greatest lower bound of B is 1.

(*iv*) The least upper bound is 5. The greatest lower bound is 1.

Note. Let B = $\{b_1, b_2, ..., b_r\}$. If a = LUB(B), then a is the first vertex that can be reached from b_1, b_2,......, b_r by upward paths. Similarly, if a = GLB(B), then a is the first vertex that can be reached from b_1, b_2,......, b_r by downward paths.

EXERCISES

1. Write the ordered pairs in the relation represented by the Hasse diagrams on the set S as shown in figure.

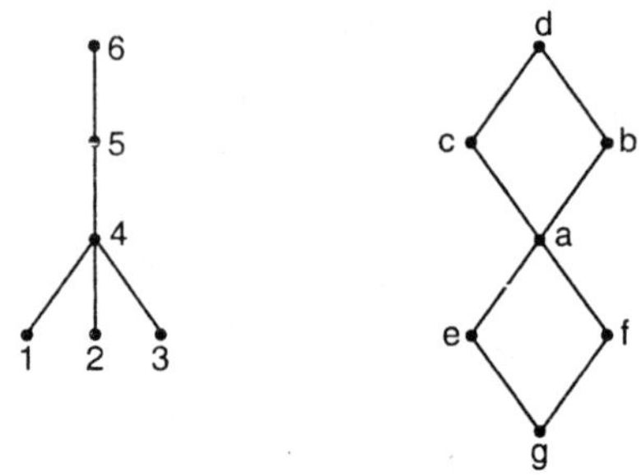

2. Draw the Hasse diagram for the relations R given below :

(*i*) A = $\{a, b, c, d\}$, R = $\{(a, a), (a, b), (a, c), (a, d), (b, b), (b, d), (c, b), (c, d)\}$

(*ii*) A = {1, 2, 3}, R = {(1,1), (1,2), (1,3), (2,2), (2,3), (3,3)}

3. Determine the Hasse diagram of the relation on A = {1, 2, 3, 4, 5} whose Matrix is shown.

$$\begin{bmatrix} 1 & 0 & 1 & 1 \\ 0 & 1 & 1 & 1 \\ 0 & 0 & 1 & 1 \\ 0 & 0 & 0 & 1 \end{bmatrix}$$

4. Determine the Hasse diagram of the partial order having the digraph as shown in figure.

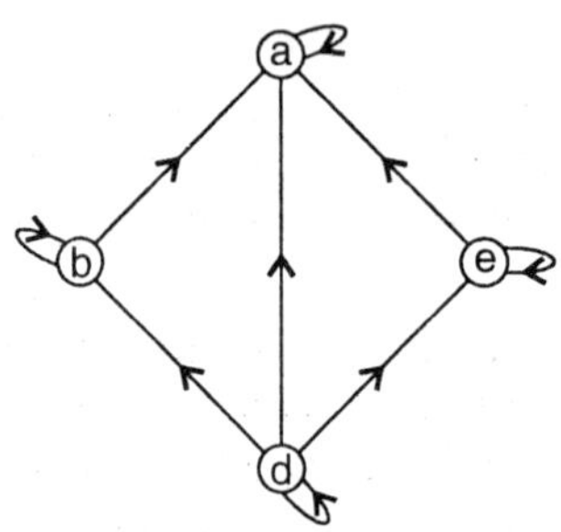

5. Determine all Maximal and Minimal elements of the posets, whose Hasse diagrams are shown in figure.

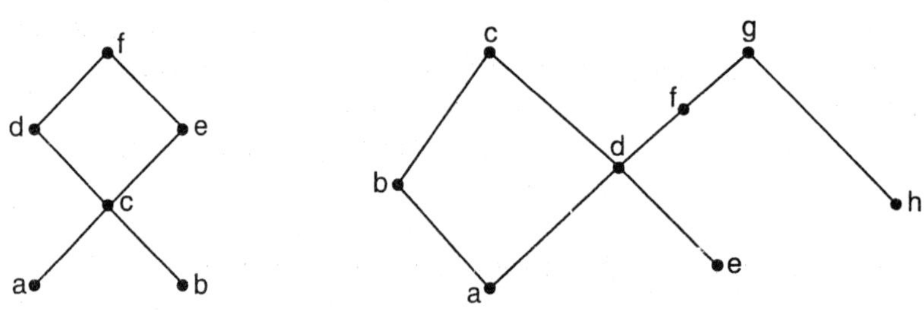

(*i*) (*ii*)

6. Determine, if possible

(*i*) All upper bounds of A (*ii*) All lower bounds of A

(*iii*) The least upper bound of A (*iv*) The greatest lower bound of A.

of the following posets whose Hasse diagrams are shown in figure.

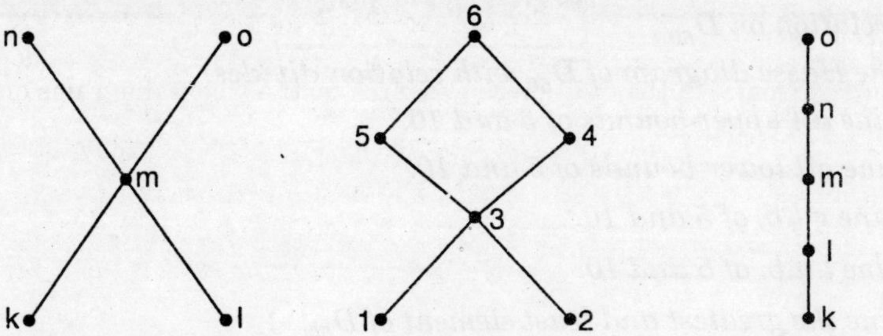

A = {*k*, *l*, *m*, *n*, *o*} A = {4, 5} A = {*m*, *n*}

(*i*) (*ii*) (*iii*)

7. Determine the following concerning the poset

({2, 4, 6, 9, 12, 18, 27, 36, 48, 60, 72}, /).

(*i*) Maximal elements (*ii*) Minimal elements

(*iii*) Greatest element, if exists (*iv*) Least element, if exists

(*v*) All upper bounds of {2, 9} (*vi*) The least upper bound of {2, 9}, if exists.

(*vii*) All lower bounds of {60, 72}

(*viii*) The greatest lower bound of (60, 72), if it exists.

8. Determine the following for the partial order represented by the Hasse diagram as shown in figure.

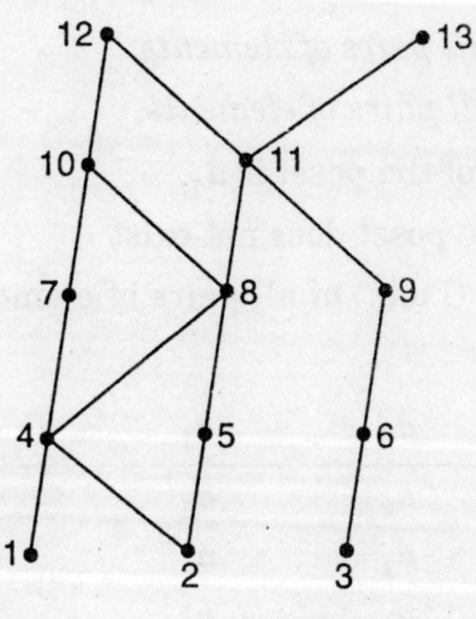

(*i*) Maximal elements (*ii*) Minimal elements

(*iii*) Greatest element, if it exists (*iv*) least element, if it exists

(*v*) All upper bounds of {1, 2, 3} (*vi*) All lower bounds of {6, 8, 9}

(*vii*) The least upper bound of {1, 2, 3}, if it exists.

(*viii*) The greatest lower bound of {6, 8, 9}, if it exists.

9. Show that a finite non empty poset has a maximal element.

10. Give a poset that has neither a maximal nor a minimal element.

11. Show that the least upper bound of a set in a poset is unique if it exists.

12. Show that the greatest upper bound of a set in a poset is unique if it exists.

SOLVED PROBLEMS

Problem 1. *Consider the set D_{50} = {1, 2, 5, 10, 25, 50} and the relation divides (/) be a partial ordering relation on D_{50}.*

(*a*) *Draw the Hasse diagram of D_{50} with relation divides.*

(*b*) *Determine all upper bounds of 5 and 10.*

(*c*) *Determine all lower bounds of 5 and 10.*

(*d*) *Determine g.l.b. of 5 and 10.*

(*e*) *Determine l.u.b. of 5 and 10.*

(*f*) *Determine the greatest and least element of D_{50}.*

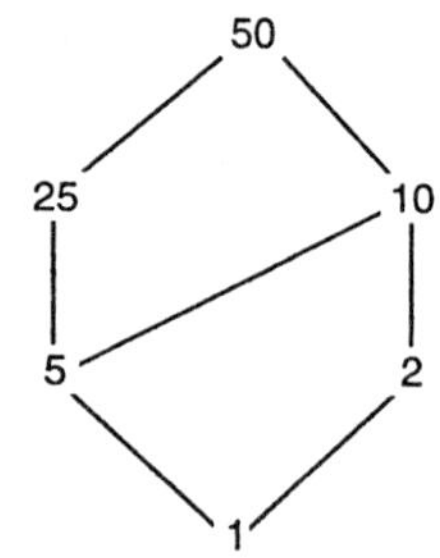

Sol. (*a*) The Hasse diagram of D_{50} is shown in figure.

(*b*) The upper bounds of 5 and 10 are 10 and 50.

(*c*) The lower bounds of 5 and 10 are 5 and 1.

(*d*) The greatest lower bound of 5 and 10 is 5.

(*e*) The least upper bound of 5 and 10 is 10.

(*f*) The greatest element of D_{50} is 50.

The least element of D_{50} is 1.

Problem 2. *Consider the Hasse diagram of the poset as shown in figure.*

(*a*) *Determine the least and greatest element of the poset, if they exist.*

(*b*) *Determine the l.u.b. of all pairs of elements.*

(*c*) *Determine the g.l.b. of all pairs of elements.*

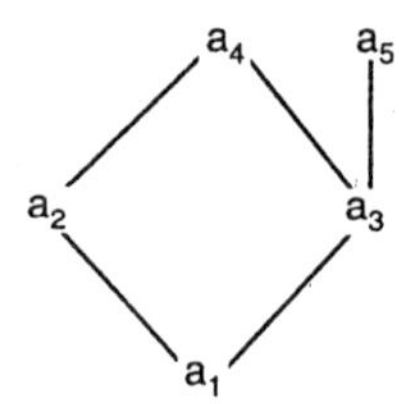

Sol. (*a*) The least element of the poset is a_1.

The greatest element of the poset does not exist.

(*b*) The least upper bound (l.u.b.) of all pairs of elements of the poset is shown in the table as follows :

l.u.b.	a_1	a_2	a_3	a_4	a_5
a_1	a_1	a_2	a_3	a_4	a_5
a_2	a_2	a_2	a_4	a_4	–
a_3	a_3	a_4	a_3	a_4	a_5
a_4	a_4	a_4	a_4	a_4	–
a_5	a_5	–	a_5	–	a_5

Note. The (–) indicates that for these pairs of elements l.u.b. does not exist.

(*c*) The greatest lower bound (g.l.b.) of all pairs of elements of the poset is shown in the table as follows :

g.l.b.	a_1	a_2	a_3	a_4	a_5
a_1	a_1	a_1	a_1	a_1	a_1
a_2	a_1	a_2	a_1	a_2	a_1
a_3	a_1	a_1	a_3	a_3	a_3
a_4	a_1	a_2	a_3	a_4	a_3
a_5	a_1	a_1	a_3	a_3	a_5

Problem 3. *Determine whether the relation represented by the following matrices is a partial order.*

$$\begin{array}{c} \\ a \\ b \\ c \\ d \end{array}\begin{array}{c} \begin{array}{cccc} a & b & c & d \end{array} \\ \begin{bmatrix} 1 & 0 & 1 & 0 \\ 0 & 1 & 1 & 0 \\ 0 & 0 & 1 & 1 \\ 1 & 1 & 0 & 1 \end{bmatrix} \end{array}$$

Sol. The relation represented by this matrix is as follows :

$$R = \{(a, a), (a, c), (b, b), (b, c), (c, c), (c, d), (d, a), (d, b)\ (d, d)\ \}$$

Now check for partial ordering.

Reflexive Relation. The relation R is reflexive since $(a, a) \in R\ \forall\ a$.

Antisymmetric. The relation R is Antisymmetric since whenever

$$(a, b) \in R \text{ and } (b, a) \in R \Rightarrow a = b.$$

Transitive. The relation R is not transitive since $(a, c) \in R$, $(c, d) \in R$, but $(a, d) \in R$.

Thus the relation R is not a partial order since it is reflexive, antisymmetric but not transitive.

Problem 4. *Determine whether the relation whose directed graph is shown in figure is a partial order.*

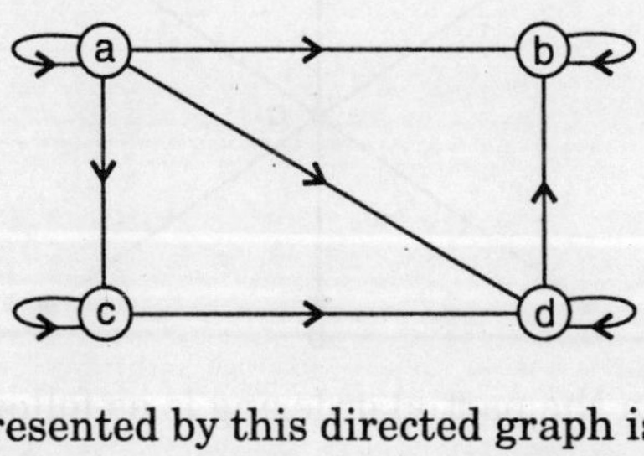

Sol. The relation R represented by this directed graph is as follows :

$$R = \{(a, a), (a, b), (a, d), (b, b), (c, c), (c, d), (d, b), (d, d)\}$$

Now check for partial ordering.

Reflexivity. The relation R is reflexive since $(a, a) \in R$, for every a.

Antisymmetric. The relation R is antisymmetric since whenever $(a, b) \in R$ and $(b, a) \in R$, we have $a = b$.

Transitive. The relation R is not transitive, since $(c, d) \in R$ and $(d, b) \in R$ but $(c, b) \notin R$.

So the relation R is not a partial order as it does not fulfill all the properties.

Problem 5. *Which of the following pairs of elements are comparable in the poset (I, /) ?*

(*i*) 5, 25 (*ii*) 8, 35 (*iii*) 3, 33 (*iv*) 4, 4

Sol. (*i*) It is comparable because 5/25.

(*ii*) It is not comparable.

(*iii*) It is comparable because 3/33.

(*iv*) It is comparable because 4/4.

Problem 6. *Determine a pair of incomparable elements from the following posets.*

(*i*) ({2, 4, 6, 8, 16}, /) (*ii*) (P {a, b, c}, $\subseteq$)

Sol. (*i*) 4, 6 and 6, 16 are pairs of incomparable elements.

(*ii*) {*a*}, {*b*} and {*b*}, {*c*} etc. are pairs of incomparable elements.

Problem 7. *Determine all the ordered pairs in the partial ordering whose Hasse diagram is shown in figure.*

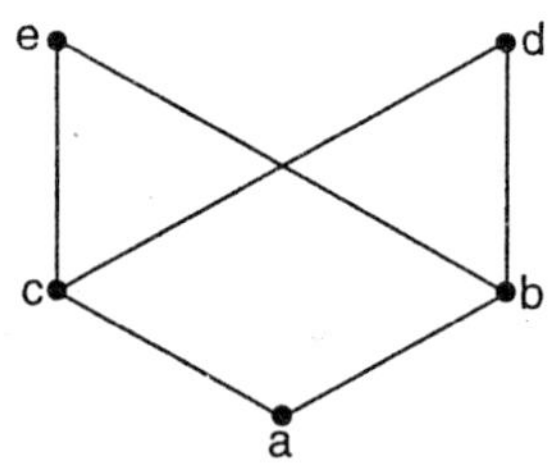

Sol. The ordered pairs in the partial ordering is as follows :

{(*a*, *a*), (*a*, *c*), (*a*, *b*), (*a*, *d*), (*a*, *e*), (*b*, *b*), (*b*, *d*), (*b*, *e*), (*c*, *c*), (*c*, *d*), (*c*, *e*), (*d*, *d*), (*e*, *e*)}

Problem 8. *Determine all the ordered pairs in the partial ordering whose Hasse diagram is shown in figure.*

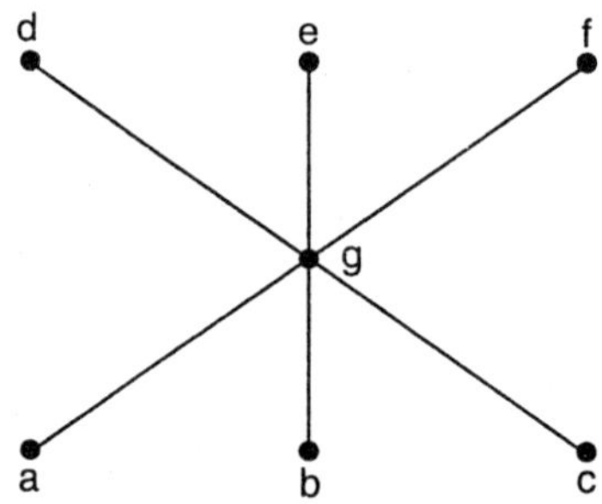

Sol. The ordered pairs in the partial ordering is as follows :

{(*a*, *a*), (*a*, *g*), (*a*, *f*), (*b*, *b*), (*b*, *g*), (*b*, *d*), ((*b*, *e*), (*b*, *f*), (*c*, *c*), (*a*, *e*), (*a*, *d*), (*c*, *g*), (*c*, *d*), (*c*, *e*), (*c*, *f*), (*g*, *d*), (*g*, *e*), (*g*, *f*)}

Problem 9. *Construct the Hasse diagram for divisibility on the set*

(*i*) { 1, 2, 3, 5, 7, 11} (*ii*) {1, 2, 3, 4, 5, 6, 7}

Sol. (*i*) The Hasse diagram for divisibility on the set {1, 2, 3, 5, 7, 11} is shown in figure.

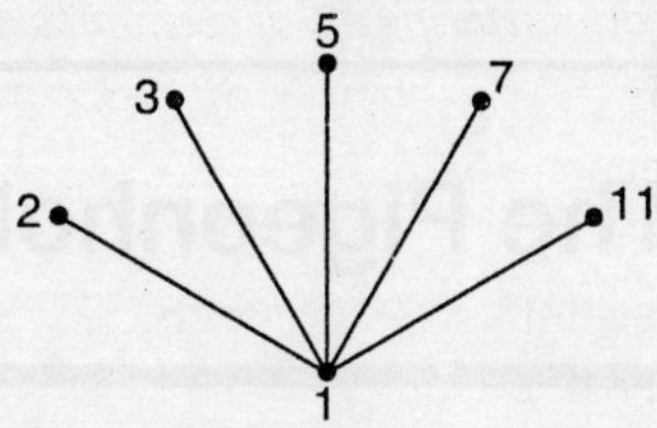

(*ii*) The Hasse diagram for divisibility on the set {1, 2, 3, 4, 5, 6, 7} is shown in figure.

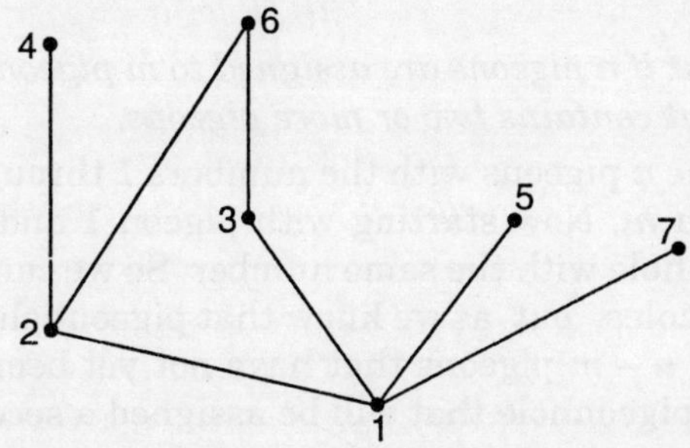

Problem 10. *Consider the set S = {a, b , c, d, e, f} be ordered as shown in figure. Determine the following :*

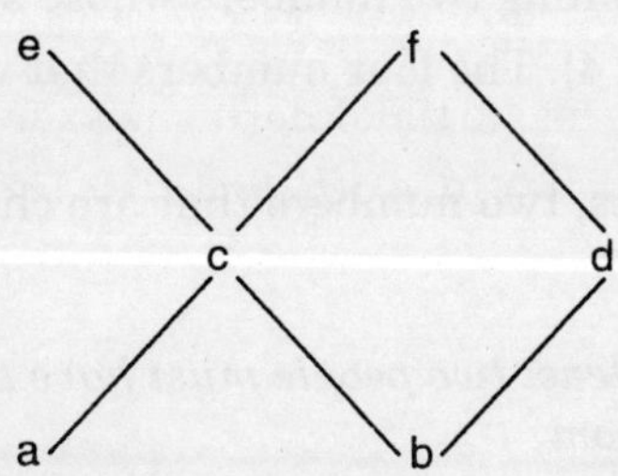

(a) All maximal and minimal elements of S.

(b) Greatest and least element.

(c) All linearly ordered subsets with three or more elements

Sol. (*a*) The maximal elements are *e* and *f*

The minimal elements are *a* and *b*

(*b*) It has neither greatest nor least element

(*c*) The linearly ordered subsets with three or more elements are

(*a*, *c*, *e*), (*a*, *c*, *f*), (*b*, *c*, *e*) (*b*, *c*, *f*), (*b*, *d*, *f*).

Chapter 4

The Pigeonhole Principle

THE PIGEONHOLE PRINCIPLE

Theorem I. *Show that if n pigeons are assigned to m pigeonholes and m < n, then there is at least one pigeonhole that contains two or more pigeons.*

Proof. Let us label the n pigeons with the numbers 1 through n and the m pigeonholes with the numbers 1 through m. Now starting with pigeon 1 and Pigeonhole 1, assign each pigeon in order to the pigeonhole with the same number. So we can assign as many pigeons as possible to distinct pigeonholes, but as we know that pigeonholes are less than pigeons *i.e.*, $m < n$. Thus, there remains $n - m$ pigeons that have not yet been assigned to a pigeonhole. Hence, there is at least one pigeonhole that will be assigned a second pigeon.

Example 1. *Show that if any four numbers from 1 to 6 are chosen, then two of them will add to 7.*

Sol. Make three sets containing two numbers whose sum is 7.

A = {1, 6}, B = {2, 5}, C = {3, 4}. The four numbers that will be chosen assigned to the set that contains it.

As there are only three sets, two numbers that are chosen is from the same set whose sum is 7.

Example 2. *Show that at least two people must have their birthday in the same month if 13 people are assembled in a room.*

Sol. We assigned each person the month of the year on which he was born. Since there are 12 months in a year.

So, according to pigeonhole principle, there must be at least two people assigned to the same month.

Example 3. *Show that if any eight + ve integers are chosen, two of them will have same remainder when divided by 7.*

Sol. Take any eight +ve integers. When these are divided by 7 each have some remainder. Since there are eight integers and only seven distinct remainders because number 7 can generate only 7 remainders, so two +ve integers must have same remainder.

EXTENDED PIGEONHOLE PRINCIPLE

It states that if n pigeons are assigned to m pigeonholes (The number of pigeons is very large than the number of pigeonholes), then one of the pigeonholes must contain at least $[(n - 1)/m] + 1$ pigeons.

Theorem II. *Prove the extended pigeonhole principle.*

Proof. We can prove this by the method of contradiction. Assume that each pigeonhole does not contain more than $[(n-1)/m]$ pigeons. Then, there will be at most $m[(n-1)/m] \le m(n-1)/m = n-1$ pigeons in all. This is in contradiction to our assumptions. Hence, for given m pigeonholes, one of these must contain at least $[(n-1)/m] + 1$ pigeons.

Example 4. *Show that if 9 colours are used to paint 100 houses, at least 12 houses will be of the same colour.*

Sol. Let us assume the colours be the pigeonholes and the houses the pigeons. Now 100 pigeons are to be assigned to 9 pigeonholes. Using the extended pigeonhole principle, $[(n-1)/m] + 1$, where $n = 100$ and $m = 9$, we have $[(100-1)/9] + 1 = 12$. Thus, there are 12 houses of the same colour.

EXERCISES

1. Ten persons have first names Ashish, Manish and Aditya and last names Goel, Gupta and Jindal. Show that at least two persons have the same first and last names.
2. Show that if we select 151 distinct computer science courses numbered between 1 and 300 inclusive, at least two are consecutively numbered.
3. An inventory consists of a list of 100 items, each marked "available" or "unavailable". There are 55 available items. Show that there are at least two available items in the list exactly nine items apart.
4. A worker is paid every other week on Friday. Show that in some month he is paid three times.
5. Show that among one Lakh people there are two who were born at exactly the same time (hour, minute, and second).
6. From the integers 1 to 200, 101 of them are chosen arbitrarily. Show that, among the chosen numbers, there exists two such that one divides another.
7. Show that among $n + 1$ arbitrarily chosen integers, there are two whose difference is divisible by n.
8. Show that one of any M consecutive integers is divisible by M.
9. Show that if five points are chosen in a square of each side of 1 inch length, then at least two of the points must be no more than 1/2 inches apart.
10. Show that if 30 dictionaries in a library contain a total of 61,327 pages , then one of the dictionaries must have at least 2045 pages. *(VTU, Karnataka, 2000)*

SOLVED PROBLEMS

Problem 1. *Seven members of a family have total Rs. 2886 in their pockets. Show that at least one of them must have at least Rs. 416 in his pocket.*

Sol. Let us assume the members be the pigeonholes and the Rupees the pigeons. Now 2886 pigeons are to be assigned to 7 pigeonholes. Using the extended pigeonhole principle, where $n = 2886$ and $m = 7$, we have $[(2886 - 1)/7] + 1 = 416$. Hence, there are 416 Rupees in one member's pocket.

Problem 2. *How many people must you have to guarantee that at least 9 of them will have birthdays in the same day of the week.*

Sol. Let us assume the days of week the pigeonholes and the people the pigeons. Now we have 7 pigeonholes and we have to find pigeons. Using the extended pigeonhole principle, we have

$$[(n-1)/7] + 1 = 9$$

$$[(n-1)/7] = 9 - 1 = 8$$

$$n - 1 = 8 \times 7$$

$$n = 56 + 1 = 57.$$

Thus, there must be 57 people to guarantee that at least 9 of them will have birthdays in the same day of the week.

Chapter 5

Probability Theory

The word 'probability' means the chance of occurring of a certain event. It is generally possible to predict the future of an event quantitatively with a certain probability of being correct. The probability is used in such cases where the outcome of trial is uncertain.

Before actually defining the probability, we will discuss some terms related with probability.

IMPORTANT TERMS RELATED WITH PROBABILITY

Trial. The performance of an experiment is called a trial.

Experiment. It is a well defined process that leads to a well defined outcome.

For example : Tossing a coin is an experiment. It may consist of one or more trials.

Sample Space. The set of all possible outcomes of an experiment is called sample space and is denoted by S.

e.g., when a die is thrown, sample space is $S = \{1, 2, 3, 4, 5, 6\}$.

It consists of six outcomes 1, 2, 3, 4, 5 and 6.

Event. The possible outcomes of a trial are called events. In other words, event is a subset of the sample space.

e.g., in the throw of a single die, getting a number less than four is an event. If this is denoted by L, then $L = \{1, 2, 3\}$ which is a subset of S.

Complement of Event. The set of all outcomes which are in sample space but not in event is called the complement of event.

e.g., in case of above example, the complement of event is $\overline{L} = \{4, 5, 6\}$.

Equally Likely Events. Events are said to be equally likely if one of them cannot be expected to occur in preference to others. In other words, it means each outcome is as likely to occur as any other outcome. *e.g.,* when a die is throw, all the six faces *i.e.,* 1, 2, 3, 4, 5 and 6 are equally likely to occur.

Mutually Exclusive or Disjoint Events. Events are called mutually exclusive if they cannot occur simultaneously.

e.g, suppose a card is drawn from a pack of cards, then the events getting a jack and getting a king are mutually exclusive because they cannot occur simultaneously.

Exhaustive Events. The total number of all possible outcomes of an experiment are called exhaustive events. *e.g.,* in the tossing of a coin, either head or tail may turn up. Therefore, there are two possible outcomes. Hence, there are two exhaustive events in tossing a coin.

Independent and Dependent Events. Events are said to be independent if occurrence of any one event does not depend on occurrence of any other event.

Events are said to be dependent if occurrence of one effect the occurrence of other events *e.g.,* the outcome of throwing a die is independent of outcome of drawing a king from pack of cards.

PROBABILITY DEFINITION

The probability of happening of an event A, denoted by P(A), is defined as

$$P(A) = \frac{\text{number of cases favourable to A}}{\text{number of possible outcomes}}.$$

Thus, if an event an happen in m ways and fails to happen in n ways and $m + n$ ways is equally likely to occur then the probability of happening of the event A is given by

$$P(A) = \frac{m}{m+n}$$

and the probability of non-happening of A is

$$P(\bar{A}) = \frac{n}{m+n}.$$

Points to be Noted

1. The probability of an event which is certain to occur is one.
2. The probability of an event which is impossible is zero.
3. If the probability of happening of an events P(A) and that of not happening is $P(\bar{A})$, then

$$P(A) + P(\bar{A}) = 1, 0 \leq P(A) \leq 1, 0 \leq P(\bar{A}) \leq 1.$$

ADDITION THEOREM

Theorem I. *If A and B are two mutually exclusive events, then*

$$P(A \cup B) = P(A) + P(B).$$

Proof. Let n = total number of exhaustive cases

n_1 = number of cases favourable to A.

n_2 = number of cases favourable to B.

Now, we have A and B two mutually exclusive events. Therefore, $n_1 + n_2$ is the number of cases favourable to A or B.

$$\therefore \quad P(A \cup B) = \frac{\text{favourable cases}}{\text{total number of exhaustive cases}} = \frac{n_1 + n_2}{n} = \frac{n_1}{n} + \frac{n_2}{n}$$

But we have, $P(A) = \frac{n_1}{n}$ and $P(B) = \frac{n_2}{n}$

Hence $P(A \cup B) = P(A) + P(B)$.

Example 1. *In a single throw of two dice find the probability of having 8 or 11.*

Sol. Number of favourable cases for throwing 8 are {(6, 2), (5, 3), (4, 4), (3, 5), (2, 6)} = 5

Number of total cases = 36

$$\therefore \quad \text{Probability of having, } 8 = \frac{5}{36}$$

Number of favourable cases for throwing, 11

$$\{(6, 5), (5, 6)\} = 2$$

$$\therefore \text{ Probability of throwing, } \quad 11 = \frac{2}{36}$$

$$\text{So, probability of throwing 8 or 11} = \frac{5}{36} + \frac{2}{36} = \frac{7}{36}.$$

Example 2. *A card is drawn at random from a well-shuffled pack of 52 cards. Find probability of getting*

(i) a jack, a queen and a king *(ii) a two of heart or two of diamond.*

Sol. (*i*) In a pack of 52 cards, we have 4 jacks, 4 queens and 4 kings.

$$\therefore \quad \text{P(a jack)} = \frac{{}^4C_1}{{}^{52}C_1} = \frac{4}{52} = \frac{1}{13}$$

$$\text{P(a queen)} = \frac{{}^4C_1}{{}^{52}C_1} = \frac{4}{52} = \frac{1}{13}$$

$$\text{P(a king)} = \frac{{}^4C_1}{{}^{52}C_1} = \frac{4}{52} = \frac{1}{13}$$

By addition theorem of probability

$$\text{P(a jack or a queen or a king)} = \frac{1}{13} + \frac{1}{13} + \frac{1}{13} = \frac{3}{13}$$

(*ii*) P(two of heart or two of diamonds) = P(two of heart) + P(two of diamonds)

$$= \frac{1}{52} + \frac{1}{52} = \frac{2}{52} = \frac{1}{26}.$$

Example 3. *Two dice are tossed once. Find the probability of getting an even number on first dice or a total of 8.*

Sol. An even number can be got on a die in 3 ways because any one of 2, 4, 6, can come. The other die can have any number. This can happen in 6 ways.

$$\therefore \text{ P(an even number on Ist die)} = \frac{3 \times 6}{36} = \frac{18}{36} = \frac{1}{2}$$

A total of 8 can be obtained in following cases {(2, 6), (3, 5), (4, 4), (5, 3), (6, 2)}

$$\therefore \quad \text{P(a total of 8)} = \frac{5}{36}$$

$$\therefore \quad \text{Total probability} = \frac{1}{2} + \frac{5}{36} = \frac{23}{36}.$$

Example 4. *Find the probability of getting a sum of 5 or 7 in toss of two dice.*

Sol. Chances for sum of 5 is {(1, 4), (2, 3), (4, 1), (3, 2)}

$\therefore \quad \text{P(sum of 5)} = \frac{4}{36} = \frac{1}{9}$

Chances for sum of 7 is {(1, 6), (6, 1), (2, 5), (5, 2), (3, 4), (4, 3)}

$\therefore \quad \text{P(sum of 7)} = \frac{6}{36} = \frac{1}{6}$

$\therefore$ Required probability $= \frac{1}{9} + \frac{1}{6} = \frac{5}{18}$.

Theorem II. *If A and B are two events that are not mutually exclusive, then*

$$P(A \cup B) = P(A) + P(B) - P(A \cap B).$$

Proof. Let n = total number of exhaustive cases

n_1 = number of cases favourable to A

n_2 = number of cases favourable to B

n_3 = number of cases favourable both A and B

But, A and B are not mutually exclusive. Therefore, A and B can occur simultaneously. So, $n_1 + n_2 - n_3$ is the number of cases favourable to A or B.

Therefore, $\quad P(A \cup B) = \frac{n_1 + n_2 - n_3}{n} = \frac{n_1}{n} + \frac{n_2}{n} - \frac{n_3}{n}$

But we have, $\quad P(A) = \frac{n_1}{n}$, $P(B) = \frac{n_2}{n}$ and $P(A \cap B) = \frac{n_3}{n}$

Hence, $\quad P(A \cup B) = P(A) + P(B) - P(A \cap B)$.

Example 5. *A card is drawn from a deck of cards. Find the probability of getting an ace or a spade card.*

Sol. Let A be event that a card drawn is a spade and B be event that a card drawn is an ace

$\therefore \quad P(A) = \frac{13}{52}$, $P(B) = \frac{4}{52}$ and $P(A \cap B) = \frac{1}{52}$

$\therefore \quad P(A \cup B) = P(A) + P(B) - P(A \cap B) = \frac{13}{52} + \frac{4}{52} - \frac{1}{52} = \frac{16}{52} = \frac{4}{13}$.

Example 6. *Two dice are tossed once. Find probability of getting an even number on first die or a total of 8.*

Sol. P(even number on Ist die or a total of 8) = P(even number on Ist die) + P(total of 8) P(even number on Ist die and total of 8)

Now, P(even number on Ist die) $= \frac{18}{36}$

Ordered pairs showing a total of 8 = {(6, 2), (5, 3), (4, 4), (3, 5), (2, 6)} = 5

$\therefore$ Probability ; P(total of 8) $= \frac{5}{36}$

$$P(\text{even number of Ist die and total of 8}) = \frac{3}{36}$$

$$\therefore \quad \text{Required probability} = \frac{18}{36} + \frac{5}{36} - \frac{3}{36} = \frac{20}{36} = \frac{5}{9}.$$

Example 7. *A dice is thrown twice. What is the probability that at least one of two throws comes up with number 3.*

Sol. Sample space contains 36 events.

Let 'A' be the event "first throw comes up with number 3" and B be the event "second throw comes up with number 3".

∴ Favourable events of

$$A = \{(3, 1), (3, 2), (3, 3), (3, 4), (3, 5), (3, 6)\} = 6$$

$$\therefore \quad P(A) = \frac{6}{36} = \frac{1}{6}$$

Favourable events of $B = \{(1, 3), (2, 3), (3, 3), (4, 3), (5, 3), (6, 3)\} = 6$

$$\therefore \quad P(B) = \frac{6}{36} = \frac{1}{6}$$

Also $\quad P(A \cap B) = \dfrac{1}{36} \quad$ [∵ Only one event will be (3, 3)]

$$\therefore \text{Required probability} = \frac{1}{6} + \frac{1}{6} - \frac{1}{36} = \frac{11}{36}.$$

Example 8. *What is the probability of getting a number greater than 2 or an even number in a single throw of a fair die.*

Sol. Let E_1 be event of getting a number greater than 2 or E_2 be the event of getting an even number.

$$\therefore \quad P(E_1) = \frac{4}{6} = \frac{2}{3}, P(E_2) = \frac{3}{6} = \frac{1}{2} \quad \text{and} \quad P(E_1 \cap E_2) = \frac{1}{3}.$$

$$\therefore \quad P(E_1 \cup E_2) = P(E_1) + P(E_2) - P(E_1 \cap E_2) = \frac{2}{3} + \frac{1}{2} - \frac{1}{3} = \frac{5}{6}.$$

Example 9. *A coin is tossed n times. What is probability that the head will present itself an odd number of times ?*

Sol. Each time tossing of coin may result in 2 ways. Hence, n throws may result in 2^n ways. The event will happen if among the result of n throws, head occupies either 1 place, 3 place, 5 place, ... etc. For these, the number of ways are ${}^nC_1, {}^nC_3, {}^nC_5, \ldots$.

Favourable number of ways $\quad = {}^nC_1 + {}^nC_3 + {}^nC_5 + \ldots = 2^{n-1}$.

$$P(\text{head occurs odd number of times}) = \frac{2^{n-1}}{2^n} = \frac{1}{2}.$$

Example 10. *A card is drawn from a deck of cards. Find probability of getting a king or a heart or a red card.*

Sol. Let A be event that a card drawn is a king

Let B be event that a card drawn is a heart

Let C be event that a card drawn is a red card

$$P(A) = \frac{4}{52}, \quad P(B) = \frac{13}{52}, \quad P(C) = \frac{26}{52}$$

$$P(A \cap B) = \frac{1}{52}, \quad P(B \cap C) = \frac{13}{52}$$

$$P(A \cap C) = \frac{2}{52}, \quad P(A \cap B \cap C) = \frac{1}{52}$$

$$P(A \cup B \cup C) = P(A) + P(B) + P(C) - P(A \cap B) - P(B \cap C) - P(A \cap C) + P(A \cap B \cap C)$$

$$= \frac{4}{52} + \frac{13}{52} + \frac{26}{52} - \frac{1}{52} - \frac{13}{52} - \frac{2}{52} + \frac{1}{52} = \frac{28}{52} = \frac{7}{13}.$$

Example 11. *From 30 tickets with the first 30 natural number one is drawn at random. Find chance that is a multiple of 3 or 7.*

Sol. $P(\text{a multiple of } 3) = \frac{10}{30}$; $P(\text{a multiple of } 7) = \frac{4}{30}$

$$P(\text{multiple of both}) = \frac{1}{30}$$

$$\therefore \quad \text{Required probability} = \frac{10}{30} + \frac{4}{30} - \frac{1}{30} = \frac{13}{30}.$$

Example 12. *Two events A and B have probability 0.25 and 0.50. The probability that both A and B occur simultaneously is 0.12. Find the probability that neither A nor B occurs.*

Sol.

$$P(A \cup B) = P(A) + P(B) - P(A \cap B) = 0.25 + 0.50 - 0.12 = 0.63$$

$$P(\overline{A \cup B}) = 1 - P(A \cup B) = 1 - 0.63 = 0.37.$$

EXERCISES

1. If three coins are tossed, represent the sample space and the event of getting
   (*i*) two heads and one tail (*ii*) at least two tails.
2. Three unbiased coins are tossed. Find the probability of getting
   (*i*) all heads (*ii*) at least two heads
   (*iii*) at most two heads.
3. The probability of a player A winning a race is 1/5 and the probability of a player B winning the race is 1/4. What is the probabilty that
   (*i*) either of them will win ? (*ii*) none of them will win ?
4. The probability that a contractor will get a road contract is 2/3 and the probability that he will not get a building contract is 5/9. If the probability of getting at least one contract is 4/5, what is the probability that he will get both the contracts ?
5. A box contains 50 nails and 150 bolts. Half of the bolts and half of the nails are rusted. If one item is chosen at random, what is the probability that it is rusted or a nail.
6. A bag contains 4 blue and 2 green balls. If a ball is taken out of this bag at random, represent the sample space and the event of this ball being blue.

7. One card is drawn from a set of 17 cards numbered 1 to 17. Find the probability that

(*i*) the number drawn is divisible by 3 or 7.

(*ii*) the number drawn is divisible by 3 or 17.

8. A card is drawn at random from the pack of 52 cards. Find the probability of drawing

(*i*) a black king (*ii*) a red card

(*iii*) a diamond or a heart (*iv*) either a red card or a king

(*v*) a card which is neither a diamond nor a king.

9. Two events A and B have probabilities 0.35 and 0.55 respectively. The probability that both A and B occur simultaneously is 0.14. Find the probability that neither A nor B occurs.

10. Two dice are thrown simultaneously. What is the probability that the sum of the numbers on the two faces is divisible by 3 or 4 ?

MULTIPLICATION THEOREM

Theorem III. *If A and B are two independent events, then the probability that both will occur is equal to product of their individual probabilities i.e., P(A ∩ B) = P(A) × P(B).*

Proof. Let event

A can happen is n_1 ways of which p are successful

B can happen in n_2 ways of which q are successful

Now, combine successful event of A with successful event of B.

Thus, total number of successful cases $= p \times q$

We have, total number of cases $= n_1 \times n_2$.

Therefore, from definition of probability

$$P(\text{A and B}) = P(A \cap B) = \frac{p \times q}{n_1 \times n_2} = \frac{p}{n_1} \times \frac{q}{n_2}.$$

We have $$P(A) = \frac{p}{n_1},\ P(B) = \frac{q}{n_2}$$

So, $$P(A \cap B) = P(A) \times P(B)$$

If, there are three independent events A, B and C, then

$$P(A \cap B \cap C) = P((A \cap B) \cap C) = P(A \cap B) \times P(C)$$
$$= P(A) \times P(B) \times P(C).$$

In general, if there are n independent events, then

$$P\left(\bigcap_{i=1}^{n} A_i\right) = \prod_{i=1}^{n} P(A_i).$$

Example 13. *Two cards are drawn from a pack of cards in succession (with replacement). Find the probability that both are jacks.*

Sol. Probability that Ist card is a jack $= \dfrac{4}{52}$

Probability that IInd card is a jack $= \dfrac{4}{52}$

Probability that both cards are jacks $= \dfrac{4}{52} \times \dfrac{4}{52} = \dfrac{1}{169}$.

Example 14. *A die is tossed twice. Find probability of having a number greater than 3 on each toss.*

Sol. Probability of having number greater than 3 in Ist toss = $\frac{3}{6}$

Probability of having number greater than 3 in IInd toss = $\frac{3}{6}$.

Probability of combined event = $\frac{3}{6} \times \frac{3}{6} = \frac{1}{4}$.

Example 15. *A bag contains 5 white and 3 black balls and four balls are successively drawn out and not replaced. What is probability that they are alternately of different colours.*

Sol. The event is W, B, W, B and B, W, B, W, where W is for white and B for black ball.
Probability of having W, B, W, B in that order

$$= \frac{5}{8} \times \frac{3}{7} \times \frac{4}{6} \times \frac{2}{5} = \frac{1}{14} \qquad \text{(Multiplication theorem)}$$

Probability of finding B, W, B, W in that order

$$= \frac{3}{8} \times \frac{5}{7} \times \frac{2}{6} \times \frac{4}{5} = \frac{1}{14}$$

∴ Probability of having balls of different colours alternately

$$= \frac{1}{14} + \frac{1}{14} = \frac{2}{14} = \frac{1}{7}. \qquad \text{(Addition theorem)}$$

Example 16. *From a bag containing 4 black and 5 white balls, a man draws four balls at random. Find the probability that there are 2 black and 2 white balls.*

Sol. Out of four black balls, 2 black balls can be drawn in $^4C_2$ways.

Out of 5 white balls, 2 white balls can be drawn in $^5C_2$ ways.

∴ Number of favourable events = $^4C_2 \times {}^5C_2$

Out of 9 balls, 4 balls can be drawn in $^9C_4$ ways

∴ Required probability = $\frac{^4C_2 \times {}^5C_2}{^9C_4} = \frac{4 \times 3}{2 \times 1} \times \frac{5 \times 4}{2 \times 1} \times \frac{4 \times 3 \times 2 \times 1}{9 \times 8 \times 7 \times 6} = \frac{10}{21}$.

Example 17. *A bag contains 5 green and 7 red balls. Two balls are drawn. Find probability that one is green and other is red.*

Sol. P(A) = P(a green ball) = $\frac{1}{5}$; P(B) = P(a red ball) = $\frac{1}{7}$

By multiplication theorem

$$P(A \text{ and } B) = P(A) \times P(B) = \frac{1}{5} \times \frac{1}{7} = \frac{1}{35}.$$

Example 18. *Two dices are thrown. Find probability of getting an odd number on one and a multiple of three on other.*

Sol. P(A) = P(an odd number) = $\frac{3}{6} = \frac{1}{2}$ [∵ 1, 3, 5 will be odd number]

$$P(B) = P(\text{a multiple of 3}) = \frac{2}{6} = \frac{1}{3} \qquad [\because \text{ 3 and 6 will be multiple of 3}]$$

$$\therefore \text{ Required probability} = \frac{1}{2} \times \frac{1}{3} = \frac{1}{6}.$$

Example 19. *A card is drawn from a well shuffled deck of 52 cards. The outcome is noted, the card is replaced and the deck reshuffled. Another card is then drawn from the deck. Find the probability of getting first card a jack and the second card a black queen.*

Sol. $$P(\text{a jack}) = \frac{4}{52} = \frac{1}{13}$$

$$P(\text{a black queen}) = \frac{2}{52} = \frac{1}{26}$$

P(first card is jack and second card is a black queen) = P(a jack) × P(a black queen)

$$= \frac{1}{13} \times \frac{1}{26} = \frac{1}{318}.$$

Example 20. *From a pack of cards, two cards are drawn at random one by one with replacement. Find the probability that first is heart and second is king.*

Sol. Probability of getting heart $= \frac{13}{52}$

Probability of getting king $= \frac{4}{52}$

Probability that Ist is heart and IInd is king $= \frac{13}{52} \times \frac{4}{52} = \frac{1}{52}.$

Example 21. *An urn contains 25 balls numbered through 1 to 25. Two balls are drawn from urn with replacement. Find probability of getting one odd and one even number.*

Sol. Let E and O represent events of drawing even and odd numbered balls. There are 12 balls having even numbers and 13 balls having odd number.

$$\therefore \qquad P(E) = \frac{12}{25},\ P(O) = \frac{13}{25}.$$

The total number of such cases consists of (first odd and second even) + (first even and second odd).

$$\therefore \text{ P(one odd and one even)} = P(O) \times P(E) + P(E) \times P(O)$$

$$= \frac{13}{25} \times \frac{12}{25} + \frac{12}{25} \times \frac{13}{25} = \frac{312}{625}.$$

EXERCISES

1. A box contains 7 red and 4 white balls. Two balls are drawn at random. Find the probability that
   (*i*) both the balls are red
   (*ii*) one ball is white and another one is red.

2. Three electric bulbs are chosen at random from 15 bulbs of which 5 are defective. Find the probability that
   (*i*) none is defective (*ii*) exactly one is defective
   (*iii*) at least one is defective.
3. A committe of 5 is to be selected from a group of 6 men and 8 women. If the selection is made randomly, find the probability that there are 3 women and 2 men.
4. A bag contains 8 red and 5 white balls. Two balls are drawn at random. Find the probability that
   (*a*) both the balls are white (*b*) both the balls are red
   (*c*) one ball is white and another one is red.
5. From a pack of 52 cards, 2 are drawn at random. Find the probability that one is king and other a queen.
6. Out of 10 outstanding students in a college, there are 4 boys and 6 girls. A team of 4 students is to be selected. Find the probability that two are boys and two are girls.
7. Four cards are drawn from a pack of 52 playing cards. Find the probability of getting :
   (*i*) all the four card of the same suit
   (*ii*) all the four card of the same number
   (*iii*) one card from each suit.
8. If from a lottery of 30 tickets, marked 1 to 30, four tickets be drawn, what is the probability that those marked 4 and 5 are among them ?
9. In shuffling a pack of cards, four cards are accidentaly dropped, find the chance that the dropped cards should be one from each suit ?

CONDITIONAL PROBABILITY

Theorem IV. *If A and B are two dependent events then the probability of occurrence of A given that B has already occurred and is denoted by P(A/B) is given by*

$$P\left(\frac{A}{B}\right) = \frac{P(A \cap B)}{P(B)} \; ; \; P(A \cap B) = P(B) \times P\left(\frac{A}{B}\right).$$

Similarly, the probability of occurrence of B given that A has already occurred is given by

$$P\left(\frac{B}{A}\right) = \frac{P(A \cap B)}{P(A)} \; ; \; P(A \cap B) = P(A) \times P\left(\frac{B}{A}\right).$$

Proof. Let S be the sample space.

Then, we have

$$P(A \cap B) = \frac{P(A \cap B)}{P(S)} = \frac{P(A \cap B)}{P(A)} \times \frac{P(A)}{P(S)}$$

$$= \frac{P(A)}{P(S)} \times \frac{P(A \cap B)}{P(A)} \qquad \because \frac{P(A \cap B)}{P(A)} = P\left(\frac{B}{A}\right)$$

$$= P(A) \times P\left(\frac{B}{A}\right) \qquad ...(i)$$

Interchange A and B in eq. (*i*), we get

$$= P(B) \times P\left(\frac{A}{B}\right).$$

Example 22. *Find the probability of drawing a heart on each of two consecutive draws from a well shuffled pack of cards if the card is not replaced after the draw.*

Sol. Let event A is heart on the first draw and event B is heart on the second draw.

Then, $$P(A \cap B) = P(A) \times P\left(\frac{B}{A}\right)$$

Now, $$P(A) = \frac{13}{52}$$

When we get a heart on the first draw, the second draw has 51 outcomes and 12 are favourable.

$$\therefore \quad P\left(\frac{B}{A}\right) = \frac{12}{51} = \frac{4}{17}$$

$$\therefore \quad P(A \cap B) = \frac{13}{52} \times \frac{4}{17} = \frac{1}{17}.$$

Example 23. *From a deck of cards, three cards are drawn one by one without replacement. Find the probability that each time it is a card of spade.*

Sol. Let event A = first card is spade

Event B = second card is spade

Event C = third card is spade

$$P(A \cap B \cap C) = P(A) \times P\left(\frac{B}{A}\right) \times P\left(\frac{C}{B \cap A}\right)$$

$$= \frac{13}{52} \times \frac{12}{51} \times \frac{11}{50} = \frac{11}{850}.$$

Example 24. *A bag contains 5 white, 7 red and 8 black balls. If four balls are drawn one by one without replacement, what is the probability that all are white.*

Sol. Let events A, B, C and D represent Ist, IInd, IIIrd and IVth white ball drawn.

$$P(A) = \frac{5}{20},\ P\left(\frac{B}{A}\right) = \frac{4}{19}$$

$$P\left(\frac{C}{A \cap B}\right) = \frac{3}{18} \quad \text{and} \quad P\left(\frac{D}{A \cap B \cap C}\right) = \frac{2}{17}$$

$$\therefore \quad P(A \cap B \cap C \cap D) = P(A) \times P\left(\frac{B}{A}\right) \times P\left(\frac{C}{A \cap B}\right) \times P\left(\frac{D}{A \cap B \cap C}\right)$$

$$= \frac{5}{20} \times \frac{4}{19} \times \frac{3}{18} \times \frac{2}{17} = \frac{1}{969}.$$

Example 25. *Two cards are drawn from a pack of 52 cards, one after the other without replacement. Find the chance that one of these cards is an ace and other is queen of opposite shade.*

Sol. There are two possibilities

(*i*) The first card drawn is an ace and second is a queen of opposite shade. Probability of this event

$$= \frac{4}{52} \times \frac{2}{51} = \frac{2}{663}.$$

(*ii*) The first card drawn is a queen and second is an ace of opposite colour. The probability of this event

$$= \frac{4}{52} \times \frac{2}{51} = \frac{2}{663}$$

Hence, required probability $= \frac{2}{663} + \frac{2}{663} = \frac{4}{663}$.

Example 26. *A bag contains 5 white and 3 red balls. Balls are drawn in succession and are not replaced. Show that the chance that Ist red ball will appear on fifth draw is* $\frac{3}{56}$.

Sol. Let W_i = event that white ball is drawn at ith draw

R_i = event that red ball is drawn at ith draw. We want that first ball should appear on fifth draw. This will happen in event W_1, W_2, W_3, W_4, R_5.

$$\therefore \quad P(W_1 \cap W_2 \cap W_3 \cap W_4 \cap R_5) = P(W_1) \times P\left(\frac{W_2}{W_1}\right) \times P\left(\frac{W_3}{W_2 \cap W_1}\right) \times P\left(\frac{W_4}{W_3 \cap W_2 \cap W_1}\right)$$

$$\times P\left(\frac{R_5}{W_4 \cap W_3 \cap W_2 \cap W_1}\right) = \frac{5}{8} \times \frac{4}{7} \times \frac{3}{6} \times \frac{2}{5} \times \frac{3}{4} = \frac{3}{56}.$$

Example 27. *There are two groups of subjects, one of which consists of 5 science and 3 engineering subjects and the other consists of 3 science and 5 engg. subjects. An unbiased die is cast. If the number 3 or number 5 turns up, a subject is selected at random from the first group. Otherwise the subject is selected at random from the second group. Find the probability that an engg. subject is selected ultimately.*

Sol. Let E_1 = event that subject is selected from the first group.

E_2 = Event that subject is selected from the second group.

Now, the probability of turning up the number 3 or 5 in the throw of an unbiased die is $\frac{2}{6} = \frac{1}{3}$.

$$\therefore \quad P(E_1) = \frac{1}{3} \quad \text{and} \quad P(E_2) = \frac{2}{3}$$

Let E is event that an engg. subject is selected.

$$\therefore \quad P\left(\frac{E}{E_1}\right) = \frac{3}{8} \quad \text{and} \quad P\left(\frac{E}{E_2}\right) = \frac{5}{8}$$

$$\therefore \quad P(E) = P(E_1) \times P\left(\frac{E}{E_1}\right) + P(E_2) \times P\left(\frac{E}{E_2}\right) = \frac{1}{3} \times \frac{3}{8} + \frac{2}{3} \times \frac{5}{8} = \frac{13}{24}.$$

Theorem V. *If $A_1, A_2, A_3, ..., A_n$ are n independent events and the probabilities of happening of these events is $P_1, P_2, P_3, ..., P_n$; then the probability that at least one of these events must happen is*

$$1 - (1 - p_1)(1 - p_2)(1 - p_3) \ldots (1 - p_n).$$

Proof. The probability that at least one of $A_1, A_2, A_3, ..., A_n$ will occur is

P(at least one of $A_1, A_2, A_3, ..., A_n$ will occur)

$$= 1 - P(\text{none of } A_1, A_2, A_3, ..., A_n \text{ will occur}) \qquad (\because \; P(A) + P(\bar{A}) = 1)$$

$$= 1 - P[A_1 \text{ will not occur}, A_2 \text{ will not occur}, ..., A_n \text{ will not occur}]$$

$$= 1 - P[\bar{A}_1 \cap \bar{A}_2 \cap \bar{A}_3 \cap ...... \cap \bar{A}_n]$$

$$= 1 - P(\bar{A}_1) \times P(\bar{A}_2) \times P(\bar{A}_3) \times ...... \times P(\bar{A}_n)$$

[$\because$ $A_1, A_2, A_3, ..., A_n$ are independent events]

$$= 1 - (1 - p_1)(1 - p_2) ... (1 - p_n)$$

$\because$ $P(\bar{A}_i) = 1 - P(A_i) = 1 - p$ for 1, 2, 3, ..., n.

Example 28. *Shanu speaks truth in 60% cases and Nanu in 70% cases. In what %age of cases are they likely to contradict each other in stating same fact.*

Sol. They will contradict each other if one of them speaks the truth and other tells a lie.
Probability that Shanu speaks the truth and Nanu tells a lie

$$= \frac{60}{100} \times \left(1 - \frac{70}{100}\right) = \frac{60}{100} \times \frac{30}{100} = \frac{9}{50}$$

Probability that Shanu tells a lie and Nanu speaks the truth

$$= \left(1 - \frac{60}{100}\right) \times \frac{70}{100} = \frac{40}{100} \times \frac{70}{100} = \frac{7}{25}$$

The percentage of cases in which they contradict each other

$$= \left(\frac{9}{50} + \frac{7}{25}\right) \times 100 = \frac{23}{50} \times 100 = 46\%.$$

Example 29. *The probability of A, B, C solving a problem are 1/2, 1/3, 1/4. Find the probability that the problem will be solved.*

Sol. We shall find the probability that the problem is not solved.
Probability that A, B, C do not solve the problem are

$$1 - \frac{1}{2}, 1 - \frac{1}{3}, 1 - \frac{1}{4} \quad \text{or} \quad \frac{1}{2}, \frac{2}{3}, \frac{3}{4}$$

By the law of multiplication

Probability that the problem is not solved $= \frac{1}{2} \times \frac{2}{3} \times \frac{3}{4} = \frac{1}{4}$

Probability that the problem is solved $= 1 - \frac{1}{4} = \frac{3}{4}$.

Example 30. *Suppose it is 11 to 6 against a person A who is now 38 years of age living till he is 73 and 6 to 3 against B now 43 living till he is 78. Find the chance that at least one of them will be alive 35 years hence.*

Sol. The chance that A will die within 35 years $= \frac{11}{17}$.

The chance that B will die within 35 years $= \frac{6}{9}$.

As the events are independent

Probability that both will die within 35 years $= \frac{11}{17} \times \frac{6}{9} = \frac{66}{153}$

$\therefore$ The chances that at least one of them will be alive $= 1 - \dfrac{66}{153} = \dfrac{87}{153}$.

Example 31. *Certain questions are given to the three independent students A, B and C to solve. A solves 25%, B solves 50%, C solves 40% of question given to them. What is probability that the question will be solved ?*

Sol. The probability that the question will not be solved = Probability that A, B and C all three will not solve it

$$= \left(1-\frac{25}{100}\right)\left(1-\frac{50}{100}\right)\left(1-\frac{40}{100}\right) = \frac{75}{100} \times \frac{50}{100} \times \frac{60}{100} = \frac{9}{40}$$

Probability that question will be solved $= 1 - \dfrac{9}{40} = \dfrac{31}{40}$.

Example 32. *An anti-aircraft gun can take a maximum of four shots at an enemy plane moving away from it. The probability of hitting the plane at first, second, third and fourth shots are 0.4, 0.3, 0.2 and 0.1. What is the probability that the gun hits the plane ?*

Sol. The probability of hitting the plane at Ist, IInd, IIIrd and IVth shots are 0.4, 0.3, 0.2 and 0.1.

Probability of not hitting the plane at Ist, IInd, IIIrd and IVth shots are (1 – 0.4), (1 – 0.3), (1 – 0.2), (1 – 0.1) *i.e.*, 0.6, 0.7, 0.8 and 0.9 respectively.

The probability of not hitting the plane at all *i.e.*, probability that none of four shots hits the plane

$$= 0.6 \times 0.7 \times 0.8 \times 0.9 = 0.3024$$

Hence, probability of hitting the plane

$$= 1 - \text{probability of not hitting the plane} = 1 - 0.3024 = 0.6976.$$

EXERCISES

1. What is the probability that a randomly generated bit string of length four contains at least two consecutive 0's, given that the first bit is a 1 ?
2. What is the probability that exactly four heads appear when a fair coin is flipped five times, given that the first flip came up heads ?
3. What is the probability that a family with two children has two boys, given that they have at least one boy ?
4. A problem in math is given to three students whose chances of solving it are 1/2, 1/4, 1/5 respectively. What is the probability that only one of them solves it correctly.
5. Ashu and Monu appeared for an interview for two vacancies. The probability of Ashu's selection is 1/3 and that of Monu's selection is 1/5. Find the probability that only one of them will be selected.
6. A problem in physics is given to three students A, B, C whose chances of solving it are 1/2, 1/3, 1/4 respectively. What is the probability

   (*i*) the problem will be solved (*ii*) the problem will not be solved.
7. In a box there are 1 black and 2 white balls, in another there are 2 black and 1 white ball. A ball is drawn from the first and put into the second and then a second ball is drawn from the second box. Find the probability that it is white.

8. A and B throw a coin alternatively till one of them gets a head and wins the game. If A starts the game, find their respective probabilities of winning.

9. In a family husband tells lies in 30% cases and wife in 35% cases, find the probability that both contradict each other on the same fact.

10. Three dice were rolled. Given that no two faces were the same, what is the probability that there was an ace ?

Theorem VI (Total Probability For Compound Events). *If $A_1, A_2, \ldots\ldots, A_n$ be subsets of a sample space S such that*

(i) Each A_i is a proper subset of S i.e., $A_i \subset S$, $i = 1, 2, \ldots\ldots, n$; $A_i \neq S$

(ii) $A_1 \cup A_2 \cup A_3 \cup \ldots\ldots \cup A_n = S$

(iii) The subsets A_i are pairwise disjoint, i.e.,

$$A_i \cap A_j = \phi \quad ; i, j = 1, 2, 3, \ldots\ldots, n. \quad i \neq j$$

If B is any event, then

$$P(B) = \sum_{i=1}^{n} P(A_i)\, P(B/A_i) \text{ provided } P(A_i) \neq 0, \quad i = 1, 2, 3, \ldots\ldots, n.$$

Proof. We have given that

$$S = \bigcup_{i=1}^{n} A_i \qquad \left\{ A_1 \cup A_2 \cup A_3 \ldots\ldots \cup A_n = \bigcup_{i=1}^{n} A_i \right\}$$

Now $$B = B \cap S = B \cap \left[\bigcup_{i=1}^{n} A_i \right] = \bigcup_{i=1}^{n} (B \cap A_i) \quad \text{(By law of distributive)}$$

Also, $A_i \cap A_j = \phi$ for $i \neq j$.

Thus $(B \cap A_i) \cap (B \cap A_j) = \phi$ for $i \neq j$

$$\therefore \quad P(B) = P\left[\bigcup_{i=1}^{n} (B \cap A_i) \right] = \sum_{i=1}^{n} P(B \cap A_i) \quad \text{(By additive law)}$$

$$= \sum_{i=1}^{n} P(A_i)\, P(B/A_i) \quad \text{(by product law)}$$

Theorem VII (Baye's Theorem). *If $A_1, A_2, A_3, \ldots\ldots, A_n$ are mutually exclusive and exhaustive events in a sample space S and B is another event in S then*

$$P(A_k/B) = \frac{P(A_k)\, P(B/A_k)}{P(A_1)\, P(B/A_1) + P(A_2)\, P(B/A_2) + \ldots\ldots + P(A_n)\, P(B/A_n)}$$

$$= \frac{P(A_k)\, P(B/A_k)}{\sum_{i=1}^{n} P(A_i)\, P(B/A_i)}$$

Proof. From the theorem of total probability for compound events, we have

$$P(B) = \sum_{i=1}^{n} P(A_i)\, P(B/A_i) \qquad \ldots(i)$$

Also, $P(B \cap A_k) = P(A_k)\, P(B/A_k)$ ...(*ii*)

Hence, $$P(A_k/B) = \frac{P(B \cap A_k)}{P(B)}$$

$$= \frac{P(A_k)\, P(B/A_k)}{\sum_{i=1}^{n} P(A_i)\, P(B/A_k)} \quad \text{(From (i) and (ii))}$$

The probabilities $P(A_1)$, $P(A_2)$, ..., $P(A_n)$ are termed as priori probabilities and there are known before conducting an experiment. The conditional probabilities $P(A_1/B)$, $P(A_2/B)$, ..., $P(A_n/B)$ are termed as posteriori probabilities and are computed after conducting the experiment.

Example 33. *Three machines A, B and C produce respectively 70%, 20% and 10% of the total number of items of a factory. The defective output of these machines are 3%, 4% and 5% respectively. An item is selected at random from the product and is found defective. Find the probability that the item was produced by machine C.*

Sol. Let A, B, C be the events that item drawn at random is produced by machines A, B, C respectively and let X be the event that the product drawn is defective.

Then $P(A_1)$ = Probability that the product drawn was manufactured by A

$$= \frac{70}{100} = .70$$

Similarly $$P(B_1) = \frac{20}{100} = .20 \quad \text{and} \quad P(C_1) = \frac{10}{100} = .10.$$

Also, $P(X/A_1)$ = Probability of drawing a defective item from the product manufactured by A

$$= \frac{3}{100} = .03$$

Similarly $$P(X/B_1) = \frac{4}{100} = .04 \quad \text{and} \quad P(X/C_1) = \frac{5}{100} = .05$$

We want to know $P(C_1/X)$ = Probability that an item is produced by machine C given that the item is defective.

Using Baye's theorem,

$$P(C_1/X) = \frac{P(C_1)\, P(X/C_1)}{P(A_1)\, P(X/A_1) + P(B_1)\, P(X/B_1) + P(C_1) P(X/C_1)}$$

$$= \frac{(0.10)\,(0.05)}{(0.70)\,(0.03) + (0.20)\,(0.04) + (0.10)\,(0.05)} = \frac{.0050}{.0340} = 0.1470588.$$

Example 34. *Suppose there are three boxes containing 3 red and 3 green balls ; 4 red and 2 green balls and 2 red and 4 green balls respectively. There is equal probability of each box being chosen. One ball is drawn from the box chosen at random. What is the probability that a red ball is drawn ?*

Sol. Let A_k be the event that *k*th box is chosen, $k = 1, 2, 3$ and $P(A_k) = 1/3$ and B the event that a red ball is drawn

Then $P(B) = P(A_1)\, P(B/A_1) + P(A_2)\, P(B/A_2) + P(A_3)\, P(B/A_3)$

We have $P(A_1) = P(A_2) = P(A_3) = 1/3$

Also $$P(B/A_1) = \frac{3}{6},\ P(B/A_2) = \frac{3}{6},\ P(B/A_3) = \frac{2}{6}$$

Hence, $$P(B) = \frac{1}{3}\left(\frac{3}{6}\right) + \frac{1}{3}\left(\frac{3}{6}\right) + \frac{1}{3}\left(\frac{2}{6}\right)$$

$$= \frac{1}{3}\left(\frac{3}{6} + \frac{3}{6} + \frac{2}{6}\right) = \frac{4}{9}.$$

RANDOM VARIABLE

A variable whose numerical value is determined by the outcome of a random experiment is called a random variable. *e.g.*,

If we toss a coin and denote the head by 1 and tail by 0, then the random variable X takes only two values 1 and 0 as shown below.

Sample Point	H	T
X	1	0

A random variable X is a real valued function, X(*a*), of the elements of the sample space S where *a* is an element of the sample space.

Note. The range of the random variable will be a set of real numbers.

Example : If a tossing of coin is performed twice and X represents the number of tails which can come up, then the X takes only the three values 2, 1, 0 corresponding to two tails one tails or no tail. It's represented symbolically as

$$X(a) = \{a : a = (2, 1, 0) \in S\}$$

Random variables are of two types

1. Discrete Random Variable
2. Continuous Random Variable.

Discrete Random Variable

A random variable X, which can take only a finite number of values in an interval of the domain, is called discrete random variable. *e.g.*,

Three dices are thrown at a time and the sum of numbers is noted which turn up on the upper face, the discrete random variable in this case is

$X(a) = \{ a : a = (a_1 + a_2 + a_3) = 3, 4, 5, ..., 18\}$

Here a_1, a_2 and a_3 are the numbers on the upper face of first, second and third die respectively.

Continuous Random Variable

A random variable X, which can take any value in the domain, or when its range is an interval or the union of intervals on the real line, is called a continuous random variable. *e.g.*,

(*i*) The maximum strength of a class is 60 students. Thus the continuous random variable,

$$x(a) = \{a : 0 \le a \le 60\}.$$

(*ii*) The life of the bridge is between 5 to 10 years. Thus the continuous random variable,

$$X(a) = \{a : 5 \le a \le 10\}.$$

Discrete Probability Distribution

The probability distribution of a random variable is the set of its possible values together with their respective probabilities. For a discrete random variable X, the possible values can be written as a list ($a_1, a_2, a_3, a_4, ...$) and corresponding probabilities $P(a_1)$, $P(a_2)$, ...

The occurrence of the value a_i with their probabilities is called the discrete probability distribution of X.

Discrete Probability Function. Generally, the probability that the random variable X assumes a specified value (a) that can be expressed in terms of general mathematical expression, as

$$f(a) = \text{Probability that X assumes the value } a$$
$$= P(a) \qquad \{x = a\}$$

The function $f(a)$ is known as probability function or probability mass function of the discrete random variable X.

Expectation. Suppose a discrete random variable X assumes the values $a_1, a_2, a_3, ... a_n$ with probabilities $p_1, p_2, p_3, ..., p_n$ respectively. Then the expectation or expected value of X, *i.e.* E(X), is defined as the sum of products of different values of a and the corresponding probabilities

It can be respresented symbolically as,

$$E(X) = \Sigma p_i a_i$$

Similarly,
$$E(X^2) = \Sigma p_i a_i^2$$
$$E(X^3) = \Sigma p_i a_i^3$$

In general, the expected value of any function b(X) of a random variable X is defined as

$$E[h(X)] = \Sigma p_i \, h(x_i)$$

SOME IMPORTANT RESULTS ON EXPECTATION

If X and Y are two variables defined over the sample space and also E(X) and E(Y) exist, then

(*i*) $E(X + Y) = E(X) + E(Y)$ [X and Y may or may not be independent variable]

(*ii*) $E(XY) = E(X)\,E(Y)$ [X and Y are independent variables]

(*iii*) $E(C) = C$

$E(CX) = CE(X)$

$E(C + X) = E(X) + C$, Here C is a constant.

(*iv*) If X and Y are two variables and C_1 and C_2 are any two constants, then

$$E(C_1 X + C_2 Y) = C_1 E(X) + C_2 E(Y)$$

(*v*) If the probability function $f(a)$ is given, the expectations are defined as

$$E(X) = \Sigma \, a f(a)$$
$$\Sigma(X^2) = \Sigma \, a^2 f(a)$$

(*vi*) The mean of a probability distribution is the expected value of X.

$$\text{Mean } (\mu) = E(X)$$

(*vii*) The variance of a probability distribution is the expected value of $(a - \mu)^2$, where μ is the mean. It is denoted as

$$\text{Variance } (\sigma^2) = E(x - u)^2$$

(*viii*) It may be shown that, the positive square root of variance gives the standard deviation (σ) *i.e.*,

$$\sigma^2 = E(a^2) - \mu^2$$

Example 35. *Find the probability distribution of the number of heads in three tosses of a coin.*

Sol. Let X be a random variable which denote the number of heads in three tosses of a coin. Then P(X = 0) = P({TTT}) = 1/8, P(X = 1) = P({HTT, THT, TTH}) = 3/8,

P(X = 2) = P({HHT, HTH, THH}) = 3/8, P(X = 3) = P({HHH}) = 1/8

Hence the required probability distribution is

X	0	1	2	3
PX	1/8	3/8	3/8	1/8

Example 36. *An entrance examination paper consists of 40 true / false questions, each of two marks and 20 multiple choice questions, each of four marks. The probability that Monu answers a true / false correctly is 0.7 and the probability that he answers multiple choice questions correctly is 0.6. What is his expected score in the examination ?*

Sol. Let the random variable X denote the score on true/false question and Y for the multiple choice question. Then

$$E(X) = 0.7 \times 2 + 0.7 \times 2 + \text{........ upto 40 times} = 56$$

$$E(Y) = 0.6 \times 4 + 0.6 \times 4 + \text{......... upto 20 times} = 48$$

The sum of the expected score is

$$E(X + Y) = E(X) + E(Y) = 56 + 48 = 104$$

Thus 104 is the expected score of Monu in the examination.

Example 37. *A box contains 3 red balls and 3 green balls, a man is to draw two balls at random without replacement and is promised Rs. 5 for each red ball be draws and Rs. 10 for each green ball. Find his expectation.*

Sol. There are three exhaustive and mutually exclusive cases for drawing two balls *i.e.*, 2 red balls 2 green balls and 1 red ball and 1 green ball. The corresponding probabilities are $\frac{3}{15}, \frac{3}{15}$ and $\frac{9}{15}$. Thus the expected value of the amount is

$$= \frac{1}{5} \times 10 + \frac{1}{5} \times 20 + \frac{3}{5} \times 15$$

$$= 2 + 4 + 9 = \text{Rs. } 15.$$

Example 38. *A random variable has the following discrete probability distribution :*

X	*2*	*4*	*6*	*8*
Probability	*0.2*	*0.2*	*0.5*	*0.1*

Determine the expectation and the standard deviation of the random variable.

Sol. We know that

$$\text{Mean } (\mu) = E(X) = \Sigma\, p_i\, a_i$$

Thus, we have

$$= (0.2 \times 2) + (0.2 \times 4) + (0.5 \times 6) + (0.1 \times 8)$$
$$= 0.4 + 0.8 + 0.3 + 0.8$$
$$\text{Mean } (\mu) = 5.0$$
$$E(X^2) = \Sigma\, p_i\, a_i^2$$
$$= (0.2 \times 2^2) + (0.2 \times 4^2) + (0.5 \times 6^2) + (0.1 \times 8^2)$$
$$= 0.8 + 3.2 + 18 + 6.4 = 28.4$$

Thus, the variance $(\sigma^2) = E(X^2) - \mu^2$

$$= 28.4 - (5.0)^2$$
$$= 28.4 - 25 = 3.4$$

and the standard deviation $(\sigma) = \sqrt{3.4}$.

BINOMIAL DISTRIBUTION

A Bernoulli's trial is an experiment having only two possible outcomes, *i.e.*, success or failure *e.g.*, the tossing of a coin (head or tail), the answer of a Multiple Choice Question (write/wrong), testing of an item (defective or not) etc. The origin of binomial distribution lies in Bernoulli's trials. Since we are considering Bernoulli's trial, certain conditions are applicable on binomial distribution, which are as follows :

(*i*) Same experiment is repeated several times.

(*ii*) The repeated trials are independent.

(*iii*) The probability of each outcome remains the same for each trial.

(*iv*) There are only two possible outcomes-success or failure.

The Binomial distribution governs the number (X) of success in n independent trials, with a probability p of success at each trial :

$$P(X = r) = C(n, r)\, p^r\, (q)^{n-r}$$

where $q = 1 - p$ and the range of possible values (r) is $\{0, 1, 2, 3, \ldots\ldots, n\}$.

The Binomial distribution can be thought of as the sum of n independent Bernoulli random variables. When n Bernoulli trials are carried out, the outcome is an n-tuple $(r_1, r_2, \ldots\ldots, r_n)$ where r_1 = S (success) or r_1 = F(failure) for $i = 1, 2, 3, \ldots\ldots, n$. Since n trials are independent, the probability of each outcome of n trials consisting of r success and $n - r$ failures is $p^r\, q^{n-r}$.

Since there are $C(n, r)$, n-tuples of S's and F's that contain r S's, the probability of r success is

$$C(n, r)\, p^r\, q^{n-r}$$

The Binomial distribution has two parameters n and p.

The mean of the Binomial distribution $= np$

and $\text{variance} = npq$.

Example 39. *If the 15% of the bulbs produced in the factory are defective, find the probability that out of 10 bulbs chosen at random (i) Exactly two will be defective (ii) at most two will be defective.*

Sol. Let us suppose that p and q be the probability of being defective and not defective respectively. Then

$$p = \frac{15}{100} = .15 \quad \text{and} \quad q = 1 - .15 = .85$$

(*i*) P(Exactly two bulbs defective in 10 bulbs) = P(2)

$$= C(10, 2)\,(.15)^2\,(.85)^8$$

(*ii*) P(at most two bulbs defectives) = P(none defective) + P(exactly one defective) + P(exactly two defective)

$$= P(0) + P(1) + P(2)$$

$$= C(10, 0)\,(.15)^0\,(.85)^{10} + C(10, 1)\,(.15)^1\,(.85)^9 + C(10, 2)\,(.15)^2\,(.85)^8$$

$$= (.85)^{10} + 10 \times (.15) \times (.85)^9 + 45 \times (.0225)\,(.85)^8$$

Example 40. *The probability that a man aged 60 will live to be 70 is 0.65. What is the probability that out of ten men now 60, at least 7 would live to be 70 ?*

Sol. The probability of success (man living upto 70) is 0.65 and the probability of failure (not living up to 70) is 1 – 0.65 = 0.35

∴ The probability of at least 7 success in 10 is

$$= {}^{10}C_7\,(.65)^7\,(.35)^3 + {}^{10}C_8\,(.65)^8\,(.35)^2 + {}^{10}C_9\,(.65)^9\,(.35)^1 + (.65)^{10}$$

$$= 0.2523 + 0.1756 + 0.0725 + 0.0135 = 0.514.$$

Example 41. *If on an average 1 ship in every 10 is wrecked, find the probability that out of 5 ships expected to arrive, at least 4 will arrive safely ?*

Sol. The probability p that a ship is wrecked $= \frac{1}{10}$

Therefore, the probability of its arriving safely $= 1 - \frac{1}{10} = \frac{9}{10}$

The chance of at least 4 arriving safely is

$$= {}^5C_4\left(\frac{9}{10}\right)^4\left(\frac{1}{10}\right) + {}^5C_5\left(\frac{9}{10}\right)^5$$

$$= 5 \times \frac{9^4}{10^5} + 1 \times \frac{9^5}{10^5} = \frac{9^4}{10^5}(5 + 9) = \frac{45927}{50000}.$$

EXERCISES

1. In a Binomial distribution consisting of 5 independent trials, probabilities of 1 and 2 success are 0.4069 and 0.2084 respectively. Determine the parameter p of the distribution.
2. The sum of the mean and the variance of a Binomial distribution for 5 trials is 1.09. Find the distribution.
3. What is the expected value of the number of success, when n Bernoulli trials are performed. Where p is probability of success on each trial ?
4. What is the variance of the random variable $X(i, j) = 2i$, where i is the number appearing on the first die and j is the number appearing on the second die, when two dice are solid ?
5. A dice is thrown 10 times. If getting an even number is a success, what is the probability of
   (*i*) At least 6 successes (*ii*) At most 5 successes
   (*iii*) At most 7 successes ?

6. If the chance that one of the ten telephonic lines is busy at an instant is 0.2. What is chance that 6 of the lines are busy ? What is the probability that all the lines are busy ?
7. Suppose that 5% of the items produced by a company are defective. If 8 items are chosen at random, what is the probability that
   (*i*) Exactly one is defective (*ii*) Exactly two are defective
   (*iii*) At least one is defective (*iv*) At least two are defective.
8. Find the probability that in a family of 4 children there will be
   (*i*) At least 1 boy
   (*ii*) At least 1 boy and 1 girl.
9. The probability of any ship of a company being destroyed on a certain Voyage is 0.03. The company owns 8 ships for the Voyage. What is the probability of
   (*i*) losing one ship (*ii*) losing at most two ships
   (*iii*) losing none.

SOLVED PROBLEMS

Problem 1. *A bag contains 8 blue and 4 red balls. Two balls are drawn at random with replacement. Find the probability of getting one blue and one red ball.*

Sol. P(a blue ball) = $P(B) = \frac{1}{8}$; P(a red ball) = $P(R) = \frac{1}{4}$

P(a blue and a red ball) = $P(B) \times P(R) + P(R) \times P(B)$

$$= \frac{1}{8} \times \frac{1}{4} + \frac{1}{4} \times \frac{1}{8} = \frac{2}{32} = \frac{1}{16}.$$

Problem 2. *An urn contains 25 balls numbered through 25. Two balls are drawn from the urn with replacement. Find probability of getting (i) both odd numbers (ii) one odd and one even number (iii) at least one odd number (iv) no odd number.*

Sol. Let E = Event of drawing even numbered ball and O = event of drawing odd numbered ball.

There are 13 balls numbered odd and 12 balls numbered even.

$$P(E) = \frac{12}{25} \quad \text{and} \quad P(O) = \frac{13}{25}$$

(*i*) Since balls are replaced back in the urn

$\therefore$ P(both odd numbers) = $P(O) \times P(O) = \frac{13}{25} \times \frac{13}{25} = \frac{169}{625}$

(*ii*) P(one odd and one even number) = $P(O) \times P(E) + P(E) \times P(O)$

$$= \frac{13}{25} \times \frac{12}{25} + \frac{12}{25} \times \frac{13}{25} = \frac{312}{625}.$$

(*iii*) Probability of getting both the balls even numbered

$$P(EE) = P(E) \,.\, P(E) = \frac{12}{25} \times \frac{12}{25} = \frac{144}{625}$$

$\therefore$ P(atleast one odd numbered ball) = $1 - P(EE) = 1 - \frac{144}{625} = \frac{481}{625}$

(*iv*) P(no odd number) = P(both even numbers) = $\frac{144}{625}$.

Problem 3. *A bag contains 3 white, 3 black and 2 red balls. One by one, three balls are drawn without replacing them. Find the probability that the third ball is red.*

Sol. The third ball is red, can be in following 3 cases

(*i*) Ist red, IInd non-red, IIIrd red = $\frac{2}{8} \times \frac{6}{7} \times \frac{1}{7}$

(*ii*) Ist non-red, IInd red, IIIrd red = $\frac{6}{8} \times \frac{2}{7} \times \frac{1}{6}$

(*iii*) Ist non-red, IInd non-red, IIIrd red = $\frac{6}{8} \times \frac{5}{7} \times \frac{2}{6}$

$$\therefore \quad \text{Required probability} = \frac{2}{8} \times \frac{6}{7} \times \frac{1}{7} + \frac{6}{8} \times \frac{2}{7} \times \frac{1}{6} + \frac{6}{8} \times \frac{5}{7} \times \frac{2}{6} = \frac{1}{4}.$$

Problem 4. *A bag contains 6 red, 4 white and 5 black balls. A person draws four balls from the box at random. Find the probability that among the balls drawn there is at least one ball of each colour ?*

Sol. The four balls may have at least one ball of each colour in the three following mutually exclusive events.

E_1 = 1 red, 1 white, 2 black balls are drawn.
E_2 = 2 red, 1 white, 1 black balls are drawn.
E_3 = 1 red, 2 white, 1 black balls are drawn.

$$\text{Total probability} = \frac{{}^6C_1 \times {}^4C_1 \times {}^5C_2}{{}^{15}C_4} + \frac{{}^6C_2 \times {}^4C_1 \times {}^5C_1}{{}^{15}C_4} + \frac{{}^6C_1 \times {}^4C_2 \times {}^5C_1}{{}^{15}C_4}$$

$$= \left\{ \frac{(6 \times 4 \times 10) + (15 \times 4 \times 5) + (6 \times 6 \times 5)}{{}^{15}C_4} \right\} = 0.5275.$$

Problem 5. *A dice is thrown. Consider two events A = {1, 2, 3, 4} and B = {4, 5, 6}. Prove that the events A and B are*

(*i*) *not independent* (*ii*) *not mutually exclusive*

(*iii*) *exhaustive.*

Sol. $P(A) = \frac{4}{6} = \frac{2}{3}$, $P(B) = \frac{3}{6} = \frac{1}{2}$; $P(AB) = \frac{1}{6}$, $P(A) \,.\, P(B) = \frac{1}{3}$

(*i*) As $P(AB) \neq P(A) \,.\, P(B)$

So, A and B are not independent *i.e.,* occurrence of one do not effect the occurrence of other.

(*ii*) $P(A \cap B) \neq$ null set

i.e., they are not mutually exclusive.

(*iii*) $A \cup B = \{1, 2, 3, 4, 5, 6\}$ *i.e.,* all the possible outcomes are existing. So, events A and B are exhaustive.

Problem 6. *If A and B are two events such that P(A + B) = 5/6, P($\overline{A}$) = 1/4, P(B) = 1/3. Show that A and B are (i) independent (ii) not exclusive.*

Sol. (*i*) $$P(A \cup B) = P(A) + P(B) - P(A \cap B)$$

or $$P(A \cap B) = P(A) + P(B) - P(A \cup B)$$

$$= \left[\left(1 - \frac{1}{4}\right)\right] + \frac{1}{3} - \frac{5}{6} = \frac{3}{4} + \frac{1}{3} - \frac{5}{6} = \frac{3}{12} = \frac{1}{4}$$

$$P(A) \,.\, P(B) = \frac{1}{3} \cdot \frac{3}{4} = \frac{1}{4}$$

As $$P(A \cap B) = P(A) \,.\, P(B)$$

So, events are independent.

(*ii*) $$P(A \cap B) \neq 0$$

So, events are not exclusive.

Problem 7. *Show that the sum of the probability of two independent events is one (certain).*

Sol. Let the probability of success of first event be p_1.

The probability of failure of first event be q_1.

Similarly, the probability of success of second event be p_2.

The probability of failure of second event be q_2.

Now, the probability of success of the first event and success of the second event $= p_1 \times p_2$.

Probability of success of first event of failure of second event $= p_1 \times q_2$.

Probability of failure of first event and success of second event $= p_2 \times q_1$.

Probability of failure of both the events $= q_1 \times q_2$.

So, there are these four possibilities.

Therefore, the sum of probabilities

$$= p_1 \times p_2 + p_1 \times q_2 + p_2 \times q_1 + q_1 \times q_2$$
$$= p_1 (p_2 + q_2) + q_1 (p_2 + q_2) = (p_1 + q_1)(p_2 + q_2)$$
$$= 1 \qquad [\because\ p_1 + q_1 = 1 \quad p_2 + q_2 = 1]$$

Problem 8. *If $P(A_1 + A_2) = \frac{5}{6}$, $P(A_1 A_2) = \frac{1}{3}$ and $P(\overline{A_2}) = \frac{1}{2}$. Show that the events A_1, A_2 are not mutually exclusive.*

Sol. $$P(\overline{A_2}) = \frac{1}{2} \quad \text{or} \quad P(A_2) = 1 - \frac{1}{2} = \frac{1}{2}\,;\ P(A_1 A_2) = \frac{1}{3}$$

$$P(A_1) \,.\, P(A_2) = \frac{1}{3} \Rightarrow P(A_1) \,.\, \frac{1}{2} = \frac{1}{3} \text{ or } P(A_1) = \frac{2}{3}$$

$$P(A_1) + P(A_2) - P(A_1 A_2) = \frac{2}{3} + \frac{1}{2} - \frac{1}{3} = \frac{5}{6} = P(A_1 + A_2)$$

So, A_1 and A_2 are independent events and hence cannot be mutually exclusive.

Problem 9. *The UPSC has a list of 150 persons. Out of these 50 are womens and 100 are men. 125 of them know Hindi and remaining do not know Hindi. 90 of them are teachers and remaining are not teachers. What is the probability of selecting a Hindi knowing woman teacher as examiner ?*

Sol. Probability of selecting a woman $= \frac{50}{150} = \frac{1}{3}$.

Probability of selecting a teacher $= \frac{90}{150} = \frac{3}{5}$.

Probability of selecting a Hindi knowing candidate $= \frac{125}{150} = \frac{5}{6}$.

As the events are independent, the probability of board selecting a Hindi knowing woman teacher

$$= \frac{1}{3} \times \frac{3}{5} \times \frac{5}{6} = \frac{1}{6}.$$

Problem 10. *In a box, there are 6 balls of which 3 are white and 3 are black. They are drawn successively (i) without replacement (ii) with replacement. What is the chance that colour are alternate ?*

Sol. (*i*) Without replacement P(that 6 balls drawn follow pattern WBWBWB)

$$= \frac{3}{6} \times \frac{3}{5} \times \frac{2}{4} \times \frac{2}{3} \times \frac{1}{2} \times \frac{1}{1} = \frac{1}{20}.$$

P(that six balls follow pattern BWBWBW)

$$= \frac{3}{6} \times \frac{3}{5} \times \frac{2}{4} \times \frac{2}{3} \times \frac{1}{2} \times \frac{1}{1} = \frac{1}{20}.$$

∴ The probability that the colours of balls are alternate = probability that either the pattern WBWBWB or BWBWBW is obtained $= \frac{1}{20} + \frac{1}{20} = \frac{1}{10}$.

(*ii*) With replacement the probability of getting pattern WBWBWB

$$= \frac{3}{6} \times \frac{3}{6} \times \frac{3}{6} \times \frac{3}{6} \times \frac{3}{6} \times \frac{3}{6} = \left(\frac{1}{2}\right)^6 = \frac{1}{64}.$$

The probability of getting the pattern BWBWBW

$$= \left(\frac{3}{6}\right)^6 = \left(\frac{1}{2}\right)^6 = \frac{1}{64}$$

Required probability = P(that the colour of balls are alternate)

$$= \frac{1}{64} + \frac{1}{64} = \frac{1}{32}.$$

Problem 11. *A box 'A' contains 2 white and 4 black balls. Another box 'B' contains 5 white and 7 black balls. A ball is transferred from the box A to box B. Then a ball is transferred from box B to box A. Find the probability that it will be white.*

Sol. We have two cases.

(*i*) When the white ball is transferred

Probability of drawing white ball from box A $= \frac{2}{6} = \frac{1}{3}$.

Now, box B contains 6 white and 7 black balls.

$\therefore$ Probability of drawing a white ball from it $= \dfrac{6}{13}$.

$\therefore$ Probability of compound event $= \dfrac{1}{3} \times \dfrac{6}{13} = \dfrac{2}{13}$.

(*ii*) When black ball is transferred

Probability of drawing black ball from box A $= \dfrac{4}{6} = \dfrac{2}{3}$.

Now, box B contains 5 white and 8 black balls.

$\therefore$ Probability of drawing a white ball from it $= \dfrac{5}{13}$.

$\therefore$ Probability of compound event $= \dfrac{2}{3} \times \dfrac{5}{13} = \dfrac{10}{39}$.

These two events are mutually exclusive.

$\therefore$ Probability of transferring a ball from box A and then drawing a white ball from box B

$$= \frac{2}{13} + \frac{10}{39} = \frac{16}{39}.$$

Problem 12. *A and B are two candidates seeking admission in PMT. Probability that A is selected is 0.5 and the probability that both A and B are selected is almost 0.3. Is it possible that the probability of B getting selected is 0.9.*

Sol.
$$P(A) = 0.5, P(A \cap B) \leq 0.3$$
$$P(A \cup B) = P(A) + P(B) - P(A \cap B)$$
or
$$P(B) = P(A \cup B) + P(A \cap B) - P(A) \leq 1 + 0.3 - 0.5 = 0.8$$
$[\because P(A \cup B) \leq 1]$

Hence, it is not possible that probability of B getting selected is 0.9.

Problem 13. *What is the probability of getting 53 Sundays in a leap year ?*

Sol. A leap year contains 366 days. Therefore, it contains 52 Sundays and two days extra.

These two extra days have the following equally likely cases

(*i*) Saturday and Sunday
(*ii*) Sunday and Monday
(*iii*) Monday and Tuesday
(*iv*) Tuesday and Wednesday
(*v*) Wednesday and Thursday
(*vi*) Thursday and Friday
(*vii*) Friday and Saturday

Hence, there are seven exhaustive cases and two favourable cases.

$\because$ Probability of getting 53 Sundays $= \dfrac{2}{7}$.

Problem 14. *A coin is biased so that the probability of heads is 2/3. What is the probability that exactly four heads come up when the coin is flipped seven times, assuming that the flips are independent.*

Sol. There are $2^7 = 128$ possible outcomes.

The number of ways for 4 heads on 7 flips = $^7C_4$

But the seven flips are independent, the probability of each of these outcomes is $(2/3)^4 (1/3)^3$.

The probability that exactly 4 heads appear is

$$= {}^7C_4 (2/3)^4 (1/3)^3 = \frac{35 \times 16}{3^7} = \frac{560}{2187}.$$

Problem 15. *What is the expected value of the sum of the numbers that appear when a pair of fair dice is rolled ?*

Sol. Let X be the random variable equal to the sum of the numbers that appear when a pair of dice is rolled.

The range of X is {2, 3, 4, 5, 6, 7, 8, 9, 10, 11, 12}.

Then $P(X = 2) = P(X = 12) = 1/36$

$P(X = 3) = P(X = 11) = 1/18$

$P(X = 4) = P(X = 10) = 1/12$

$P(X = 5) = P(X = 9) = 1/9$

$P(X = 6) = P(X = 8) = 5/36$

$P(X = 7) = 6/36 = 1/6.$

Putting the value in the formula, we have

$$E(X) = 2.\frac{1}{36} + 3.\frac{1}{18} + 4.\frac{1}{12} + 5.\frac{1}{9} + 6.\frac{5}{36} + 7.\frac{1}{6} + 8.\frac{5}{36} + 9.\frac{1}{9} + 10.\frac{1}{12} + 11.\frac{1}{18} + 12.\frac{1}{36}$$

$$= 7.$$

Problem 16. *What is the Variance of the random variable X (a, b) where a is the number appearing on the first die and b is the number appearing on the second die, when two dice are rolled.*

Sol. We have to find the expected values of X and X^2.

Since $P(X = k) = 1/6$ for $k = 2, 4, 6, 8, 10, 12$ and 0 otherwise.

Thus $E(X) = (2 + 4 + 6 + 8 + 10 + 12)/6 = 7$

and $E(X^2) = (2^2 + 4^2 + 6^2 + 8^2 + 10^2 + 12^2)/6 = 182/3$

The variance of the random variable X is

$$V(X) = E(X^2) - E(X)^2$$

$$= 182/3 - 49 = 35/3.$$

Problem 17. *A and B in turn toss an ordinary die for a prize Rs. 110. The first to toss a 'six' wins . If A has the first throw, what is his expectation ?*

Sol. Chance of throwing 6 with a die = 1/6

Chance of not throwing a 6 with a die = $1 - \frac{1}{6} = \frac{5}{6}$

A will win if he throws a six in Ist, 3rd, 5th, ... attempts.

His chances of throwing '6' in Ist, 3rd, 5th, ... throws are successively.

$$\frac{1}{6};\frac{5}{6}\times\frac{5}{6}\times\frac{1}{6}\quad i.e.\quad \left(\frac{5}{2}\right)^2\times\frac{1}{6};\left(\frac{5}{6}\right)^4\times\frac{1}{6}\ \text{....... etc.}$$

$\therefore$ Chance of A's success is

$$= \frac{1}{6}+\left(\frac{5}{2}\right)^2\times\frac{1}{6}+\left(\frac{5}{4}\right)^4\times\frac{1}{6}+\text{.......}\infty$$

$$= \frac{1/6}{1-(5/6)^2} = \frac{6}{11}$$

So A expectation is $= \frac{6}{11}\times 110 = 60.$

Problem 18. *There are three coins, identical in appearance one of which is ideal and the other two biased with probabilities 1/3 and 2/3 respectively for a head. One coin is taken at random and tossed twice. If a head appears both the times, what is the probability that the ideal coin was chosen.*

Sol. Let B as the event of obtaining 2 heads in two tosses of the selected coin and A_1, A_2, A_3 respectively. The events of choosing the ideal, second and third coins. Then,

Probability of getting a head in a toss $= \frac{1}{2}$

$$P(A_1) = P(A_2) = P(A_3) = \frac{1}{3} \quad \text{and} \quad P\left(\frac{B}{A_1}\right) = \left(\frac{1}{2}\right)^2 = \frac{1}{4}$$

The probabilities of turning head up with second and third coins respectively being $\frac{1}{3}$ and $\frac{2}{3}$, we have

$$P\left(\frac{B}{A_2}\right) = \left(\frac{1}{3}\right)^2 = \frac{1}{9} \quad \text{and} \quad P\left(\frac{B}{A_3}\right) = \left(\frac{2}{3}\right)^2 = \frac{4}{9}$$

Using Baye's formula the required probability is

$$P\left(\frac{A_1}{B}\right) = \frac{P(A_1)\,P(B/A_1)}{P(A_1)\,P(B/A_1)+P(A_2)\,P(B/A_2)+P(A_3)\,P(B/A_3)}$$

$$= \frac{\frac{1}{3}\cdot\frac{1}{4}}{\frac{1}{3}\cdot\frac{1}{4}+\frac{1}{3}\cdot\frac{1}{9}+\frac{1}{3}\cdot\frac{4}{9}} = \frac{\frac{1}{12}}{\frac{1}{12}+\frac{1}{27}+\frac{4}{27}}$$

$$= \frac{1/12}{29/108} = \frac{9}{29}.$$

Problem 19. *In the coming year, there will be three candidates for the post of principal. Mr. Gupta, Mr. Chopra and Mr. Verma whose chances of getting appointment are in the proportion 4 : 2 : 3 respectively The probability that Mr. Gupta if selected will abolish co-education in college is 0.3. The probability of Mr. Chopra and Mr. Verma doing the same are respectively 0.5 and 0.8. What is the probability that co-education will be abolished in the college next year.*

Sol. Let A_1, A_2, A_3 are the events that Mr. Gupta, Mr. Chopra and Mr. Verma are selected as principal and B is the even that co-education is abolished. Then,

$$P(A_1) = \frac{4}{4+2+3} = \frac{4}{9} \quad P(A_2) = \frac{2}{4+2+3} = \frac{2}{9}$$

$$P(A_3) = \frac{3}{4+2+3} = \frac{3}{9}.$$

We have

$$P(B/A_1) = 0.3, \quad P(B(A_2)) = 0.5, P(B/A_3) = 0.8$$

$$\begin{aligned} \therefore \quad P(B) &= \text{Probability that co-education will be abolished} \\ &= P[(B \cap A_1) \cup (B \cap A_2) \cup (B \cap A_3)] \\ &= P(B \cap A_1) + P(B \cap A_2) + P(B \cap A_3) \\ &= P(A_1)\, P\left(\frac{B}{A_1}\right) + P(A_2)\, P\left(\frac{B}{A_2}\right) + P(A_3)\, P\left(\frac{B}{A_3}\right) \\ &= \frac{4}{9} \times 0.3 + \frac{2}{9} \times 0.5 + \frac{3}{9} \times 0.8 = \frac{23}{45}. \end{aligned}$$

Problem 20. *The contents of 3 urns 1, 2, 3 are*

1 white, 2 black, 3 red balls ;

2 white, 1 black, 1 red ball ;

4 white, 5 black, 3 red balls.

One urn is selected at random and two balls drawn. They happen to be white and red. What is the probability that they come from urn 2 or 3 ? Also find the probability that the balls come from urn 1.

Sol. Let A_1, A_2 and A_3 respectively as the events that the urn 1, 2 or 3 is chosen and B is the event that two balls taken from selected urn are white and red, we have

$$P(A_1) = P(A_2) = P(A_3) = \frac{1}{3}$$

$$P\left(\frac{B}{A_1}\right) = \frac{{}^1C_1 \times {}^3C_1}{{}^6C_2} = \frac{1}{5}$$

$$P\left(\frac{B}{A_2}\right) = \frac{{}^2C_1 \times {}^1C_1}{{}^4C_2} = \frac{1}{3}$$

$$P\left(\frac{B}{A_3}\right) = \frac{{}^4C_1 \times {}^3C_1}{{}^{12}C_2} = \frac{2}{11}$$

Applying Baye's formula

$$P\left(\frac{A_2}{B}\right) = \frac{P(A_2)\, P(B/A_2)}{P(A_1)\, P(B/A_1) + P(A_2)\, P(B/A_2) + P(A_3)\, P(B/A_3)}$$

$$= \frac{\frac{1}{3} \cdot \frac{1}{3}}{\frac{1}{3} \cdot \frac{1}{5} + \frac{1}{3} \cdot \frac{1}{3} + \frac{1}{3} \cdot \frac{2}{11}} = \frac{55}{118}$$

Similarly, $P\left(\frac{A_3}{B}\right)=\frac{30}{118}$ $P\left(\frac{A_1}{B}\right)=\frac{33}{118}$.

Problem 21. *In a bolt factory machines A, B, C manufacture 25%, 35% and 40% respectively of the total of their output 5%, 4% and 2% are defective bolts. A bolt is drawn at random from the produce and is found defective. What are the probabilities that it was manufactured by machine A, B or C ?*

Sol. $P(A)=\frac{25}{100}, P(B)=\frac{35}{100}, P(C)=\frac{40}{100}$

Denoting by x, the event of drawing a defective bolt

$$P\left(\frac{X}{A}\right)=\frac{5}{100},\quad P\left(\frac{X}{B}\right)=\frac{4}{100},\quad P\left(\frac{X}{C}\right)=\frac{2}{100}$$

We have to determine $P\left(\frac{A}{X}\right), P\left(\frac{B}{X}\right), P\left(\frac{C}{X}\right)$

Using Baye's formula, we have

$$P\left(\frac{A}{X}\right)=\frac{P(A)\,P(X/A)}{P(A)\,P\left(\frac{X}{A}\right)+P(B)\,P\left(\frac{X}{B}\right)+P(C)\,P\left(\frac{X}{C}\right)}$$

$$=\frac{\frac{25}{100}\cdot\frac{5}{100}}{\frac{25}{100}\cdot\frac{5}{100}+\frac{35}{100}\cdot\frac{4}{100}+\frac{40}{100}\cdot\frac{2}{100}}$$

$$=\frac{125}{125+140+80}=\frac{125}{345}$$

Similarly $P\left(\frac{B}{X}\right)=\frac{140}{345}$ $\quad P\left(\frac{C}{X}\right)=\frac{80}{345}$

Problem 22. *In a university, the geographical distribution of male students is as follows– 50% come from East, 30% come from west and 20% come from North. The following proportion of male candidates wear ties—80% of Easterners, 60% of westerners and 40% of Northeners. What is the probability that a student who wears a tie comes from East ?*

Sol. Let A_1, A_2, A_3 as the events that male students come from East, West and North respectively and B is event that wear tie, then

$$P(A_1)=\frac{50}{100},\quad P(A_2)=\frac{30}{100},\quad P(A_3)=\frac{20}{100}$$

$$P\left(\frac{B}{A_1}\right)=\frac{80}{100},\quad P\left(\frac{B}{A_2}\right)=\frac{60}{100},\quad P\left(\frac{B}{A_3}\right)=\frac{40}{100}$$

Using Baye's formula.

$$P\left(\frac{A_1}{B}\right)=\frac{P(A_1)\,P(B/A_1)}{P(A_1)\,P(B/A_1)+P(A_2)\,P(B/A_2)+P(A_3)\,P(B/A_3)}$$

$$= \frac{\frac{50}{100}\cdot\frac{80}{100}}{\frac{50}{100}\cdot\frac{80}{100}+\frac{30}{100}\cdot\frac{60}{100}+\frac{20}{100}\cdot\frac{40}{100}} = \frac{40}{40+18+8} = \frac{40}{66} = \frac{20}{33}.$$

Problem 23. *A perfect cubic die is thrown a large number of times in sets of 8. The occurence of a 5 or 6 is called a success. In what proportion of the sets would you expect 3 successes.*

Sol. p = probability of getting a '5' or a '6' with six faced die

$$= \frac{2}{6} = \frac{1}{3}, \text{ so that } q = 1 - p = \frac{2}{3} \text{ and } n = 8$$

The binomial distribution is therefore $N\left(\frac{2}{3}+\frac{1}{3}\right)^8$

Probability of 3 successes in one set of 8 = ${}^8C_3 p^3 q^5$

$$= \frac{8.7.6}{1.2.3}\left(\frac{1}{3}\right)^3\left(\frac{2}{3}\right)^5 = \frac{1792}{6561} = 0.2731$$

$\therefore$ Percentage *i.e.*, probability of 3 successes in 100 sets = 0.2731 × 100 = 27.31%.

Problem 24. *Assuming that half the population are consumers of biscuits, so that chance of an individual being a consumer is $\frac{1}{2}$ and assuming that 100 investigators each take ten individuals to see whether they are consumers, how many investigators would you expect to report that 3 people or less were consumers ?*

Sol. $p = q = \frac{1}{2}$, N = 100, n = 10

$\therefore$ Binomial distribution is $N(q+p)^n$ *i.e.*, $100\left(\frac{1}{2}+\frac{1}{2}\right)^{10}$

No. of investigators reporting that there are 3 persons or less are consumers–

$$= N[p(0) + p(1) + p(2) + p(3)]$$

$$= 100\left[\left(\frac{1}{2}\right)^{10} + {}^{10}C_1\left(\frac{1}{2}\right)^1\left(\frac{1}{2}\right)^9 + {}^{10}C_2\left(\frac{1}{2}\right)^2\left(\frac{1}{2}\right)^8 + {}^{10}C_3\left(\frac{1}{2}\right)^3\left(\frac{1}{2}\right)^7\right]$$

$$= \frac{100}{2^{10}}[1 + 10 + 45 + 120]$$

$$= \frac{17600}{2^{10}} = 17 \text{ (approx.)}$$

Problem 25. *Determine the binomial distribution for which the mean is 4 and variance 3 and also find its mode.*

Sol. We have $E(X) = np = 4$

$$\text{Var.}(X) = npq = 3$$

giving $p = \frac{1}{4}, \quad q = \frac{3}{4} \quad$ and $\quad n = 16$

The parameters of binomial distribution under consideration are $n = 16$ and $p = \frac{1}{4}$ and distribution is $(q + p)^n = \left(\frac{3}{4}+\frac{1}{4}\right)^{16}$

Now, mode = $np = 4$ as np is a whole number.

Problem 26. *An irregular six-faced die is thrown and the expectation that in 100 throws it will give 5 even numbers is twice the expectation that it will give 4 even numbers. How many times in 10,000 sets of 10 throws would you expect it to give no even numbers.*

Sol. If p be the expectation of getting an even number and given that the probability of 5 even numbers in 10 throws of a die is twice that of 4 even numbers in 10 throws of the die, then we have

$$^{10}C_5\, p^5 q^5 = 2 \times {}^{10}C_4\, p^4\, q^6 = \frac{3}{5} p$$

$$= q = 1 - p \quad \Rightarrow \quad p = \frac{5}{8} \text{ and } q = \frac{3}{8}$$

$\therefore$ Frequency for no even number in 10,000 sets of 10 throws = $10{,}000\, q^{10}$

$$= 10{,}000 \left(\frac{3}{8}\right)^{10} = 1 \text{ (approx.)}$$

SECTION II

Chapter 6

Recurrence Relations and Generating Functions

DEFINITION

A recurrence relation is a functional relation between the independent variable x, dependent variable $f(x)$ and the differences of various order of $f(x)$. A recurrence relation is also called a difference equation and we will use these two terms interchangeably.

For example : The equation $f(x + 3h) + 3f(x + 2h) + 6f(x + h) + 9f(x) = 0$ is a recurrence relation.

It can also be written as

$$a_{r+3} + 3a_{r+2} + 6a_{r+1} + 9a_r = 0$$

or

$$y_{k+3} + 3y_{k+2} + 6y_{k+1} + 9y_k = 0.$$

For example : The Fibonacci sequence is defined by the recurrence relation $a_r = a_{r-2} + a_{r-1}$, $r \geq 2$, with the initial conditions $a_0 = 1$ and $a_1 = 1$.

ORDER OF THE RECURRENCE RELATION

The order of the recurrence relation or difference equation is defined to be the difference between the highest and lowest subscripts of $f(x)$ or a_r or y_k.

For example : The equation $13a_r + 20a_{r-1} = 0$
is a first order recurrence relation.

For example : The equation $8f(x) + 4f(x + 1) + 8f(x + 2) = k(x)$
is a second order difference equation.

DEGREE OF THE DIFFERENCE EQUATION

The degree of a difference equation is defined to be the highest power of $f(x)$ or a_r or y_k.

For example : The equation $y_{k+3}^3 + 2y_{k+2}^2 + 2y_{k+1} = 0$
has the degree 3, as the highest power of y_k is 3.

For example : The equation $a_r^4 + 3a_{r-1}^3 + 6a_{r-2}^2 + 4a_{r-3} = 0$
has the degree 4, as the highest power of a_r is 4.

For example : The equation $y_{k+3} + 2y_{k+2} + 4y_{k+1} + 2y_k = k(x)$
has the degree 1, because the highest power of y_k is 1 and its order is 3.

For example : The equation $f(x + 2h) - 4f(x + h) + 2f(x) = 0$ has the degree 1 and its order is 2.

SEQUENCE

A sequence is a list of objects in some particular order. It means it has a first element, a second element, a third element and so on. Two types of sequences

(*i*) Finite sequence

(*ii*) Infinite sequence

(*i*) **Finite sequence.** A sequence is called a finite, if the list of elements stops after n steps, where $n \in N$.

e.g., The sequence 2, 4, 8, 16, 32, 64 is a finite sequence.

(*ii*) **Infinite sequence.** A sequence is called an Infinite, if the list of elements does not stop or we can say it may go on forever.

e.g., 1. The list 2, 4, 6, 8, 10, 12, 14, ... is an infinite sequence.

2. The list 1, 4, 9, 16, 25, ... is an infinite sequence of the squares of all +ve integers.

Representation of Sequence

The sequence may be represented either by explicitly describing the first few terms of the sequence followed by a dotted line in case of infinite sequence and describing all the elements in the sequence when it is finite.

Another compact way of describing the sequence is describe it through a formula.

Two types of formulas are used to describe the sequence.

(*i*) Recursive formula

(*ii*) Explicit formula

(*i*) **Recursive formula.** A formula which refers to previous terms to define the next term, is called a recursive formula. Every recursive formula must include a starting place in it.

e.g., 1. The recursive formula for the finite sequence 2, 4, 8, 16, 32, 64 is $a_1 = 2, a_n = 2a_{n-1}$ and $2 \le n \le 6$.

2. The recursive formula for the infinite sequence 3, 7, 11, 15, 19, 23, ... is $a_1 = 3$, $a_n = a_{n-1} + 4$.

(*ii*) **Explicit formula.** A formula which tells us exactly what value any particular term has, is called an explicit formula.

e.g., 1. The explicit formula for the infinite sequence 1, – 1, 1, – 1, 1, – 1, ... is $a_n = (-1)^{n+1}$.

2. The finite sequence 65, 61, 57, 53, 49, 45 can be defined by the explicit formula $a_n = 69 - 4n, 1 \le n \le 6$.

Example 1. *Write the recursive formula for the fibbonacci sequence i.e., 1, 1, 2, 3, 5, 8, 13, 21, 34, 55, ...*

Sol. The sequence can be represented as

$$a_n = a_{n-1} + a_{n-2}$$

with $a_0 = 1$ and $a_1 = 1$.

Note 1. If the characteristic equation $x^2 - r_1x - r_2 = 0$ of the recurrence relation $a_n = r_1 a_{n-1} + r_2 a_{n-2}$ has two distinct roots, S_1 and S_2, then $a_n = uS_1^{\ n} + vS_2^{\ n}$, where u and v depend on the initial conditions, is the explicit formula for the sequence.

2. If the characteristic equation $x^2 - r_1x - r_2 = 0$ has a single root S, then the explicit formula is $a_n = uS^n + vnS^n$, where u and v depend on the initial conditions.

Example 2. *Find an explicit formula for the fibonacci sequence defined by $a_n = a_{n-1} - a_{n-2}$ with the initial conditions $a_0 = 1$ and $a_1 = 1$.*

Sol. We need the roots of the equation associated with the fibonacci sequence to describe the explicit formula for the sequence.

The fibonacci sequence is defined by a linear homogeneous recurrence relation of degree 2 *i.e.*, $a_n = a_{n-1} + a_{n-2}$

The corresponding equation of the recurrence relation is

$$x^2 - x - 1 = 0$$

The roots of this quadratic equation are $\frac{1+\sqrt{5}}{2}$ and $\frac{1-\sqrt{5}}{2}$

Now we have to determine u and v, so solve the following

$$u\left(\frac{1+\sqrt{5}}{2}\right) + v\left(\frac{1-\sqrt{5}}{2}\right) = 1 \text{ and } u\left(\frac{1+\sqrt{5}}{2}\right)^2 + v\left(\frac{1-\sqrt{5}}{2}\right)^2 = 1$$

For the given initial conditions, $u = \frac{1}{\sqrt{5}}$ and $v = -\frac{1}{\sqrt{5}}$

Therefore, the explicit formula for the fionacci sequence is

$$a_n = \frac{1}{\sqrt{5}}\left(\frac{1+\sqrt{5}}{2}\right)^n - \frac{1}{\sqrt{5}}\left(\frac{1-\sqrt{5}}{2}\right)^n$$

Example 3. *Find the explicit formula for the recurrence relation $a_n = 2a_{n-1} - a_{n-2}$ with initial conditions $a_1 = 1.5$ and $a_2 = 3$.*

Sol. We need to find the roots of the associated equation. The equation for this difference equation is

$$x^2 - 2x + 1 = 0$$

The equation's two roots 1, 1 *i.e.*, one multiple root, which is 1.

Thus, $a_n = u(1)^n + vn(1)^n$ is equation to find the values of u and v.

From this equation and using the initial conditions solve the following

$$d_1 = 1.5 = u(1)^1 + v.1(1)^1 \; ; \; d_2 = 3 = u(1)^2 + v.2(1)^2$$

so the value of $u = 0$ and $v = 1.5$

Hence the explicit formula becomes

$$a_n = 1.5\, n.$$

EXERCISES

1. Give an explicit formula for S_n, where $S_0 = 3$ and $S_n = -5\, S_{n-1}$, $n \geq 1$
2. Give a recursive formula for the sequence

   2, 5, 7, 12, 19, 31, ...
3. Find out the first four terms of the sequence whose general term is given below.

   (*i*) $a_n = 3n^2 + 3n - 6$ (*ii*) $a_n = -3\, a_{n-1} + 1$, $a_1 = -3$

4. Give recursive as well as explicit formula for the following sequence
(*i*) 0, 2, 0, 2, 0, 2, ... (*ii*) 1, – 1, 1, – 1, 1, – 1,......
(*iii*) 1, 1/2, 1/4, 1/8, 1/16, ...
5. Give the recursive formula for tower of Hanoi problem.
[**Hint.** $a_n = 2a_{n-1} + 1, n > 1$ with $a_1 = 1$]
6. Determine the first five terms of the sequence defined by the following recurrence relations.
(*i*) $a_n = n(a_{n-1})^2, \quad n \geq 1, \quad a_0 = 1$
(*ii*) $a_n = n\, a_{n-1} + n^2 a_{n-2}, \quad n \geq 2, a_0 = 1, a_1 = 1.$

LINEAR RECURRENCE RELATIONS WITH CONSTANT COEFFICIENTS

A recurrence relation is called linear if its degree is one.

The general form of linear recurrence relation with constant coefficient is

$$C_0 y_{n+r} + C_1 y_{n+r-1} + C_2 y_{n+r-2} + \ldots + C_r y_n = R(n)$$

where $C_0, C_1, C_2, \ldots, C_n$ are constants and $R(n)$ is same function of independent variable n.

A solution of a recurrence relation in any function which satisfies the given equation.

LINEAR HOMOGENEOUS RECURRENCE RELATION WITH CONSTANT COEFFICIENTS

The equation is said to be linear homogeneous difference equation if and only if $R(n) = 0$ and it will be of order n.

The equation is said to be linear non-homogeneous difference equation if $R(n) \neq 0$.

For example : The equation $a_{r+3} + 6a_{r+2} + 12a_{r+1} + 8a_r = 0$ is a linear homogeneous equation of order 3.

For example : The equation $a_{r+2} - 4a_{r+1} + 4a_r = 3r + 2^r$ is a linear non-homogeneous equation of order 2.

A linear homogeneous difference equation with constant coefficients is given by

$$C_0 y_n + C_1 y_{n-1} + C_2 y_{n-2} + \ldots + C_r y_{n-r} = 0 \qquad \ldots(i)$$

where $C_0, C_1, C_2, \ldots, C_r$ are constants.

The solution of the equation (*i*) is of the form $A\alpha_1^K$, where α_1 is the characteristic root and A is constant.

Substitute the values of $A\alpha^K$ for y_n in equation (*i*), we have

$$C_0 A\alpha^K + C_1 A\alpha^{K-1} + C_2 A\alpha^{K-2} + \ldots + C_r A\alpha^{K-r} = 0 \qquad \ldots(ii)$$

After simplifying equation (*ii*), we have

$$C_0 \alpha^r + C_1 \alpha^{r-1} + C_2 \alpha^{r-2} + \ldots + C_r = 0 \qquad \ldots(iii)$$

The equation (*iii*) is called the characteristic equation of the difference equation.

If α_1 is one of the roots of the characteristic equation, then $A\alpha_1^K$ is a homogeneous solution to the difference equation.

To find the solution of the linear homogeneous difference equations, we have the four cases, that are discussed as follows.

Case I. If the characteristic equation has n distinct real roots $\alpha_1, \alpha_2, \alpha_3, \ldots, \alpha_n$.

Thus, $\alpha_1^K, \alpha_2^K, \ldots\ldots, \alpha_n^K$ are all solutions of equation (*i*).

Also, we have, $A_1 \alpha_1^K, A_2 \alpha_2^K, \ldots\ldots, A_n \alpha_n^K$ are all solutions of equation (*i*). The sums of solutions are also solutions.

Hence, the homogeneous solution of the difference equation is

$$y_K = A_1\alpha_1^K + A_2\alpha_2^K + \ldots + A_n \alpha_n^K.$$

Case II. If the characteristic equation has repeated real roots.

If $\alpha_1 = \alpha_2$, then $(A_1 + A_2 K) \alpha_1^K$ is also a solution.

If $\alpha_1 = \alpha_2 = \alpha_3$, then $(A_1 + A_2 K + A_3 K^2) \alpha_1^K$ is also a solution.

Similarly, if root α_1 is repeated n times, then

$$(A_1 + A_2 K + A_3 K^2 + \ldots + A_n K^{n-1}) \alpha_1^K$$

is the solution of the homogeneous equation.

Case III. If the characteristic equation has one imaginary root.

If $\alpha + i\beta$ is the root of the characteristic equation, then $\alpha - i\beta$ is also the root, where α and β are real.

Thus, $(\alpha + i\beta)^K$ and $(\alpha - i\beta)^K$ are solutions of the equations. This implies

$$(\alpha + i\beta)^K A_1 + (\alpha - i\beta)^K A_2$$

is also a solution of the characteristic equation, where A_1 and A_2 are constants which are to be determined.

Case IV. If the characteristic equation has repeated imaginary roots.

When the characteristic equation has repeated imaginary roots,

$$(C_1 + C_2 K) (\alpha + i\beta)^K + (C_3 + C_4 K)(\alpha - i\beta)^K$$

is the solution of the homogeneous equation.

Example 4. *Solve the difference equation* $a_r - 3a_{r-1} + 2a_{r-2} = 0$.

Sol. The characteristic equation is given by

$$s^2 - 3s + 2 = 0 \quad \text{or} \quad (s - 1)(s - 2) = 0$$

$\Rightarrow \quad s = 1, 2$

Therefore, the homogeneous solution of the equation is

$$a_r = C_1 + C_2 \,.\, 2^r.$$

Example 5. *Solve the difference equation* $a_r - 6a_{r-1} + 8a_{r-2} = 0$.

Sol. The characteristic equation is

$$s^2 - 6s + 8 = 0 \quad \text{or} \quad (s - 2)(s - 4) = 0$$

$\Rightarrow \quad s = 2, 4$

Therefore, the homogeneous solution of the equation is

$$a_r = C_1 \,.\, 2^r + C_2 \,.\, 4^r.$$

Example 6. *Solve the difference equation* $9y_{K+2} - 6y_{K+1} + y_K = 0$.

Sol. The characteristic equation is

$$9s^2 - 6s + 1 = 0 \quad \text{or} \quad (3s - 1)^2 = 0$$

$\Rightarrow \quad s = \frac{1}{3}$ and $\frac{1}{3}$

Therefore, the homogeneous solution is given by

$$y_K = (C_1 + C_2 K) \,.\, \left(\frac{1}{3}\right)^K.$$

Example 7. *Solve the difference equation* $a_r - 4a_{r-1} + 4a_{r-2} = 0.$

Sol. The characteristic equation is given by

$$s^2 - 4s + 4 = 0 \quad \text{or} \quad (s-2)^2 = 0$$

$\Rightarrow \quad s = 2$ and 2

Therefore, the homogeneous solution of equation is $a_r = (C_1 + C_2 r) . 2^r$.

Example 8. *Solve the difference equation* $a_r + a_{r-1} + a_{r-2} = 0.$

Sol. The characteristic equation is $s^2 + s + 1 = 0$

The roots of this characteristic equation are imaginary, *i.e.*, $s = \dfrac{-1+i\sqrt{3}}{2}$ and $\dfrac{-1-i\sqrt{3}}{2}$.

Therefore, the homogeneous solution of the equation is

$$a_r = \left[\frac{-1+i\sqrt{3}}{2}\right]^r C_1 + \left[\frac{-1-i\sqrt{3}}{2}\right]^r C_2.$$

Example 9. *Solve the difference equation* $y_K - y_{K-1} - y_{K-2} = 0.$

Sol. The characteristic equation is $s^2 - s - 1 = 0$

$$s = \frac{1 \pm \sqrt{1+4}}{2} = \frac{1 \pm \sqrt{5}}{2}$$

Therefore, the homogeneous solution of the equation is

$$y_K = C_1\left[\frac{1+\sqrt{5}}{2}\right]^K + C_2\left[\frac{1-\sqrt{5}}{2}\right]^K.$$

Example 10. *Solve the difference equation* $a_{r+4} + 2a_{r+3} + 3a_{r+2} + 2a_{r+1} + a_r = 0.$

Sol. The characteristic equation is $s^4 + 2s^3 + 3s^2 + 2s + 1 = 0$

or $$(s^2 + s + 1)(s^2 + s + 1) = 0$$

or $$s = \frac{-1 \pm i\sqrt{3}}{2}, \frac{-1 \pm i\sqrt{3}}{2}$$

Therefore, the homogeneous solution is given by

$$a_r = (C_1 + C_2 r)\left(\frac{-1+i\sqrt{3}}{2}\right)^r + (C_3 + C_4 r)\left(\frac{-1-i\sqrt{3}}{2}\right)^r.$$

Example 11. *Solve the difference equation* $y_{K+4} + 4y_{K+3} + 8y_{K+2} + 8y_{K+1} + 4y_K = 0.$

Sol. The characteristic equation is $s^4 + 4s^3 + 8s^2 + 8s + 4 = 0$

or $$(s^2 + 2s + 2)(s^2 + 2s + 2) = 0$$

or $$s = -1 \pm i, -1 \pm i$$

Therefore, the homogeneous solution is given by

$$y_K = (C_1 + C_2 K)(-1+i)^K + (C_3 + C_4 K)(-1-i)^K.$$

PARTICULAR SOLUTION

(a) Homogeneous Linear Difference Equations

We can find the particular solution of the difference equation, when the equation is of homogeneous linear type by putting the values of the initial conditions in the homogeneous solution.

Example 12. *Solve the difference equation $2a_r - 5a_{r-1} + 2a_{r-2} = 0$ and find particular solution such that $a_0 = 0$ and $a_1 = 1$.*

Sol. The characteristic equation is $2s^2 - 5s + 2 = 0$

or $$(2s - 1)(s - 2) = 0$$

or $$s = \frac{1}{2} \quad \text{and} \quad 2.$$

Therefore, the homogeneous solution is given by

$$a_{r(h)} = C_1 \left(\frac{1}{2}\right)^r + C_2 \cdot 2^r \qquad ...(i)$$

Putting $r = 0$ and $r = 1$ in equation (*i*), we get

$$a_0 = C_1 + C_2 = 0 \qquad ...(a)$$

$$a_1 = \frac{1}{2} C_1 + 2C_2 = 1 \qquad ...(b)$$

Solving eq. (*a*) and (*b*), we have

$$C_1 = -\frac{2}{3} \quad \text{and} \quad C_2 = \frac{2}{3}$$

Hence, the particular solution is

$$a_{r(P)} = -\frac{2}{3} \cdot \left(\frac{1}{2}\right)^r + \frac{2}{3} \cdot (2)^r.$$

Example 13. *Solve the difference equation $a_r - 4a_{r-1} + 4a_{r-2} = 0$ and find the particular solution, given that $a_0 = 1$ and $a_1 = 6$.*

Sol. The characteristic equation is

$$s^2 - 4s + 4 = 0 \quad \text{or} \quad (s - 2)^2 = 0 \quad \text{or} \quad s = 2, 2$$

Therefore, the homogeneous solution is given by

$$a_{r(n)} = (C_1 + C_2 r) \cdot 2^r \qquad ...(i)$$

Putting $r = 0$ and $r = 1$ in equation (*i*), we get

$$a_0 = (C_1 + 0) \cdot 2^0 = 1 \quad \therefore \quad C_1 = 1$$

$$a_1 = (C_1 + C_2) \cdot 2 = 6 \quad \therefore \quad C_1 + C_2 = 3 \Rightarrow C_2 = 2$$

Hence, the particular solution is $a_{r(P)} = (1 + 2r) \cdot 2^r$.

Example 14. *Solve the difference equation $9a_r - 6a_{r-1} + a_{r-2} = 0$ satisfying the conditions $a_0 = 0$ and $a_1 = 2$.*

Sol. The characteristic equation is

$$9s^2 - 6s + 1 = 0 \quad \text{or} \quad (3s - 1)^2 = 0$$

or $$s = \frac{1}{3}, \frac{1}{3}$$

Therefore, the homogeneous solution is given by

$$a_{r(h)} = (C_1 + C_2 r) \cdot \left(\frac{1}{3}\right)^r \quad \text{...(i)}$$

Putting $r = 0$ and $r = 1$ in equation (*i*), we get

$$a_0 = C_1 = 0 \quad \text{...(a)}$$

$$a_1 = (C_1 + C_2) \cdot \frac{1}{3} = 2. \quad \therefore \quad C_1 + C_2 = 6 \Rightarrow C_2 = 6$$

Hence, the particular solution is

$$a_{r(P)} = 6r \cdot \left(\frac{1}{3}\right)^r.$$

Example 15. *Solve the difference equation $a_r - 7a_{r-1} + 10a_{r-2} = 0$ satisfying the conditions $a_0 = 0$ and $a_1 = 6$.*

Sol. The characteristic equation is

$$s^2 - 7s + 10 = 0 \quad \text{or} \quad (s - 5)(s - 2) = 0 \quad \text{or } s = 2, 5$$

Therefore, the homogeneous solution is given by

$$a_{r(h)} = C_1 \cdot 2^r + C_2 \cdot 5^r. \quad \text{...(i)}$$

Putting $r = 0$ and $r = 1$ in equation (*i*), we get

$$a_0 = C_1 + C_2 = 0 \quad \text{...(a)}$$

$$a_1 = 2C_1 + 5C_2 = 6 \quad \text{...(b)}$$

From eq. (*a*) $\quad C_1 = - C_2$

From eq. (*b*) $\quad 3C_2 = 6 \quad \therefore \quad C_2 = 2$ and $C_1 = - 2.$

Hence, the particular solution is $a_{r(P)} = - 2 \cdot 2^r + 2 \cdot 5^r.$

EXERCISES

1. Solve the following recurrence relations :

   (*i*) $a_n = 4a_{n-1} + 5a_{n-2}, a_1 = 2, a_2 = 6$ (*ii*) $a_n = - 3\, a_{n-1} - 2a_{n-2}, a_1 = - 2, a_2 = 4.$
2. Solve the following recurrence relations :

   (*i*) $a_n - a_{n-1} - a_{n-2} = 0, a_0 = 1$ and $a_1 = 1$ (*ii*) $a_n - 4a_{n-1} + 4a_{n-2} = 0, a_0 = 1$ and $a_1 = 6.$
3. Show that $a_n = c_1 2^n + c_2 4^n$ is a solution of the difference equation $a_n - 6a_{n-1} + 8a_{n-2} = 0.$
4. Solve the following recurrence relations :

   (*i*) $a_n = 4\, a_{n-1} - 4a_{n-2}$ (*ii*) $a_n = - 6a_{n-1} - 9\, a_{n-2}$
5. Solve the following recurrence relations :

   (*i*) $a_r - 3a_{r-1} + 2a_{r-2} = 0$ (*ii*) $a_r + a_{r-1} + a_{r-2} = 0$

   (*iii*) $a_r + a_{r-1} + a_{r-2} + a_{r-3} = 0.$

(*b*) **Non-Homogeneous Linear Difference Equations.** There are two methods to find the particular solution of non-homogeneous linear difference equations. These are as follows :

1. Undetermined coefficients method
2. E and Δ operator method.

1. Undetermined Coefficients Method. This method is used to find particular solution of non-homogeneous linear difference equations, whose R.H.S. term R(n) consists of terms of special forms.

In this method, firstly we assume the general form of the particular solution according to the form of R(n) containing a number of unknown constant coefficients, which have to be determined. Then according to the difference equation, we will determine the exact solution.

The general form of particular solution to be assumed for the special forms of R(n), to find the exact solution is shown in the table.

Form of R(n)	General form to be assumed
Z, here Z is a constant	A
Z^r, here Z is a constant	A . Z^r
P(r), a polynomial of degree n	$A_0 r^n + A_1 r^{n-1} + ... + A_n$
Z^r. P(r), here P(r) is a polynomial of nth degree in r. Z is a constant.	$[A_0 r^n + A_1 r^{n-1} + ... + A_n]. Z^r$

Example 16. *Find the particular solution of the difference equation*

$$a_{r+2} - 3a_{r+1} + 2a_r = Z^r, \quad ...(i)$$

where Z is some constant.

Sol. The general form of solution is = A . Z^r

Now putting this solution on L.H.S. of equation (i), we get

$$= AZ^{r+2} - 3AZ^{r+1} + 2AZ^r = (Z^2 - 3Z + 2)AZ^r \quad ...(ii)$$

Equating equation (ii) with R.H.S. of equation (i), we get

$$(Z^2 - 3Z + 2)A = 1$$

or

$$A = \frac{1}{Z^2 - 3Z + 2} = \frac{1}{(Z-1)(Z-2)} \quad (Z \neq 1, Z \neq 2)$$

Therefore, the particular solution is $\dfrac{Z^r}{(Z-1)(Z-2)}$.

Example 17. *Find the particular solution of the difference equation*

$$a_{r+2} - 5a_{r+1} + 6a_r = 5^r. \quad ...(i)$$

Sol. Let us assume the general form of the solution = A . 5^r.

Now to find the value of A, put this solution on L.H.S. of the equation (i), then this becomes

$$= A.\ 5^{r+2} - 5\ .\ A5^{r+1} + 6\ .\ A5^r$$
$$= 25A\ .\ 5^r - 25A\ .\ 5^r + 6A\ .\ 5^r$$
$$= 6A\ .\ 5^r \quad ...(ii)$$

Equating equation (ii) to R.H.S. of equation (i), we get

$$A = \frac{1}{6}$$

Therefore, the particular solution of the difference equation is = $\dfrac{1}{6} . 5^r$.

Example 18. *Find the particular solution of the difference equation*

$$a_{r+2} - 4a_r = r^2 + r - 1. \quad ...(i)$$

Sol. The homogeneous solution of the difference equation is given by

$$a_{r(h)} = C_1(2)^r + C_2(-2)^r$$

To find the particular solution, let us assume the general form of the solution is

$$= A_1 r^2 + A_2 r + A_3.$$

Putting this solution in L.H.S. of equation (*i*), we get

$$= A_1(r+2)^2 + A_2(r+2) + A_3 - 4A_1 r^2 - 4A_2 r - 4A_3$$

$$= -3A_1 r^2 + (4A_1 - 3A_2)r + (4A_1 + 2A_2 - 3A_3) \quad ...(ii)$$

Equating equation (*ii*) with R.H.S. of equation (*i*), we get

$$-3A_1 = 1$$

$$4A_1 - 3A_2 = 1$$

$$4A_1 + 2A_2 - 3A_3 = -1$$

After solving these three equations, we get

$$A_1 = -\frac{1}{3}, A_2 = -\frac{7}{9}, A_3 = -\frac{17}{27}$$

Therefore, the particular solution is $= -\frac{r^2}{3} - \frac{7}{9}r - \frac{17}{27}$.

Example 19. *Find the particular solution of the difference equation*

$$a_{r+2} - 2a_{r+1} + a_r = 3r + 5. \quad ...(i)$$

Sol. The homogeneous solution of the difference equation is given by

$$a_{r(h)} = C_1 + C_2 r \quad ...(ii)$$

Corresponding to the term $3r + 5$, we assume the general form of the solution as $A_1 r + A_2$ but due to occurrence of these terms in equation (*ii*), we multiply this by suitable power of r so that none of the term will occur in equation (*ii*). Thus multiply by r^2.

Hence, the general form of the solution becomes

$$= A_1 r^3 + A_2 r^2 \;......$$

Putting this solution in L.H.S. of equation (*i*), we get

$$= A_1(r+2)^3 + A_2(r+2)^2 - 2A_1(r+1)^3 - 2A_2(r+1)^2 + A_1 r^3 + A_2 r^2$$

$$= A_1(r^3 + 8 + 6r^2 + 12r) + A_2(r^2 + 4 + 4r) - 2A_1(r^3 + 1 + 3r^2 + 3r) - 2A_2(r^2 + 1 + 2r) + A_1 r^3 + A_2 r^2$$

$$= (12A_1 + 4A_2 - 6A_1 - 4A_2)r + (8A_1 + 4A_2 - 2A_1 - 2A_2)$$

$$= (6A_1)r + (6A_1 + 2A_2) \quad ...(iii)$$

Equating equation (*iii*) with R.H.S. of equation (*i*), we get

$$6A_1 = 3 \quad \therefore \quad A_1 = \frac{1}{2}$$

$$6A_1 + 2A_2 = 5 \quad \therefore \quad A_2 = 1$$

Therefore, the particular solution is $\frac{1}{2}r^3 + r^2$.

Example 20. *Find the particular solution of the difference equation*

$$a_{r+2} + a_{r+1} + a_r = r \cdot 2^r. \quad ...(i)$$

Sol. Let us assume the general form of the solution = $(A_0 + A_1 r) . 2^r$

Now, put this solution in the L.H.S. of equation (*i*), we get

$$= 2^{r+2} [A_0 + A_1 (r + 2)] + 2^{r+1} [A_0 + A_1 (r + 1)] + 2^r (A_0 + A_1 r)$$
$$= 4. 2^r (A_0 + A_1 r + 2A_1) + 2 . 2^r (A_0 + A_1 r + A_1) + 2^r (A_0 + A_1 r)$$
$$= r. 2^r(7A_1) + 2^r (7A_0 + 10A_1) \quad ...(ii)$$

Equating equation (*ii*) with R.H.S. of equation (*i*), we get

$$7A_1 = 1 \quad \therefore \quad A_1 = \frac{1}{7}$$

$$7A_0 + 10A_1 = 0 \quad \therefore \quad A_0 = \frac{-10}{49}$$

Therefore, the particular solution is $2^r \left(\frac{-10}{49} + \frac{1}{7} r\right)$.

Example 21. *Find the particular solution of the difference equation*

$$a_r - 4a_{r-1} + 4a_{r-2} = (r + 1) . 2^r. \quad ...(i)$$

Sol. The homogeneous solution of the difference equation is given by

$$a_{r(h)} = (C_1 + C_2 r) . 2^r \quad ...(ii)$$

because it has two real and equal roots *i.e.,* 2 and 2.

To find the particular solution, let us assume the general form of the solution is $= 2^r (A_1 r + A_0)$, but due to occurrence of there terms in equation (*ii*), we multiply this by suitable power of r so that none of the terms will occur in equation (*ii*). Thus multiply by r^2.

Hence, the general form of the solution becomes = $2^r (A_1 r + A_0) . r^2$

Putting this solution in L.H.S. of equation (*i*), we get

$$= 2^r. (A_1 r + A_0) . r^2 - 4 . 2^{r-1} [A_1(r - 2) + A_0] . (r - 1)^2 + 4 . 2^{r-2} [A_1 (r - 2) + A_0] . (r - 2)^2$$
$$= 2^r .(A_1 r + A_0) . r^2 - 2(r^2 + 1 - 2r) . 2^r (A_1 r - A_1 + A_0) + (r^2 + 4 - 4r) . 2^r. (A_1 r - 2A_1 + A_0)$$
$$= r . 2^r (6A_1) + 2^r (- 6A_1 + 2A_0) \quad ...(iii)$$

Equating equation (*iii*) with R.H.S. of equation (*i*), we get

$$6A_1 = 1 \quad \therefore \quad A_1 = \frac{1}{6}$$

$$- 6A_1 + 2A_0 = 1 \quad \therefore \quad A_0 = 1$$

Therefore, the particular solution is $= r^2. 2^r\left(\frac{r}{6} + 1\right)$.

EXERCISES

1. Find the particular solution of the recurrence relation

$$a_n - 3a_{n-1} + 4a_{n-2} = 1.$$

2. Find the particular solution of the following recurrence relations :

(*i*) $a_n - 4a_{n-1} + 3a_{n-2} = 2^n$ (*ii*) $a_{n+2} - 2a_{n+1} + a_n = 2^n$

(*iii*) $a_{n+2} - 2a_{n+1} + a_n = 1 + n.$

3. Find the particular solution of the following recurrence relations :

(i) $a_n - a_{n-1} - a_{n-2} = n + 3$ (ii) $a_n + 5a_{n-1} + 6a_{n-2} = 42.4^n$

(iii) $a_n - 7a_{n-1} + 16a_{n-2} = n.4^n$.

4. Solve the following recurrence relations :

(i) $a_{n+2} - 2a_{n+1} + a_n = 3n + 5$ (ii) $a_n + 4a_{n-1} + 4a_{n-2} = n^2 - 3n + 5$

(iii) $a_{n+2} - 5a_{n+1} + 6a_n = n^2$ (iv) $a_n - 6a_{n-1} + 8a_{n-2} = 3$.

2. E and Δ Operator Method. Before discussing this method to find the particular solution of the non-homogeneous linear difference equation, we will discuss the meaning of the operator E and Δ.

Meaning of Operator E. The operation of E on $f(x)$ means that give an increment to the value of x in the function. The operation of E is, put $(x + h)$ in the function wherever there is x. Here h is increment quantity. So

$$\mathrm{E}f(x) = f(x + h)$$

Here, E is operated on $f(x)$, therefore, E is a symbol known as shift operator.

Meaning of Operator Δ. The operator Δ is an operation of two steps.

Firstly, x in the function is incremented by a constant and then former is subtracted from the later *i.e.*,

$$\Delta f(x) = f(x + h) - f(x).$$

Theorem I. *Prove that* $E \cong 1 + \Delta$.

Proof. The operation of Δ on $f(x)$ is of two steps. First, increment the value of x in the function. So, whenever, there is x in $f(x)$ put $x + h$ (here h is constant increment), which means operation of E on $f(x)$ *i.e.*,

$$f(x + h) = \mathrm{E}f(x).$$

Second, subtract the original function from the value obtained in first step, hence

$$\Delta f(x) = \mathrm{E}f(x) - 1\, f(x) = (\mathrm{E} - 1)\, f(x)$$

[Where E and I are operators and 1 is an identity operator]

So, the operation of Δ on $f(x)$ is equivalent to the operation of (E – 1) on $f(x)$.

Therefore, we have

$$\mathrm{E} \cong 1 + \Delta.$$

Theorem II. *Show that* $E^n f(x) = f(x + nh)$.

Proof. We know that $\mathrm{E}\, f(x) = f(x + h)$

Now

$$\begin{aligned} \mathrm{E}^n f(x) &= \mathrm{E}\,.\,\mathrm{E}\,.\,\mathrm{E}\,.\,\mathrm{E}.\ \ldots\ n \text{ times } f(x) \\ &= \mathrm{E}^{n-1}[\mathrm{E}f(x)] = \mathrm{E}^{n-1} f(x + h) \\ &= \mathrm{E}^{n-2}[\mathrm{E}\, f(x + h)] = \mathrm{E}^{n-2} f(x + 2h) \\ &\ldots\ldots\ldots\ldots\ldots\ldots \\ &\ldots\ldots\ldots\ldots\ldots\ldots \\ &= \mathrm{E}\, f\,[x + (n - 1)h] = f(x + nh). \end{aligned}$$

Theorem III. *Show that* $E\ Cf(x) = CE\ f(x)$.

Proof. We know that E $Cf(x) = Cf(x + h) = \mathrm{CE}\ f(x + h)$. Hence proved.

There is no effect of operation of E on any constant. Therefore, the operation of E on any constant will be equal to the constant itself.

By E and Δ operator method, we will find the solution of

$$C_0 y_{n+r} + C_1 y_{n+r-1} + C_2 y_{n+r-2} + \ldots\ldots + C_n y_n = R(n) \quad \ldots(i)$$

Equation (*i*) can be written as

$$C_0 E^r y_n + C_1 E^{r-1} y_n + C_2 E^{r-2} y_n + \ldots + C_n y_n = R(n)$$

i.e.,
$$(C_0 E^r + C_1 E^{r-1} + C_2 E^{r-2} + \ldots\ldots + C_n) y_n = R(n)$$

Putting
$$C_0 E^r + C_1 E^{r-1} + C_2 E^{r-2} + \ldots + C_n = P(E)$$

So
$$P(E) y_n = R(n)$$

$$\therefore \quad y_n = \frac{R(n)}{P(E)} \quad \ldots(ii)$$

To find the particular solution of (*ii*) for different forms of R(*n*), we have the following cases.

Case I. When R(*n*) is some constant A

We know that, the operation of E on any constant will be equal to the constant itself *i.e.,*

$$EA = A$$

Therefore,
$$P(E)\,A = (C_0 E^r + C_1 E^{r-1} + C_2 E^{r-2} + \ldots + C_n)A$$
$$= (C_0 + C_1 + C_2 + \ldots + C_n)A$$
$$= P(1)A$$

i.e.,
$$\frac{A}{P(E)} = \frac{1}{P(1)} . A$$

Therefore, using equation (*ii*), the particular solution of (*i*) is

$$y_n = \frac{A}{P(1)},\ P(1) \neq 0$$

P(1) is obtained by putting E = 1 in P(E).

Case II. When R(*n*) is of the form $A . Z^n$, where A and Z are constants

We have,
$$P(E)(A . Z^n) = \{C_0 E^r + C_1 E^{r-1} + \ldots + C_n\}(A . Z^n)$$
$$= A\{C_0 Z^{r+n} + C_1 Z^{r+n-1} + \ldots + C_n Z^n\}$$
$$= A\{C_0 Z^r + C_1 Z^{r-1} + \ldots + C_n\}. Z^n$$
$$= AP(Z) . Z^n$$

To get P(Z) put E = Z in P(E)

Therefore,
$$\frac{A . Z^n}{P(E)} = \frac{A . Z^n}{P(Z)},\ \text{provided } P(Z) \neq 0$$

Thus,
$$y_n = \frac{A . Z^n}{P(Z)},\ P(Z) \neq 0$$

If A = 1, then
$$y_n = \frac{Z^n}{P(Z)}$$

When P(Z) = 0 then for equation

(*i*)
$$(E - Z) y_n = A . Z^n$$

For this, the particular solution becomes $= A. \dfrac{1}{(E - Z)} . Z^n = A . n\, Z^{n-1}$

(*ii*) $(E - Z)^2 y_n = A \cdot Z^n$

For this, the particular solution becomes $= A. \dfrac{1}{(E-Z)^2} . Z^n = \dfrac{A . n(n-1)}{2!} . Z^{n-2}$

(*iii*) $(E - Z)^3 y_n = A \cdot Z^n$

For this, the particular solution becomes $= A . \dfrac{1}{(E-Z)^3} . Z^n = \dfrac{A . n(n-1)(n-2)}{3!} . Z^{n-3}$ and so on.

Case III. When R(*n*) be a polynomial of degree *m* in *n*

We know that $E \equiv 1 + \Delta$

So, $P(E) = P(1 + \Delta)$

or $$\frac{1}{P(E)} = \frac{1}{P(1+\Delta)},$$

which can be expanded in ascending power of Δ as far as upto Δ^m

$$\Rightarrow \quad \frac{1}{P(E)} = \frac{1}{P(1+\Delta)} = (b_0 + b_1\Delta + b_2\Delta + ... + b_m \Delta^m + ...)$$

$$\Rightarrow \quad \frac{1}{P(E)} . R(n) = (b_0 + b_1\Delta + b_2\Delta + ... + b_m\Delta^m + ...) . R(n)$$

$$= b_0 R(n) + b_1\Delta R(n) + ... + b_m \Delta^m R(n)$$

All other higher terms will be zero because R(*n*) is a polynomial of degree *m*.

Thus, the particular solution of equation (*i*), in this case will be

$$y_n = b_0 R(n) + b_1 \Delta R(n) + ... + b_m \Delta^m R(n).$$

Case IV. When R(*n*) is of the form R(*n*) . Z^n, where R(*n*) is a polynomial of degree *m* and Z is some constant

We have $E^r [Z^n R(n)] = Z^{r+n} R(n + r) = Z^r. Z^n. E^r. R(n) = Z^n (ZE)^r R(n)$

Similarly, we have

$$\frac{1}{P(E)} [Z^n R(n)] = Z^n \frac{1}{P(ZE)} . (R(n)) = Z^n [P(Z + Z\Delta)]^{-1}. R(n)$$

Thus, the particular solution of equation (*i*), in this case will be

$$y_n = Z^n [P(Z + Z\Delta)]^{-1} R(n).$$

Example 22. *Find the particular solution of the difference equation*

$$2a_{r+1} - a_r = 12.$$

Sol. The above equation can be written as

$$(2E - 1) a_r = 12$$

The particular solution is given by

$$a_r = \frac{1}{(2E-1)} . 12$$

Put E = 1, in the equation. The particular solution is $a_r = 12$.

Example 23. *Find the particular solution of difference equation* $a_r - 3a_{r-1} + 2a_{r-2} = 2^r$.

Sol. The above equation can be written as

$$(E^2 - 3E + 2) a_r = 2^r$$

Therefore, $P(E) = E^2 - 3E + 2$ or $P(E) = (E - 2)(E - 1)$

Thus, the particular solution is given by

$$a_r = \frac{1}{(E-2)(E-1)} \cdot 2^r$$

$$= r.\ \frac{1}{(E-1)} \cdot 2^{r-1} = r.\ \frac{1}{(2-1)} \cdot 2^{r-1}$$

$$a_r = r.\ 2^{r-1}.$$

Example 24. *Find the particular solution of difference equation*

$$a_r - 4a_{r-1} + 4a_{r-2} = 2^r.$$

Sol. The above equation can be written as

$$(E^2 - 4E + 4)\ a_r = 2^r$$

Therefore, $P(E) = E^2 - 4E + 4 = (E-2)^2$

Thus, the particular solution is given by

$$a_r = \frac{1}{(E-2)^2} \cdot 2^r = \frac{r(r-1)}{2!} \cdot 2^{r-2}$$

$$a_r = r(r-1).\ 2^{r-3}.$$

Example 25. *Find the particular solution of the difference equation*

$$a_r - 2a_{r-1} = 7r^2.$$

Sol. The above equation can be written as $(E-2)\ a_r = 7r^2$

Thus, the particular solution is given by

$$a_r = \frac{1}{(E-2)} \cdot 7r^2 = \frac{1}{(\Delta-1)} \cdot 7r^2 = -7.\frac{1}{(1-\Delta)} \cdot r^2$$

$$= -7(1-\Delta)^{-1}.\ (r^2) = -7(1+\Delta+\Delta^2)\ (r^2)$$

By factorial notation, $[r]^n = r(r-1)(r-2) \ldots (r-n+1)$.

Hence $(r^2) = [r]^2 + [r]$

$$a_r = -7(1+\Delta+\Delta^2)\ ([r]^2 + [r]) = -7([r]^2 + [r] + 2[r] + 1 + 2 + 0)$$

The operation of Δ on $[r]^n$ is analogous to its differentiation

$$a_r = -7[r]^2 - 21[r] - 21 = -7r(r-1) - 21r - 21$$

$$a_r = -7r^2 - 14r - 21.$$

Example 26. *Find the particular solution of the difference equation*

$$a_r - 2a_{r-1} = 7r.$$

Sol. The above equation can be written as $(E-2)\ a_r = 7r$

Thus, the particular solution is given by

$$a_r = \frac{1}{(E-2)} \cdot 7r = 7.\ \frac{1}{(\Delta-1)} \cdot r = -7\ \frac{1}{(1-\Delta)}\ [r]$$

(By factorial notation $r = [r]$)

$$= -7(1-\Delta)^{-1}\ [r] = -7(1+\Delta+\Delta^2+\ldots)\ [r]$$

By neglecting higher powers

$$a_r = -7(1+\Delta)\ [r] = -7([r]+1)$$

$$a_r = -7(r+1).$$

Example 27. *Find the particular solution of the difference equation*

$$a_r + a_{r-1} + a_{r-2} = r.\ 2^r.$$

Sol. The above equation can be written as $(E^2 + E + 1)\, a_r = r\ .\ 2^r$

Thus, the particular solution is given by

$$a_r = \frac{1}{(E^2 + E + 1)}\ .\ r.\ 2^r = 2^r\ \frac{1}{(2E)^2 + 2E + 1}\ .\ r$$

$$= 2^r\ (4(1+\Delta)^2 + 2(1+\Delta) + 1)^{-1}.\ r = 2^r\ (7 + 10\Delta + 4\Delta^2)^{-1}.\ r$$

$$= \frac{2^r}{7}\left(1 + \frac{10\Delta + 4\Delta^2}{7}\right)^{-1} .\ [r] = \frac{2^r}{7}\left(1 - \frac{10\Delta + 4\Delta^2}{7}\right).\ [r] = \frac{2^r}{7}\left([r] - \frac{10}{7}\right)$$

$$a_r = \frac{2^r}{7}\left(r - \frac{10}{7}\right).$$

EXERCISES

Solve all the questions in the previous EXERCISES using E and Δ method.

TOTAL SOLUTION

The total solution or the general solution of a non-homogeneous linear difference equation with constant coefficients is the sum of the homogeneous solution and particular solution. If no initial conditions are given, then our work is finished. If n initial conditions are given, obtain n linear equations in n unknowns and solve them, if possible to get a total solution.

As we already know that the homogeneous solution of the recurrence relation is obtained by putting the right hand side of the equation equal to zero and the particular solution is obtained with R(n) on the right hand side of the equation.

If $y_{(h)}$ denotes the homogeneous solution of the recurrence relation and $y_{(p)}$ denotes the particular solution of the recurrence relation then, the total solution or the general solution y of the recurrence relation is given by

$$y = y_{(h)} + y_{(p)}.$$

Example 28. *Solve the difference equation*

$$a_r - 4a_{r-1} + 4a_{r-2} = 3r + 2^r. \qquad \text{...}(i)$$

Sol. The homogeneous solution of this equation is obtained by putting R.H.S. equal to zero *i.e.,*

$$a_r - 4a_{r-1} + 4a_{r-2} = 0$$

The homogeneous solution is $a_{r(h)} = (C_1 + C_2\, r)\ .\ 2^r$

as shown in example 7.

The equation (i) can be written as $(E^2 - 4E + 4)\, a_r = 3r + 2^r$

The particular solution is given as

$$a_{r(p)} = \frac{1}{(E^2 - 4E + 4)}.\ (3r + 2^r) = \frac{1}{(E-2)^2}\ .\ (3r) + \frac{1}{(E-2)^2}\ .\ 2^r$$

$$= 3 \cdot \frac{1}{(1-\Delta)^2}(r) + \frac{r(r-1)}{2!} \cdot 2^{r-2} = 3(1-\Delta)^{-2}(r) + \frac{r(r-1)}{2!} \cdot 2^{r-2}$$

$$= 3(1+2\Delta) \cdot [r] + \frac{r(r-1)}{2!} \cdot 2^{r-2} = 3(r+2) + r(r-1) \cdot 2^{r-3}$$

$$a_{r(p)} = 3(r+2) + r(r-1) \cdot 2^{r-3}$$

Therefore, the total solution is $a_r = (C_1 + C_2 r) \cdot 2^r + 3(r+2) + r(r-1) \cdot 2^{r-3}$.

Example 29. *Solve the difference equation*

$$a_r + 4a_{r-1} + 4a_{r-2} = r^2 - 3r + 5. \quad \ldots(i)$$

Sol. Put the R.H.S. of the equation equal to zero *i.e.,*

$$a_r + 4a_{r-1} + 4a_{r-2} = 0$$

This can be written as

$$(E^2 + 4E + 4)\, a_r = 0$$

or $$(E+2)^2 = 0 \quad \text{or} \quad E = -2, -2$$

The homogeneous solution is given by

$$a_{r(h)} = (C_1 + C_2 r) \cdot (-2)^r$$

The equation (i) can be written as

$$(E^2 + 4E + 4)\, a_r = r^2 - 3r + 5$$

The particular solution is given by

$$a_{r(p)} = \frac{1}{E^2+4E+4} \cdot [(r)^2 - 3(r) + 5] = \frac{1}{(E+2)^2}(r^2 - 3r + 5)$$

$$= \frac{1}{(3+\Delta)^2} \cdot ([r]^2 - 2[r] + 5)$$

{By factorial notation, $[r]^2 = r(r-1)$ and $[r] = r$}

$$= \frac{1}{9} \cdot \frac{1}{\left(1+\frac{\Delta}{3}\right)^2} \cdot ([r]^2 - 2[r] + 5)$$

$$= \frac{1}{9}\left(1+\frac{\Delta}{3}\right)^{-2} \cdot ([r]^2 - 2[r] + 5)$$

$$= \frac{1}{9}\left(1-\frac{2\Delta}{3}+\frac{3\Delta^2}{9}\right) \cdot ([r]^2 - 2[r] + 5)$$

$$= \frac{1}{9}\left([r]^2 - \frac{10}{3}[r] + 7\right)$$

$$a_{r(p)} = \frac{1}{9}\left(r^2 - \frac{13}{3}r + 7\right)$$

Therefore, the total solution is

$$a_r = a_{r(h)} + a_{r(p)}$$

$$a_r = (C_1 + C_2 r) \cdot (-2)^r + \frac{1}{9}\left(r^2 - \frac{13}{3}r + 7\right).$$

Example 30. *Solve the difference equation*

$$a_r + a_{r-1} + a_{r-2} = r.\ 2^r. \qquad ...(i)$$

Sol. Put the R.H.S. of the equation to zero *i.e.,*

$$a_r + a_{r-1} + a_{r-2} = 0$$

The homogeneous solution is

$$a_{r(h)} = \left[\frac{-1+i\sqrt{3}}{2}\right]^r C_1 + \left[\frac{-1-i\sqrt{3}}{2}\right]^r C_2$$

as shown in example 8.

The equation (*i*) can be written as $(E^2 + E + 1)\, a_r = r.\ 2^r$

The particular solution is given $a_{r(p)} = \dfrac{2^r}{7}\left(r - \dfrac{10}{7}\right)$

as shown in example.

Therefore, the total solution is

$$a_r = a_{r(h)} + a_{r(p)}$$

$$a_r = \left(\frac{-1+i\sqrt{3}}{2}\right)^r C_1 + \left(\frac{-1-i\sqrt{3}}{2}\right)^r C_2 + \frac{2^r}{7}\left(r - \frac{10}{7}\right).$$

Example 31. *Solve the difference equation*

$$a_r + 6a_{r-1} + 9a_{r-2} = 3 \qquad ...(i)$$

with initial conditions $a_0 = 0$ *and* $a_1 = 1$.

Sol. Put the R.H.S. of the equation (*i*) equal to zero *i.e.,*

$$a_r + 6a_{r-1} + 9a_{r-2} = 0$$

This can be written as $\quad E^2 + 6E + 9 = 0$

or $\quad (E + 3)(E + 3) = 0$

or $\quad E = -3, -3$

The homogeneous solution is $a_{r(h)} = (C_1 + C_2\, r)\,.\,(-3)^r$

Now, we have $\quad P(E) = (E + 3)(E + 3)$

Putting E = 1,we get P(1) = 16

Hence, the particular solution is $a_{r(p)} = \dfrac{3}{16}$

Therefore, the total solution is $\quad a_r = (C_1 + C_2\, r)(-3)^r + \dfrac{3}{16}$

Now, putting $r = 0$, we have $\quad a_0 = (C_1 + 0)(1) + \dfrac{3}{16} = 0$

$\therefore \quad C_1 = -\dfrac{3}{16}$

Now, putting $r = 1$, we have $\quad a_1 = (C_1 + C_2)(-3) + \dfrac{3}{16} = 1$

Substitute the value of $C_1 = -\frac{3}{16}$ in the equation, we have

$$\left(-\frac{3}{16} + C_2\right)(-3) + \frac{3}{16} = 1$$

$$\left(-\frac{3}{16} + C_2\right)(-3) = \frac{13}{16}$$

$$C_2 = -\frac{1}{12}$$

Therefore, the total solution satisfying initial condtions is

$$a_r = \left(-\frac{3}{16} - \frac{r}{12}\right)(-3)^r + \frac{3}{16}.$$

EXERCISES

1. Solve the following recurrence relations :
   (*i*) $a_{n+2} + 2a_{n+1} - 15a_n = 6n + 10$; $a_0 = 1, a_1 = 2$
   (*ii*) $a_n - 4a_{n-1} + 3a_{n-2} = 2^n$; $a_1 = 1, a_2 = 1$.
2. Solve the following recurrence relations :
   (*i*) $a_r - 7a_{r-1} + 10a_{r-2} = 5^r$; $a_0 = 0$ and $a_1 = 1$
   (*ii*) $a_r + 6a_{r-1} + 9a_{r-2} = 5$; $a_0 = 0$ and $a_1 = 1$.
3. Find the solution of the recurrence relation
   $$a_n = -5a_{n-1} - 6a_{n-2} + 42.4^n$$
   with initial conditions $a_1 = 56$ and $a_2 = 278$.
4. Solve the recurrence relation
   $$a_n = 7a_{n-1} - 16a_{n-2} + 12a_{n-3} + n4^n$$
   with initial conditions $a_0 = -2$, $a_1 = 0$ and $a_2 = 5$.
5. Solve the recurrence relation
   $$a_{n+2} - 5a_{n+1} + 6a_n = 2$$
   with initial conditions $a_0 = 1, a_1 = 2$.
6. Solve the recurrence relation
   $$a_r - 5a_{r-1} + 6a_{r-2} = 2^r + r, r \geq 2$$
   with initial conditions $a_0 = 1, a_1 = 1$.
7. Solve the recurrence relation
   $$a_r - 7a_{r-1} + 10a_{r-2} = 2r^2 + 2$$
   with initial conditions $a_0 = 1, a_1 = 1$.

DISCRETE NUMERIC FUNCTION

A function whose domain is the set of whole number and range is the set of real number is called a discrete numeric function or simply a numeric function. For a function a, we use a_0, $a_1, a_2, \ldots, a_r, \ldots$ to denote the values of the function at 0, 1, 2, ..., r, ... Thus a_r is the value of the numeric function at r.

e.g., 1. The function $a = 5r^2 + 10, r \geq 0$ is a numeric function. We can also write this function as

$$a_r = 5r^2 + 10, r \geq 0.$$

2. $$a_r = \begin{cases} 5r, & 0 \le r \le 10 \\ 2^r, & r \ge 11 \end{cases}$$

3. $$a_r = \begin{cases} -5, & r = 2, 4, 6 \\ 1, & \text{otherwise} \end{cases}$$

Example 32. *A person deposits 1000 Rupees in the saving account at an interest rate of 4% per annum compounded annually. Determine the amount in the account at the end of r years. Also write the numeric function.*

Sol. The amount after one year is

$$a_1 = 1000\left(1 + \frac{4}{100}\right)^1 = 1000\,(1.04)^1$$

The amount after two years is

$$a_2 = 1000\left(1 + \frac{4}{100}\right)^2 = 1000\,(1.04)^2$$

Similarly, the amount after r years is

$$a_r = 1000\left(1 + \frac{4}{100}\right)^r = 1000\,(1.04)^r$$

Hence, the required numeric function.

Example 33. *A ball is dropped to the floor from a height of 50 m. Assume that the ball always rebounds to reach half of the height from which it falls.*

(a) Let a_r denotes the height it reaches in the rth rebound. Write the numeric function a

(b) Let b_r denotes the loss in height during the rth rebound. Express b_r in terms of a_r and also write the numeric function b.

Sol. (*a*) The height attained by the ball in the first rebound is

$$a_1 = \frac{50}{2}$$

The height attained by the ball in the second rebound is

$$a_2 = \frac{1}{2} \times \frac{50}{2} = \frac{50}{2^2}$$

and in the same way, the height attained by the ball in the rth rebound is

$$a_r = \frac{50}{2^r}$$

which is the required numeric function.

(*b*) The loss in height during the first rebound is

$$b_1 = \left(50 - \frac{50}{2}\right)$$

The loss in height during the second rebound is

$$b_2 = \left(50 - \frac{50}{2^2}\right)$$

and similarly, the loss in height during the rth rebound is

$$b_r = 50 - \frac{50}{2^r} \text{ m}$$

$$b_r = 50 - a_r \qquad \Big|\because\ a_r = \frac{50}{2^r}$$

which is the required numeric function.

Sum and Product of Two Numeric Functions

The sum of two numeric functions a_r and b_r is a numeric function k_r whose value at k is equal to the sum of the values of the given numeric function at r.

The product of two numeric functions a_r and b_r is a numeric function l_r whose value at r is equal to the product of the values of the given numeric function at r.

Example 34. *Consider the two functions a and b where*

$$a_r = \begin{cases} 0, & 0 \le r \le 3 \\ 2^{-r} + 3, & r \ge 5 \end{cases}$$

and

$$b_r = \begin{cases} 1 - 2^r, & 0 \le r \le 2 \\ r + 4, & r \ge 3 \end{cases}$$

Sol. Let c be the sum of the two functions a and b, denoted by $c = a + b$. Then

$$c_r = a_r + b_r = \begin{cases} 1-2^r, & 0 \le r \le 2 \\ 7, & r = 3 \\ 2^{-r} + r + 7, & r \ge 4 \end{cases}$$

Let d denotes the product of the two functions, given as $d = ab$, then

$$d_r = a_r\, b_r = \begin{cases} 0, & 0 \le r \le 3 \\ r.2^{-r} + 2^{-r+2} + 3r + 7, & r \ge 4 \end{cases}$$

Modulus of a Numeric Function

Let a be a discrete numeric function. Then modulus of a denoted by $|\ a\ |$ is a numeric function defined as

$$|\ a\ | = a_r \text{ if } a_r \text{ is positive}$$
$$= -\,a_r \text{ if } a_r \text{ is negative.}$$

Example 35. *Consider the numeric function a defined as*

$$a_r = (-1)^r \left(\frac{2}{r^3}\right), r \ge 0.$$

Determine $|\ a\ |$.

Sol. Let us assume that $c = |\ a\ |$

So $$c_r = \frac{2}{r^3}, r \ge 0.$$

The Accumulated Sum of a Numeric Function

The accumulated sum of a numeric function a is a numeric function whose value at r is equal to $\sum_{i=0}^{r} a_i$.

Example 36. *A person deposits Rs. 1000 in a saving account at an annual compounded interest rate of 4%. Determine the total amount after r years.*

Sol. Let $a_r = 1000\,(1.04)^r, \quad r \geq 0$

Let b be the accumulated sum of a. Then

$$b_r = \sum_{i=0}^{r} a_i = \sum_{i=0}^{r} 1000\,(1.04)^i = \frac{10,0000}{4}((1.04)^{r+1} - 1), r \geq 0.$$

$S^i a$ for a Numeric Function

Consider a numeric function a and a +ve integer i. Then $S^i a$ is used to denote a numeric function such that its value at r is 0 for $r = 0, 1, 2, 3, \ldots\ldots i - 1$ and is a_{r-i} for $r \geq i$.

e.g., Let us consider the numeric function

$$a_r = \begin{cases} 3, & 0 \leq r \leq 20 \\ 5, & r \geq 21 \end{cases}$$

Now, let us assume that $b = S^4 a$. Then

$$b_r = \begin{cases} 0, & 0 \leq r \leq 3 \\ 3, & 4 \leq r \leq 24 \\ 5, & r \geq 25 \end{cases}$$

Again, let us assume that $c = S^7 a$. Then

$$c_r = \begin{cases} 0, & 0 \leq r \leq 6 \\ 3, & 7 \leq r \leq 27 \\ 5, & r \geq 28 \end{cases}$$

$S^{-i} a$ for a Numeric Function

Consider a numeric function a and a +ve integer i. Then $S^{-i} a$ is used to denote a numeric function such that its value at r is a_{r+i} for $r \geq 0$.

e.g., Let us consider the numeric function

$$a_r = \begin{cases} 3, & 0 \leq r \leq 20 \\ 5, & r \geq 21 \end{cases}$$

Now, Let us assume that $b = S^{-4}\, a$. Then

$$b_r = \begin{cases} 3, & 0 \leq r \leq 16 \\ 5, & r \geq 17 \end{cases}$$

Again, let us assume that $c = S^{-7}\, a$. Then

$$c_r = \begin{cases} 3, & 0 \leq r \leq 13 \\ 5, & r \geq 14 \end{cases}$$

Example 37. *Let a be a numeric function such that*

$$a_r = \begin{cases} 1, & 0 \leq r \leq 3 \\ 2^r + 3, & r \geq 4 \end{cases}$$

Determine S^2 a and S^{-2} a.

Sol. Let us assume that $b = S^2 a$. Then

$$b_r = \begin{cases} 0, & 0 \le r \le 1 \\ 1, & 2 \le r \le 5 \\ 2^{r+2} + 3, & r \ge 6 \end{cases}$$

Now, let us assume that $c = S^{-2} a$. Then

$$c_r = \begin{cases} 1, & 0 \le r \le 1 \\ 2^{r-2} + 3, & r \ge 2 \end{cases}$$

The Forward difference of a Numeric Function (Δa)

The forward difference of a numeric function a is a numeric function denoted by Δa, whose value at r is given by

$$\Delta a_r = a_{r+1} - a_r, \quad r \ge 0$$

The Backward difference of a Numeric Function (∇a)

The backward difference of a numeric function a is a numeric function denoted by ∇a, whose value at r is given by

$$\nabla a_r = a_r - a_{r-1}, \quad r \ge 1.$$

Example 38. *Consider the numeric function given below :*

$$a_r = \begin{cases} 0, & 0 \le r \le 3 \\ 2^r + 5, & r \ge 4 \end{cases}$$

Determine the forward difference Δa_r and backward difference ∇a_r.

Sol. Let us assume that $b_r = \Delta a_r$ be the forward difference of the numeric function. Then

$$b_r = a_{r+1} - a_r$$

Thus
$$b_r = \begin{cases} 0, & 0 \le r \le 2 \\ 21, & r = 3 \\ 2^r, & r \ge 4 \end{cases}$$

Now let us assume that $c_r = \Delta a_r$ be the backward difference of the numeric function. Then

$$c_r = a_r - a_{r-1}$$

Thus
$$c_r = \begin{cases} 0, & 0 \le r \le 3 \\ 21, & r = 4 \\ 2^{r-1}, & r \ge 5 \end{cases}$$

Convolution of Two Numeric Functions

If a and b are two numeric functions then their convolution denoted by $a * b$ is a numeric function c defined as

$$a * b = a_0 b_r + a_1 b_{r-1} + a_2 b_{r-2} + \ldots\ldots + a_r b_0$$

$$= \sum_{i=0}^{r} a_i \, b_r - i$$

Example 39. *Let a and b be two numeric functions defined as*

$$a_r = 2^r,\ r \geq 0$$

and

$$b_r = 5^r,\ r \geq 0.$$

*Determine a * b.*

Sol. Let k be the convolution of a and b *i.e.*, $k_r = a_r * b_r$

Then
$$k_r = \sum_{i=0}^{r} 2^i \cdot 5^{r-i},\ r \geq 0.$$

Example 40. *Determine a * b, where a and b are two numeric functions such that*

$$a_r = \begin{cases} 1, & 0 \leq r \leq 2 \\ 0, & r \geq 3 \end{cases}$$

$$b_r = \begin{cases} 1, & 0 \leq r \leq 2 \\ 0, & r \geq 3 \end{cases}$$

Sol. Let us suppose that $c = a * b$

Then, we know, that

$$c_r = a_0\, b_r + a_1\, b_{r-1} + a_2\, b_{r-2} + \ldots.. + a_r\, b_0$$

Hence
$$c_r = \begin{cases} 1, & r = 0 \\ 2, & r = 1 \\ 3, & r = 2 \\ 2, & r = 3 \\ 1, & r = 4 \\ 0, & r \geq 5 \end{cases}$$

EXERCISES

1. Let a be a numeric function such that
$$a_r = \begin{cases} 2, & 0 \leq r \leq 3 \\ 2^{-r} + 5, & r \geq 4 \end{cases}$$
Determine S^2a and $S^{-2}a$, Δa and ∇a.
2. Let a be a numeric function such that
$$a_r = r^3 - 2r^2 + 3r + 2$$
Determine Δa, $\Delta^2 a$ and $\Delta^3 a$.
3. Let a and b be two numeric functions such that $a_r = r + 1$ and $b_r = 2^r$ for all $r \geq 0$. Determine $\Delta(ab)$.
4. Let $a_r = \begin{cases} 1, & r = 0 \\ 3, & r = 1 \\ 2, & r = 2 \\ 0, & r \geq 3 \end{cases}$
$$b_r = 5^r \text{ for all } r$$
Determine c so that $a * c = b$.
5. Every particle inside a nuclear reactor splits into two particles in each second. Suppose one particle is injected into the reactor every second beginning at $t = 0$. How many particles are there in the reactor at the nth second ?

GENERATING FUNCTIONS

Generating functions is a method to solve the recurrence relations.

Let us consider, the sequence $a_0, a_1, a_2, ..., a_r, ...$ of real numbers. For some interval of real numbers containing zero whose value at t is given, the function G(t) is defined by the series

$$G(t) = a_0 + a_1 t + a_2 t^2 + ... + a_r t^r + ... \qquad ...(i)$$

This function G(t) is called the generating function of the sequence a_r.

Now, for the constant sequence 1, 1, 1, 1, ... the generating function is

$$G(t) = \frac{1}{(1-t)}$$

because it can be expressed as

$$G(t) = (1-t)^{-1} = 1 + t + t^2 + t^3 + t^4 + ... \qquad \text{[By Binomial expansion]}$$

Comparing, this with equation (i), we get

$$a_0 = 1, a_1 = 1, a_2 = 1 \text{ and so on.}$$

For, the constant sequence 1, 2, 3, 4, 5, ... the generating function is

$$G(t) = \frac{1}{(1-t)^2} \text{ because it can be expressed as}$$

$$G(t) = (1-t)^{-2} = 1 + 2t + 3t^2 + 4t^3 + ... + (r+1)\,t^r$$

Comparing this with equation (i), we get

$$a_0 = 1, a_1 = 2, a_2, = 3, a_3 = 4 \text{ and so on.}$$

The generating function of Z^r, ($Z \neq 0$ and Z is a constant) is given by

$$G(t) = 1 + Zt + Z^2 t^2 + Z^3 t^3 + ... + Z^r t^r$$

or
$$G(t) = \frac{1}{(1-Zt)} \qquad \text{[Assume } |Zt| < 1]$$

So
$$G(t) = \frac{1}{(1-Zt)} \text{ generates } Z^r, Z \neq 0.$$

Also, If $a_r^{(1)}$ has the generating function $G_1(t)$ and $a_r^{(2)}$ has the generating function $G_2(t)$, then $\lambda_1 a_r^{(1)} + \lambda_2 a_r^{(2)}$ has the generating function $\lambda_1 G_1(t) + \lambda_2 G_2(t)$. Here λ_1 and λ_2 are constants.

Theorem IV. *If we know the generating function for the sequence a_r, then the generating function of sequences a_{r+1}, a_{r+2} can be find easily.*

Proof. Let us consider, that G(t) be the generating function for the sequence, having the general term a_r, then we have

$$G(t) = a_0 + a_1 t + a_2 t^2 + ... + a_r t^r + ...$$

or
$$\frac{G(t) - a_0}{t} = a_1 + a_2 t + a_3 t^2 + ... + a_r t^{r-1} + a_{r+1} t^r + ...$$

Thus $\frac{G(t) - a_0}{t}$ is the generating function for the sequence a_{r+1}.

In the same way,

$$\frac{G(t) - a_0 - a_1 t}{t^2} = a_2 + a_3 t + a_4 t^2 + ...$$

is the generating function for a_{r+2}.

The summary of results, which we have deduced from the above discussion, is shown in Fig. 1.

S. No.	General term of sequence a_r	Generating function G(t)
1.	1	$1/(1-t)$
2.	$r+1$	$1/(1-t)^2$
3.	r	$t/(1-t)^2$
4.	$r(r+1)$	$2t/(1-t)^3$
5.	Z^r, here Z is a constant	$1/(1-Zt)$
6.	${}^nC_r A^{n-r}. Z^r$ $(r = 0, 1, 2, 3, ..., n)$ $r > n$	$(A+Zt)^n$
7.	a_r	$G(t)$
8.	a_{r+1}	$\frac{G(t)-a_0}{t}$
9.	a_{r+2}	$\frac{G(t)-a_0-a_1 . t}{t^2}$
10.	a_{r+3}	$\frac{G(t)-a_0-a_1 . t-a_2 t^2}{t^3}$
11.	a_{r+n}	$\frac{G(t)-a_0-a_1 t-......-a_{n-1} t^{n-1}}{t^n}$
12.	$\lambda_1 a_r^{(1)} + \lambda_2 a_r^{(2)}$	$\lambda_1 G_1(t) + \lambda_2 G_2(t)$

Fig. 1

Example 41. *Solve the recurrence relation*

$$a_{r+2} - 3a_{r+1} + 2a_r = 0 \quad ...(i)$$

by the method of generating functions with the initial conditions $a_0 = 2$ and $a_1 = 3$.

Sol. Let us assume that

$$G(t) = \sum_0^\infty a_r t^r$$

Multiply equation (*i*) by t^r and summing from $r = 0$ to ∞, we have

$$\sum_0^\infty a_{r+2} t^r - 3\sum_0^\infty a_{r+1} t^r + 2\sum_0^\infty a_r t^r = 0$$

or $$(a_2 + a_3 t + a_4 t^2 + ...) - 3(a_1 + a_2 t + a_3 t^2 + ...) + 2(a_0 + a_1 t + a_2 t^2 + ...) = 0$$

$$[\because \; G(t) = a_0 + a_1 t + a_2 t^2 + ...]$$

$$\therefore \quad \frac{G(t)-a_0-a_1 t}{t^2} - 3\left(\frac{G(t)-a_0}{t}\right) + 2G(t) = 0 \quad ...(ii)$$

Now, put $a_0 = 2$ and $a_1 = 3$ in equation (*ii*) and solving, we get

$$G(t) = \frac{2-3t}{1-3t+2t^2} \quad \text{or} \quad G(t) = \frac{2-3t}{(1-t)(1-2t)}$$

Now, Let $\frac{2-3t}{(1-t)(1-2t)} = \frac{A}{1-t} + \frac{B}{1-2t}$

i.e., $$2 - 3t = A(1-2t) + B(1-t) \quad ...(iii)$$

Put $t = 1$ on both sides of equation (*iii*) to find A. Hence

$$-1 = -A \quad \therefore \quad A = 1$$

Put $t = \frac{1}{2}$ on both sides of equation (*iii*) to find B. Hence

$$\frac{1}{2} = \frac{1}{2}B \quad \therefore \quad B = 1$$

Thus $G(t) = \frac{1}{1-t} + \frac{1}{1-2t}$. Hence, $a_r = 1 + 2^r$.

Example 42. *Solve the recurrence relation*

$$a_r - 7a_{r-1} + 10a_{r-2} = 0 \qquad ...(i)$$

by the method of generating functions with the initial conditions $a_0 = 3$ and $a_1 = 3$.

Sol. Let us assume that

$$G(t) = \sum_0^{\infty} a_r t^r$$

Multiply equation (*i*) by t^r and summing from $r = 2$ to ∞, we have

$$\sum_2^{\infty} a_r t^r - 7\sum_2^{\infty} a_{r-1} t^r + 10 \sum_2^{\infty} a_{r-2} t^r = 0$$

or $$(a_2 t^2 + a_3 t^3 + ...) - 7(a_1 t^2 + a_2 t^3 + ...) + 10(a_0 t^2 + a_1 t^3 + ...) = 0$$

$$G(t) - a_0 - a_1 t - 7t[G(t) - a_0] + 10t^2 G(t) = 0 \qquad ...(ii)$$

Now, put $a_0 = 3$ and $a_1 = 3$ in equation (*ii*) and solving, we get

$$G(t) = \frac{3 + 24t}{10t^2 - 7t + 1} = \frac{3 + 24t}{(5t-1)(2t-1)}$$

By partial fractions

$$G(t) = \frac{10}{(2t-1)} - \frac{13}{(5t-1)} \text{ or } G(t) = \frac{13}{(1-5t)} - \frac{10}{(1-2t)}$$

Therefore, $$a_r = 13(5)^r - 10(2)^r.$$

Example 43. *Solve the recurrence relation*

$$a_{r+2} - 2a_{r+1} + a_r = 2^r \qquad ...(i)$$

by the method of generating functions with the initial conditions $a_0 = 2$ and $a_1 = 1$.

Sol. By taking generating functions of equation (*i*) both the sides, we have

$$\frac{G(t) - a_0 - a_1 t}{t^2} - 2\left(\frac{G(t) - a_0}{t}\right) + G(t) = \frac{1}{1-2t} \qquad ...(ii)$$

Now, put $a_0 = 2$ and $a_1 = 1$ in equation (*ii*) and solving, we get

$$G(t) - 2 - t - 2t\, G(t) - 2t + t^2 G(t) = \frac{t^2}{1-2t}$$

$$(t^2 - 2t + 1)\, G(t) = 2 + 3t + \frac{t^2}{1-2t}$$

$$(1-t)^2\, G(t) = 2 + 3t + \frac{t^2}{1-2t}$$

$$G(t) = \frac{2}{(1-t)^2} + \frac{3t}{(1-t)^2} + \frac{t^2}{(1-2t)(1-t)^2}$$

By partial fraction $\dfrac{t^2}{(1-2t)(1-t)^2} = \dfrac{1}{(1-2t)} - \dfrac{1}{(1-t)^2}$

Hence, $G(t) = \dfrac{1}{(1-t)^2} + \dfrac{3t}{(1-t)^2} + \dfrac{1}{(1-2t)}$

Therefore, $a_r = (r+1) + 3r + 2^r$

$$a_r = 1 + 4r + 2^r.$$

Example 44. *Solve the recurrence relation*

$$a_{r+2} - 5a_{r+1} + 6a_r = 2 \quad \text{...(i)}$$

by the method of generating functions satisfying the initial conditions $a_0 = 1$ and $a_1 = 2$.

Sol. Let us assume that

$$G(t) = \sum_{0}^{\infty} a_r t^r$$

Now, by taking generating functions of equation (*i*), we have

$$\frac{G(t) - a_0 - a_1 t}{t^2} - 5\left(\frac{G(t) - a_0}{t}\right) + 6G(t) = \frac{2}{1-t}$$

Put $a_0 = 1$ and $a_1 = 3$ in the above equation and solving, we get

$$G(t) = \frac{5t^2 - 4t + 1}{(1-t)(1-2t)(1-3t)}$$

By partial fractions

$$G(t) = \frac{1}{(1-t)} + \frac{-1}{(1-2t)} + \frac{1}{(1-3t)}$$

Therefore, the solution after applying inverse transformations

$$a_r = 1 - 2^r + 3^r.$$

EXERCISES

1. Solve the following recurrence relation
$$a_r - 5a_{r-1} + 6a_{r-2} = 2^r + r, r \geq 2$$
with $a_0 = 1$, $a_1 = 1$, by the method of generating function.
2. Determine the generating function for the sequence
$$B = \{a_n\}, \qquad \text{where } a_n = \begin{cases} 1 & \text{if } 0 \leq n \leq 3 \\ 5 & \text{if } 4 \leq n \leq 7 \\ 0 & \text{if } n \geq 8 \end{cases}$$
3. Solve the following recurrence relations using the generating function method :
(*i*) $a_{n+2} - 2\,a_{n+1} + a_n = 2^n$ (*ii*) $a_{n+2} - 2a_{n+1} + a_n = 1 + n.$

4. Solve the following recurrence relations using the generating function method :
   (*i*) $a_n - 4a_{n-2} = 0$; $a_0 = 0$ and $a_1 = 1$
   (*ii*) $a_{n+2} + 4a_{n+1} + 4a_n = 0$; $a_0 = 0$ and $a_1 = 1$
   (*iii*) $a_n - 7a_{n-1} + 10a_{n-2} = 0$; $a_0 = 3$ and $a_1 = 3$

SOLVED PROBLEMS

Problem 1. *Solve the following difference equation*

$$y_{K+3} + y_{K+2} - 8y_{K+1} - 12y_K = 0.$$

Sol. The characteristic equation is given by

$$s^3 + s^2 - 8s - 12 = 0$$

or $$(s-3)(s+2)(s+2) = 0$$

or $$s = 3, -2, -2$$

Therefore, the homogeneous solution of the equation is

$$y_K = C_1 . 3^K + (C_2 + C_3 K) . (-2)^K.$$

Problem 2. *Solve the difference equation*

$$y_{K+3} + 6y_{K+2} + 12y_{K+1} + 8y_K = 0.$$

Sol. The characteristic equation is

$$s^3 + 6s^2 + 12s + 8 = 0$$

or $$(s+2)(s+2)(s+2) = 0$$

or $$s = -2, -2, -2.$$

Therefore, the homogeneous solution of the equation is

$$y_K = (C_1 K^2 + C_2 K + C_3) . (-2)^K.$$

Problem 3. *Solve the difference equation* $a_r - 4a_{r-1} + 6a_{r-2} - 4a_{r-3} + a_{r-4} = 0.$

Sol. The characteristic equation is given by

$$s^4 - 4s^3 + 6s^2 - 4s + 1 = 0$$

or $$(s-1)(s^3 - 3s^2 + 2s - 1) = 0$$

or $$(s-1)(s-1)(s^2 - 2s + 1) = 0$$

or $$(s-1)(s-1)(s-1)(s-1) = 0$$

or $$s = 1, 1, 1, 1$$

Therefore, the homogeneous solution of the equation is

$$y_K = (C_1 r^3 + C_2 r^2 + C_3 r + C_4) . (1)^r.$$

Problem 4. *Solve the difference equation*

$$a_r - 2a_{r-1} + 2a_{r-2} - a_{r-3} = 0.$$

Sol. The characteristic equation is given by

$$s^3 - 2s^2 + 2s - 1 = 0$$

or $$(s-1)(s^2 - s + 1) = 0$$

or $$s = 1, \frac{1+i\sqrt{3}}{2}, \frac{1-i\sqrt{3}}{2}$$

Therefore, the homogeneous solution of the equation is

$$a_r = C_1 + C_2\left[\frac{1+i\sqrt{3}}{2}\right]^r + C_3\left[\frac{1-i\sqrt{3}}{2}\right]^r.$$

Problem 5. *Solve the difference equation* $a_{r+2} - 2a_{r+1} + a_r = 3r + 5.$ ...(*i*)

Sol. Put R.H.S. of the equation equal to zero *i.e.,*

$$a_{r+2} - 2a_{r+1} + a_r = 0$$

This can be written as $(E^2 - 2E + 1)\, a_r = 0$

or $(E-1)^2 = 0 \Rightarrow E = +1, 1$

Therefore, the homogeneous solution is

$$a_{r(h)} = C_1 + C_2 r$$

The particular solution of equation (*i*) is

$$a_{r(p)} = \frac{1}{2}r^3 + r^2$$

as shown in example.

Therefore, the total solution of the difference equation is

$$a_r = C_1 + C_2 r + \frac{1}{2}r^3 + r^2.$$

Problem 6. *Solve the difference equation*

$$y_K - y_{K-1} - 6y_{K-2} = -30 \qquad \text{...}(i)$$

Given that $y_0 = 20$, $y_1 = -5$.

Sol. The equation (*i*) can be written as

$$(s^2 - s - 6)\, y_K = -30$$

Its homogeneous solution is given as

$$(s^2 - s - 6) y_K = 0$$

or $(s-3)(s+2) = 0$

or $s = 3, -2$

Thus, the homogeneous solution is

$$y_{K(h)} = C_1 \cdot 3^K + C_2 \cdot (-2)^K$$

For the particular solution of equation (*i*), consider the general form of solution = A

Put in the general form in L.H.S. of equation (*i*), we get

$$A - A - 6A = -30 \Rightarrow A = 5$$

Therefore, the particular solution = 5

The total solution is $y_K = C_1 \cdot 3^K + C_2(-2)^K + 5$

Now, putting K = 0, we have

$$y_0 = C_1 + C_2 + 5 = 20$$

$\therefore$ $C_1 + C_2 = 15$ ...(*ii*)

Now, putting K = 1, we have

$$y_1 = 3C_1 - 2C_2 + 5$$

$$-5 = 3C_1 - 2C_2 + 5 = 3C_1 - 2C_2 = -10 \qquad \text{...}(iii)$$

Solving (*ii*) and (*iii*), we have

$$C_1 = 4, C_2 = 11$$

Therefore, the total solution satisfying the initial conditions is

$$y_K = 4 . 3^K + 11(-2)^K + 5.$$

Problem 7. *Solve the recurrence relation*

$$a_{r+2} - 2a_{r+1} + a_r = r^2 . 2^r. \quad ...(i)$$

Sol. Put R.H.S. of the equation (*i*) equal to zero *i.e.,*

$$a_{r+2} - 2a_{r+1} + a_r = 0$$

This can be written as $(s^2 - 2s + 1)\, a_r = 0$

or $$(s-1)^2 = 0$$

or $$s = 1, 1$$

Therefore, the homogeneous solution is

$$a_{r(h)} = C_1 + C_2\, r$$

For, finding particular solution of equation (*i*), consider its general form

$$a_r = 2^r (A_0 + A_1 r + A_2 r^2)$$

Put this solution in the L.H.S. of equation (*i*), we get

$$= 2^{r+2} [A_2 (r+2)^2 + A_1 (r+2) + A_0] - 2 . 2^{r+1} [A_2 (r+1)^2 + A_1(r+1) + A_0] + 2^r (A_2 r^2 + A_1 r + A_0)$$

$$= r^2 . 2^r (A_2) + r. 2^r (8A_2 + A_1) + 2^r (12A_2 + 4A_1 + A_0) \quad ...(ii)$$

Equating equation (*ii*) with R.H.S. of equation (*i*), we get

$$A_2 = 1$$

$$8A_2 + A_1 = 0$$

$$12A_2 + 4A_1 + A_0 = 0$$

On solving these equations, we get

$$A_0 = 20,\ A_1 = -8, A_2 = 1$$

Therefore, the particular solution is

$$a_{r(p)} = 2^r (r^2 - 8r + 20)$$

Hence, the total solution is

$$a_r = C_1 + C_2 r + 2^r (r^2 - 8r + 20).$$

Problem 8. *Solve the recurrence relation*

$$a_r + a_{r-1} + a_{r-2} = 0 \quad ...(i)$$

Satisfying the initial conditions $a_0 = 0$ and $a_1 = 2$.

Sol. This equation can be written as $(s^2 + s + 1)\, a_r = 0$

or $$s = \frac{-1 + i\sqrt{3}}{2}, \frac{-1 - i\sqrt{3}}{2}$$

Therefore, the solution is given by

$$a_r = C_1 \left[\frac{-1 + i\sqrt{3}}{2}\right]^r + C_2 \left[\frac{-1 - i\sqrt{3}}{2}\right]^r \quad ...(ii)$$

Put $r = 0$, in equation (*ii*), we have

$$a_0 = C_1 + C_2 = 0 \quad ...(iii)$$

Put $r = 1$, in equation (*ii*), we have

$$a_1 = C_1\left[\frac{-1+i\sqrt{3}}{2}\right] + C_2\left[\frac{-1-i\sqrt{3}}{2}\right] = 2 \quad ...(iv)$$

On solving equations (*iii*) and (*iv*), we have

$$C_1 = \frac{2}{i\sqrt{3}}$$

$$C_2 = \frac{-2}{i\sqrt{3}}$$

Therefore, the solution satisfying initial conditions is

$$a_r = \frac{2}{i\sqrt{3}}\left[\frac{-1+i\sqrt{3}}{2}\right]^r - \frac{2}{i\sqrt{3}}\left[\frac{-1-i\sqrt{3}}{2}\right]^r.$$

Problem 9. *Solve the recurrence relation*

$$a_r - a_{r-1} - a_{r-2} = 0 \quad ...(i)$$

satisfying the initial conditions $a_0 = 1$ *and* $a_1 = 1$.

Sol. The equation (*i*) can also be written as

$$(s^2 - s - 1)\, a_r = 0 \quad \text{or} \quad s = \frac{1+\sqrt{5}}{2}, \frac{1-\sqrt{5}}{2}$$

Therefore, the solution is given by

$$a_r = C_1\left[\frac{1+\sqrt{5}}{2}\right]^r + C_2\left[\frac{1-\sqrt{5}}{2}\right]^r \quad ...(ii)$$

Put $r = 0$, in equation (*ii*), we have

$$a_0 = C_1 + C_2 = 1 \quad ...(iii)$$

Put $r = 1$, in equation (*ii*), we have

$$a_1 = C_1\left[\frac{1+\sqrt{5}}{2}\right] + C_2\left[\frac{1-\sqrt{5}}{2}\right] = 1 \quad ...(iv)$$

On solving equations (*iii*) and (*iv*), we have

$$C_1 = \frac{\sqrt{5}+1}{2\sqrt{5}}$$

$$C_2 = \frac{\sqrt{5}-1}{2\sqrt{5}}$$

Therefore, the solution satisfying initial conditions is

$$a_r = \left(\frac{\sqrt{5}+1}{2\sqrt{5}}\right)\left(\frac{1+\sqrt{5}}{2}\right)^r + \left(\frac{\sqrt{5}-1}{2\sqrt{5}}\right)\left(\frac{1-\sqrt{5}}{2}\right)^r.$$

Problem 10. *Solve the recurrence relation*

$$a_{r+1} - 3a_r = 2^r. \quad ...(i)$$

Sol. Put R.H.S. of the equation (*i*) equal to zero *i.e.*,

$$a_{r+1} - 3a_r = 0$$

This can be written as

$$(s-3)a_r = 0 \quad \text{or} \quad s = 3.$$

Therefore, the homogeneous solution is

$$a_{r(h)} = C_1 \,.\, 3^r$$

For the particular solution of the equation (*i*), consider the general form of the solution

$$= A \,.\, 2^r$$

Put this solution on L.H.S. of equation (*i*), we have

$$= A \,.\, 2^{r+1} - 3A2^r = (2A - 3A) \,.\, 2^r$$

$$= -A \,.\, 2^r \qquad \text{...}(ii)$$

Equating equation (*ii*) with R.H.S. of equation (*i*), we get

$$A = -1$$

Therefore, the particular solution is

$$a_{r(p)} = -(2^r)$$

Hence, the total solution is $a_r = C_1 \,.\, 3^r - (2^r)$.

Problem 11. *Solve the recurrence relation*

$$a_{r+2} - 5a_{r+1} + 6a_r = 5^r. \qquad \text{...}(i)$$

Sol. Put the R.H.S. of equation equal to zero *i.e.*,

$$a_{r+2} - 5a_{r+1} + 6a_r = 0$$

This can also be written as

$$(s^2 - 5s + 6)a_r = 0$$

or $$(s-3)(s-2) = 0$$

or $$s = 3, 2$$

Therefore, the homogeneous solution is

$$a_{r(h)} = C_1 \,.\, 2^r + C_2 \,.\, 3^r$$

For, finding the particular solution consider the general form of the solution $= A \,.\, 5^r$.

Now, put this solution on L.H.S. of equation (*i*), we get

$$= A \,.\, 5^{r+2} - 5 \,.\, A \,.\, 5^{r+1} + 6 \,.\, A. \, 5^r$$

$$= 25A \,.\, 5^r - 25A \,.\, 5^r + 6A \,.\, 5^r$$

$$= 6A \,.\, 5^r \qquad \text{...}(ii)$$

Equating equation (*ii*) to R.H.S. of equation (*i*), we get

$$A = \frac{1}{6}$$

Therefore, the particular solution is

$$a_{r(p)} = \frac{1}{6} \,.\, 5^r$$

Hence, the total solution is

$$a_r = C_1 \,.\, 2^r + C_2 \,.\, 3^r + \frac{1}{6} \,.\, 5^r.$$

Problem 12. *Solve the recurrence relation*

$$a_r - a_{r-1} = 7r. \quad ...(i)$$

Sol. Put the R.H.S. of equation (*i*) equal to zero *i.e.,*

$$a_r - a_{r-1} = 0$$

This equation can also be written as

$$(s-1)a_r = 0 \quad \text{or} \quad s = 1$$

Therefore, the homogeneous solution is

$$a_{r(h)} = C_1$$

For finding, the particular solution, consider the general form of the solution $= A_0 r + A_1$. Since one of the terms of the solution occurs in the homogeneous solution, so multiply the solution by r, we get $A_0 r^2 + A_1 . r$.

Now, put this solution in L.H.S. of equation (*i*), we have

$$= A_0 r^2 + A_1 r - A_0(r-1)^2 - A_1 (r-1) = 2A_0 r + A_1 - A_0 \quad ...(ii)$$

Equating equation (*ii*) with R.H.S. of equation (*i*), we get

$$A_0 = \frac{7}{2}$$

$$A_1 - A_0 = 0 \quad \Rightarrow \quad A_1 = \frac{7}{2}$$

Therefore, the particular solution is $a_{r(p)} = \frac{7}{2}r^2 + \frac{7}{2}r$

Hence, the total solution is

$$a_r = C_1 + \frac{7}{2}r^2 + \frac{7}{2}r.$$

Problem 13. *Solve the recurrence relation*

$$a_{r+2} - 5a_{r+1} + 6a_r = r^2. \quad ...(i)$$

Sol. The homogeneous solution of the equation (*i*) is

$$a_{r(h)} = C_1 . 2^r + C_2 . 3^r$$

For finding, the particular solution of equation (*i*), consider the general form of the solution

$$= A_1 r^2 + A_2 r + A_3$$

Put this solution in L.H.S. of equation (*i*), we have

$$= A_1 (r+2)^2 + A_2 (r+2) + A_3 - 5A_1 (r+1)^2 - 5A_2 (r+1) - 5A_3$$
$$+ 6A_1 r^2 + 6A_2 r + 6A_3$$

$$= (2A_1)r^2 + (-6A_1 + 2A_2)r + (-A_1 - 3A_2 + 2A_3) \quad ...(ii)$$

Equating equation (*ii*) with R.H.S. of equation (*i*), we get

$$2A_1 = 1 \quad ...(iii)$$

$$-6A_1 + 2A_2 = 0 \quad ...(iv)$$

$$-A_1 - 3A_2 + 2A_3 = 0 \quad ...(v)$$

On solving equation (*iii*), (*iv*) and (*v*), we get

$$A_1 = \frac{1}{2}, A_2 = \frac{3}{2}, A_3 = \frac{5}{2}$$

Therefore, the particular solution is

$$a_{r(p)} = \frac{1}{2}r^2 + \frac{3}{2}r + \frac{5}{2}$$

Hence, the total solution is

$$a_r = C_1 \,.\, 2^r + C_2 \,.\, 3^r + \frac{1}{2}r^2 + \frac{3}{2}r + \frac{5}{2}.$$

Problem 14. *Solve the recurrence relation*

$$y_{K+1} - 5y_K = 5^K \qquad ...(i)$$

satisfying the initial condition $y_0 = 3$.

Sol. Put the R.H.S. of the equation (*i*) equal to zero *i.e.,*

$$y_{K+1} - 5y_K = 0$$

This can also be written as

$$(s-5)y_K = 0 \quad \text{or} \quad s = 5$$

Therefore, the homogeneous solution is $y_{K(h)} = C_1 \,.\, 5^K$.

For finding, the particular solution, consider the general form of the solution = A . 5^K.

Since, this term occurs in the homogeneous solution, so multiply the solution by K.

Hence, the general form of the solution becomes = AK . 5^K.

Now, put this solution on L.H.S. of equation (*i*), we have

$$= A(K+1) \,.\, 5^{K+1} - 5AK \,.\, 5^K = A \,.\, 5^{K+1}$$

$$= 5A \,.\, 5^K \qquad ...(ii)$$

Equating equation (*ii*) with R.H.S. of equation (*i*), we get

$$A = \frac{1}{5}$$

Therefore, the particular solution is

$$y_{K(p)} = \frac{1}{5}K \,.\, 5^K$$

Hence, the total solution is

$$y_K = C_1 \,.\, 5^K + \frac{1}{5}K \,.\, 5^K \quad \text{or} \quad y_K = \left(C_1 + \frac{K}{5}\right) .\, 5^K \qquad ...(iii)$$

Put K = 0, in equation (*iii*), we have

$$y_0 = C_1 = 3.$$

The total solution satisfying the initial condition is

$$y_K = \left(3 + \frac{K}{5}\right) .\, 5^K.$$

Problem 15. *Solve the recurrence relation*

$$a_r - 3a_{r-1} + 2a_{r-2} = 0 \qquad ...(i)$$

satisfying the initial conditions $a_0 = 1$ *and* $a_1 = 4$.

Sol. The equation (*i*) can also be written as

$$(s^2 - 3s + 2)a_r = 0 \quad \text{or} \quad (s-2)(s-1) = 0$$

or

$$s = 1, 2$$

Therefore, the solution is $a_r = C_1 + C_2 . 2^r$ ...(*ii*)

Put $r = 0$, in equation (*ii*), we have

$$a_0 = C_1 + C_2 = 1 \qquad ...(iii)$$

Put $r = 1$, in equation (*ii*), we have

$$a_1 = C_1 + 2C_2 = 4 \qquad ...(iv)$$

On solving equations (*iii*) and (*iv*), we get

$$C_1 = -2$$
$$C_2 = 3$$

Therefore, the solution satisfying the initial conditions is

$$a_r = -2 + 3 . 2^r.$$

Problem 16. *Solve the recurrence relation*

$$y_K = r . y_{K-1} + Z \qquad ...(i)$$

satisfying the initial conditions $y_0 = 0$. $r, z > 0, r \neq 1$.

Sol. The equation (*i*) can be written as

$$(E - r)\, y_K = Z \qquad ...(ii)$$

Put the R.H.S. of the equation (*ii*) equal to zero *i.e.*,

$$(E - r) y_K = 0$$

or $$(E - r) = 0 \quad \text{or} \quad E = r$$

Therefore, the homogeneous solution is

$$y_{K(h)} = C_1 . r^K$$

The particular solution of equation (*ii*), is given by

$$y_{K(p)} = \frac{1}{(E - r)} . z$$

Put E = 1, because z is a constant. Hence

$$y_{K(p)} = \frac{z}{(1 - r)}$$

Hence, the total solution is

$$y_K = C_1 . r^K + \frac{z}{(1 - r)} \qquad ...(iii)$$

Put K = 0, in equation (*iii*), we have

$$y_0 = C_1 + \frac{z}{(1 - r)} = 0$$

$$\therefore \qquad C_1 = -\frac{z}{1 - r}$$

The total solution satisfying the initial condition is

$$y_K = -\frac{z}{1 - r} . r^K + \frac{z}{(1 - r)} \quad \text{or} \quad y_K = \frac{z(1 - r^K)}{(1 - r)}.$$

Problem 17. *Solve the recurrence relation*

$$y_{K+3} + y_{K+2} - 8y_{K+1} - 12y_K = 2K^2 + 5. \qquad ...(i)$$

Sol. Put the R.H.S. of equation (*i*) equal to zero *i.e.,*

$$y_{K+3} + y_{K+2} - 8y_{K+1} - 12y_K = 0$$

i.e.,
$$(E^3 + E^2 - 8E - 12)y_K = 0$$

or
$$E = 3, -2, -2$$

Therefore, the homogeneous solution is

$$y_{K(h)} = C_1 . 3^K + (C_2 + C_3 K)(-2)^K$$

The particular solution of equation (*i*) is given as

$$y_{K(p)} = \frac{1}{(E^3 + E^2 - 8E - 12)} . (2K^2 + 5)$$

$$= \frac{1}{(E-3)(E+2)^2} . (2K^2 + 5) \qquad \text{[Put } E = 1 + \Delta]$$

$$= \frac{1}{(\Delta - 2)(\Delta + 3)^2} . (2K^2 + 5)$$

$$= (\Delta^3 + 4\Delta^2 - 3\Delta - 18)^{-1} . (2K^2 + 5)$$

$$= -\frac{1}{18}\left(1 + \frac{3\Delta - 4\Delta^2 - \Delta^3}{18}\right)^{-1} . (2K^2 + 5)$$

$$y_{K(p)} = -\frac{1}{18}\left(1 - \frac{3\Delta - 4\Delta^2 - \Delta^3}{18} + \frac{(3\Delta - 4\Delta^2 - \Delta^3)^2}{(18)^2} + \ldots\right) . (2K^2 + 5)$$

$$= -\frac{1}{18}\left(1 - \frac{\Delta}{6} + \frac{\Delta^2}{4}\right) . (2K^2 + 5) \qquad \text{[By neglecting higher degrees]}$$

By factorial notation $2K^2 + 5 = 2[K]^2 + 2[K] + 5$.

Hence,
$$y_{K(p)} = -\frac{1}{18}\left(1 - \frac{\Delta}{6} + \frac{\Delta^2}{4}\right) . (2[K]^2 + 2[K] + 5)$$

$$= -\frac{1}{18}\left(2[K]^2 + 2[K] + 5 - \frac{1}{6}(4[K] + 2) + \frac{1}{4}(4)\right)$$

$$= -\frac{1}{18}\left(2[K]^2 - \frac{4}{3}[K] + \frac{17}{3}\right)$$

$$= -\frac{1}{18}\left(2K(K-1) - \frac{4}{3}K + \frac{17}{3}\right)$$

$$y_{K(p)} = -\frac{K^2}{9} + \frac{1}{27}K - \frac{17}{54}$$

Hence, the total solution is

$$y_K = C_1 . 3^K + (C_2 + C_3 K)(-2)^K + \left(\frac{-K^2}{9} + \frac{1}{27}K - \frac{17}{54}\right).$$

Problem 18. *Solve the recurrence relation*

$$y_{K+4} + 4y_{K+3} + 8Y_{K+2} + 8y_{K+1} + 4y_K = 10. \qquad \ldots(i)$$

Sol. Put the R.H.S. of equation (*i*) equal to zero *i.e.*,

$$y_{K+4} + 4y_{K+3} + 8y_{K+2} + 8y_{K+1} + 4y_K = 0$$

i.e.,
$$(E^4 + 4E^3 + 8E^2 + 8E + 4)\, y_K = 0$$

Its homogeneous solution is

$$y_{K(h)} = (C_1 + C_2 K)(-1 + i)^K + (C_3 + C_4 K)(-1 - i)^K$$

as shown before.

The particular solution of equation (*i*) is given as

$$y_{K(p)} = \frac{1}{(E^4 + 4E^3 + 8E^2 + 8E + 4)} . 10$$

By putting E = 1, we have

$$y_{K(p)} = \frac{10}{1 + 4 + 8 + 8 + 4} = \frac{10}{25} = \frac{2}{5}$$

Hence, the total solution is

$$y_K = (C_1 + C_2 K)(-1 + i)^K + (C_3 + C_4 K)(-1 - i)^K + \frac{2}{5}.$$

Problem 19. *Obtain the numeric function c made up of the letters {a, b, c, d, α, β, Δ} with the first portion made up of English letters and the second portion made up of Greek letters.*

Sol. Let a_r denote the number of sequences of length *r* made from the letters *a, b, c, d,* and is defined as

$$a_r = 4^r$$

Also let b_r denote the number of sequences of length *r* made from the letters {α, β, Δ} and is defined as

$$a_r = 3^r$$

Then $c_r = a_r * b_r$ is the convolution of *a* and *b*, and is given by

$$c_r = \sum_{i=0}^{r} 4^i . 3^{r-i}, r \geq 0.$$

Problem 20. *Let a be a numeric function given by*

$$a_r = \begin{cases} 3, & 0 \leq r \leq 3 \\ 2^{-2r} + 4, & r \geq 4 \end{cases}$$

Determine $S^2 a$ *and* $S^{-2} a$.

Sol. Let us assume that $b = S^2 a$. Then

$$b_r = \begin{cases} 0 & \text{if } 0 \leq r \leq 1 \\ 3 & \text{if } 2 \leq r \leq 5 \\ 2^{-2r+2} + 4 & \text{if } r \geq 6 \end{cases}$$

Again, let $c = S^{-2} a$. Then

$$c_r = \begin{cases} 2, & 0 \leq r \leq 1 \\ 2^{2r-2} + 4, & r \geq 2 \end{cases}$$

Problem 21. *Let a be a numeric function given by*

$$a_r = \begin{cases} 2, & 0 \le r \le 3 \\ 2^{-r} + 3, & r \ge 4 \end{cases}$$

Determine Δa and ∇a.

Sol. Let us assume that $b = \Delta a$, then

$$b_r = \begin{cases} 0, & 0 \le r \le 2 \\ 17/16, & r = 3 \\ -2^{-(r+1)}, & r \ge 4 \end{cases}$$

Again, let us assume that $c = \nabla a$, then.

$$c_r = \begin{cases} 0, & 0 \le r \le 3 \\ 17/16, & r = 4 \\ -2^{-r}, & r \ge 5 \end{cases}$$

Problem 22. *Let a and b be two numeric functions such that $a_r = r + 1$ and $b_r = 2^r$, $r \ge 0$. Determine Δab.*

Sol. Let us assume that $c = ab$, then

$$\begin{aligned} c_r &= a_r\, b_r,\ r \ge 0 \\ &= (r + 1) \,.\, 2^r \end{aligned}$$

$$\therefore \qquad \begin{aligned} \Delta c_r &= c_{r+1} - c_r \\ &= (r + 2) \,.\, 2^{r+1} - (r + 1) \,.\, 2^r \\ &= 2((r + 2) - (r + 1))\, 2^r = (r + 3) \,.\, 2^r. \end{aligned}$$

Problem 23. *Determine $a * b$, where a and b are two numeric functions such that*

$$a_r = \begin{cases} 1, & 0 \le r \le 2 \\ 0, & r \ge 3 \end{cases}$$

$$b_r = \begin{cases} 1, & 0 \le r \le 2 \\ 0, & r \ge 3 \end{cases}$$

Sol. Let us assume that $c = a * b$, then

$$c_r = a_0 b_r + a_1\, b_{r-1} + \ldots\ldots + a_r\, b_0$$

Thus

$$c_r = \begin{cases} 1, & r = 0 \\ 2, & r = 1 \\ 3, & r = 2 \\ 2, & r = 3 \\ 1, & r = 4 \\ 0, & r \ge 5 \end{cases}$$

Problem 24. *Determine $a * b$, where a and b are two numeric functions such that*

$$a_r = \begin{cases} 0, & 0 \le r \le 2 \\ 2^r, & r \ge 3 \end{cases} \qquad b_r = 2^r, \quad r \ge 0.$$

Sol. Let us assume that $c = a * b$, then

$$c_r = a_0\, b_r + a_r\, b_{r-1} + \ldots\ldots + a_r\, b_0$$

Hence,

$$c_r = \begin{cases} 0, & 0 \le r \le 2 \\ (r - 2) \,.\, 2^r, & r \ge 3 \end{cases}$$

Chapter 7

Propositions

PROPOSITION

A proposition is a statement which is either true or false. It is a declarative sentence.

For example : The following statements are all propositions :

(*i*) Jawahar Lal Nehru is the first prime minister of India.

(*ii*) It rained yesterday.

(*iii*) If x is an integer, then x^2 is a + ve integer.

For example : The following statements are not propositions :

(*i*) Please report at 11 a.m. sharp.

(*ii*) What is your name ?

(*iii*) $x^2 = 13$.

PROPOSITIONAL VARIABLES

The lower case letters starting from P onwards are used to represent propositions *e.g.*,

p : India is in Asia

$q : 2 + 2 = 4$.

Example 1. *Classify the following statements as propositions or non-propositions.*

(*i*) *The population of India goes upto 100 million in year 2000.*

(*ii*) $x + y = 30$

(*iii*) *Come here*

(*iv*) *The Intel Pentium-III is a 64-bit computer.*

Sol. (*i*) Proposition (*ii*) Not a proposition

(*iii*) Not a proposition (*iv*) Proposition.

COMBINATION OF PROPOSITIONS

We can combine the propositions to produce new propositions. There are three fundamental and three derived connectors to combine the propositions. These are explained as follows one by one.

(*a*) Fundamental Connectors

1. **Conjunction.** It means ANDing of two statements. Assume p and q be two propositions. Conjunction of p and q to be a proposition which is true when both p and q are true, otherwise false. It is denoted by $p \wedge q$. (Fig. 1)

Truth tables are used to determine the truth or falsity of the combined proposition.

p	q	$p \wedge q$
T	T	T
T	F	F
F	T	F
F	F	F

Fig. 1. Truth Table of $p \wedge q$

2. **Disjunction.** It means ORing of two statements. Assume p and q be two propositions. Disjunction of p and q to be a proposition which is true when either one or both p and q are true and is false when both p and q are false. It is denoted by $p \vee q$. (Fig. 2)

p	q	$p \vee q$
T	T	T
T	F	T
F	T	T
F	F	F

Fig. 2. Truth Table of $p \vee q$

3. **Negation.** It means opposite of original statement. Assume p be a proposition. Negation of p to be a proposition which is true when p is false, and is false when p is true. It is denoted by $\sim p$. (Fig. 3)

p	$\sim p$
T	F
F	T

Fig. 3. Truth Table of $\sim p$

Example 2. *Consider the following :*

p : He is rich

q : He is Generous.

Write the proposition which combines the proposition p and q using conjunction ($\wedge$), *disjunction* ($\vee$), *and negation* ($\sim$).

Sol. Conjunction. He is rich and generous *i.e.,* $p \wedge q$.

Disjunction. He is rich or generous *i.e.,* $p \vee q$.

Negation. He is not rich *i.e.,* $\sim p$

He is not generous *i.e.,* $\sim q$.

It is false that he is rich or generous *i.e.,* $\sim (p \vee q)$.

He is neither rich nor generous *i.e.,* $\sim p \wedge \sim q$.

It is false that he is not rich *i.e.,* $\sim (\sim p)$.

Example 3. *Let p be "It is hot day" and q be "The temperature is 45°C". Write in simple sentences the meaning of following :*

(i) $\sim p$ (ii) $\sim (p \vee q)$ (iii) $\sim (p \wedge q)$
(iv) $\sim (\sim p)$ (v) $p \vee q$ (vi) $p \wedge q$
(vii) $\sim p \wedge \sim q$ (viii) $\sim (\sim p \vee \sim q)$.

Sol. (i) It is not a hot day.
(ii) It is false that it is hot day or temperature is 45°C.
(iii) It is not true that it is hot day and temperature is 45°C.
(iv) It is false that it is not a hot day.
(v) It is hot day or temperature is 45°C.
(vi) It is hot day and temperature is 45°C.
(vii) It is neither a hot day nor temperature is 45°C.
(viii) It is false that it is not a hot day or temperature is not 45°C.

Example 4. *Consider the following statements :*

p : He is coward.
q : He is lazy.
r : He is rich.

Write the following compound statements in the symbolic form.
(i) *He is either coward or poor.*
(ii) *He is neither coward nor lazy.*
(iii) *It is false that he is coward but not lazy.*
(iv) *He is coward or lazy but not rich.*
(v) *It is false that he is coward or lazy but not rich.*
(vi) *It is not true that he is not rich.*
(vii) *He is rich or else he is both coward and lazy.*

Sol. (i) $p \wedge \sim r$ (ii) $\sim p \wedge \sim q$ (iii) $\sim (p \wedge \sim q)$
(iv) $(p \vee q) \wedge \sim r$ (v) $\sim((p \vee q) \wedge \sim r)$ (vi) $\sim (\sim r)$
(vii) $r \vee (p \wedge q)$.

(b) **Derived Connectors**

1. **NAND.** It means negation after ANDing of two statements. Assume p and q be two propositions. Nanding of p and q to be a proposition which is false when both p and q are true, otherwise true. It is denoted by $p \uparrow q$. (Fig. 4)

p	q	$p \uparrow q$
T	T	F
T	F	T
F	T	T
F	F	T

Fig. 4

2. **NOR or Joint Denial.** It means negation after ORing of two statements. Assume p and q be two propositions. NOring of p and q to be a proposition which is true when both p and q are false, otherwise false. It is denoted by $p \downarrow q$. (Fig. 5)

p	q	$p \downarrow q$
T	T	F
T	F	F
F	T	F
F	F	T

Fig. 5

3. **XOR.** Assume p and q be two propositions. XORing of p and q is true if p is true or if q is true but not both and vice-versa. It is denoted by $p \oplus q$. (Fig. 6)

p	q	$p \oplus q$
T	T	F
T	F	T
F	T	T
F	F	F

Fig. 6

Example 5. *Generate the truth table for following :*

(*i*) $A \oplus B \oplus C$ (*ii*) $A \uparrow B \uparrow C$.

Sol. The -1 truth table for above formulas are as shown in Figs. 7 and 8.

(*i*)

A	B	C	A ⊕ B	A ⊕ B ⊕ C
T	T	T	F	T
T	T	F	F	F
T	F	T	T	F
T	F	F	T	T
F	T	T	T	F
F	T	F	T	T
F	F	T	F	T
F	F	F	F	F

Fig. 7

(*ii*) Truth table for (*ii*) is

A	B	C	A ↑ B	A ↑ B ↑ C
T	T	T	F	T
T	T	F	F	T
T	F	T	T	F
T	F	F	T	T
F	T	T	T	F
F	T	F	T	T
F	F	T	T	F
F	F	F	T	T

Fig. 8

Example 6. *Prove that* $X \oplus Y \equiv (X \wedge \sim Y) \vee (\sim X \wedge Y)$.

Sol. Construct the truth table for both the propositions. (Fig. 9)

X	Y	X ⊕ Y	~ Y	~ X	X ∧ ~ Y	~ X ∧ Y	(X ∧ ~ Y) ∨ (~ X ∧ Y)
T	T	F	F	F	F	F	F
T	F	T	T	F	T	F	T
F	T	T	F	T	F	T	T
F	F	F	T	T	F	F	F

Fig. 9

As the truth table for both the proposition are same.

$$X \oplus Y \equiv (X \wedge \sim Y) \vee (\sim X \wedge Y). \text{ Hence proved.}$$

Example 7. *Show that* $(p \oplus q) \vee (p \downarrow q)$ *is equivalent to* $p \uparrow q$.

Sol. Construct the truth table for both the propositions.

p	q	$(p \oplus q)$	$(p \downarrow q)$	$(p \oplus q) \vee (p \downarrow q)$	$p \uparrow q$
T	T	F	F	F	F
T	F	T	F	T	T
F	T	T	F	T	T
F	F	F	T	T	T

Fig. 10

Since the values of $(p \oplus q) \vee (p \downarrow q)$ is same as $p \uparrow q$ as in Fig. 10. Hence, they are equivalent.

Example 8. *Show that* $(p \uparrow q) \oplus (p \uparrow q)$ *is equivalent to* $(p \vee q) \wedge (p \downarrow q)$.

Sol. Construct the truth table for both the propositions

p	q	$p \uparrow q$	$(p \uparrow q) \oplus (p \uparrow q)$	$p \vee q$	$p \downarrow q$	$(p \vee q) \wedge (p \downarrow q)$
T	T	F	F	T	F	F
T	F	T	F	T	F	F
F	T	T	F	T	F	F
F	F	T	F	F	T	F

Fig. 11

Since, the values of $(p \uparrow q) \oplus (p \uparrow q)$ and $(p \vee q) \wedge (p \downarrow q)$ are same as in Fig. 11. Hence, they are equivalent.

(c) Some Other Connectors

1. **Conditional.** Statements of the form "If p then q" are called conditional statements.

It is denoted as $p \to q$ and read as "p implies q" or "q is necessary for p" or "p is sufficient for q".

Conditional statement is true if both p and q are true or if p is false. It is false if p is true and q is false. The proposition p is called hypothesis and the proposition q is called conclusion. The truth table of conditional statement is

p	q	$p \to q$
T	T	T
T	F	F
F	T	T
F	F	T

Fig. 12. Truth Table of $p \to q$

For example : The followings are conditional statements :

1. If $a = b$ and $b = c$, then $a = c$.
2. If I will get money, then I will purchase computer.

VARIATIONS IN CONDITIONAL STATEMENT

Contrapositive. The proposition $\sim q \rightarrow \sim p$ is called contrapositive of $p \rightarrow q$

Converse. The proposition $q \rightarrow p$ is called the converse of $p \rightarrow q$.

Inverse. The proposition $\sim p \rightarrow \sim q$ is called the inverse of $p \rightarrow q$.

Example 9. *Show that $p \rightarrow q$ and its contrapositive $\sim q \rightarrow \sim p$ are logically equivalent.*

Sol. Construct truth table for both the propositions. (as in Fig. 13)

p	q	$\sim p$	$\sim q$	$p \rightarrow q$	$\sim q \rightarrow \sim p$
T	T	F	F	T	T
T	F	F	T	F	F
F	T	T	F	T	T
F	F	T	T	T	T

Fig. 13

As, the values in both cases are same, hence both propositions are equivalent.

Example 10. *Show that proposition $q \rightarrow p$ and $\sim p \rightarrow \sim q$ is not equivalent to $p \rightarrow q$.*

Sol. Construct truth table for all the above propositions :

p	q	$\sim p$	$\sim q$	$p \rightarrow q$	$q \rightarrow p$	$\sim p \rightarrow \sim q$
T	T	F	F	T	T	T
T	F	F	T	F	T	T
F	T	T	F	T	F	F
F	F	T	T	T	T	T

Fig. 14

As the values of $p \rightarrow q$ in table is not equal to $q \rightarrow p$ and $\sim p \rightarrow \sim q$ as in Fig. 14. So both of them are not equal to $p \rightarrow q$ but they are themselves logically equivalent.

Example 11. *Prove that the following propositions are equivalent to $p \rightarrow q$.*

(i) $\sim (p \wedge \sim q)$ *(ii)* $\sim p \vee q$ *(iii)* $\sim q \rightarrow \sim p$.

Sol. Construct the truth table for all the above propositions :

p	q	$\sim p$	$\sim q$	$\sim p \vee q$	$\sim q \rightarrow \sim p$	$(p \wedge \sim q)$	$\sim (p \wedge \sim q)$	$p \rightarrow q$
T	T	F	F	T	T	F	T	T
T	F	F	T	F	F	T	F	F
F	T	T	F	T	T	F	T	T
F	F	T	T	T	T	F	T	T

Fig. 15

In the above table (Fig. 15) the values of $p \rightarrow q$ is equivalent to (*i*), (*ii*) and (*iii*), hence they are equivalent to $p \rightarrow q$. Hence proved.

2. **Biconditional.** Statements of the form "if and only if" are called biconditional statements.

It is denoted as $p \leftrightarrow q$ and read as "p if and only if q". The proposition $p \leftrightarrow q$ is true if p and q have the same truth values and is false if p and q do not have the same truth values. The name of biconditional comes from the fact that $p \leftrightarrow q$ is equivalent to $(p \to q) \wedge (q \to p)$.

The truth table of $p \leftrightarrow q$ is

p	q	$p \leftrightarrow q$
T	T	T
T	F	F
F	T	F
F	F	T

Fig. 16. Truth Table of $p \leftrightarrow q$

For example : (*i*) Two lines are parallel if and only if they have same slope.

(*ii*) You will pass the exam if and only if you will work hard.

Example 12. *Prove that $p \leftrightarrow q$ is equivalent to $(p \to q) \wedge (q \to p)$.*

Sol. Construct the truth tables of both propositions :

p	q	$p \leftrightarrow q$
T	T	T
T	F	F
F	T	F
F	F	T

Fig. 17

p	q	$p \to q$	$q \to p$	$(p \to q) \wedge (q \to p)$
T	T	T	T	T
T	F	F	T	F
F	T	T	F	F
F	F	T	T	T

Fig. 18

Since, the truth tables are same, hence they are logically equivalent. (Fig. 17 and Fig. 18). Hence proved.

PRINCIPLE OF DUALITY

Two formulas A_1 and A_2 are said to be duals of each other if either one can be obtained from the other by replacing $\wedge$ (AND) by $\vee$ (OR) and $\vee$ (OR) by $\wedge$ (AND). Also if the formula contains T (True) or F (False), then we replace T by F and F by T to obtain the dual.

Note 1. The two connectives $\wedge$ and $\vee$ are called dual of each other.

2. Like AND and OR, $\uparrow$ (NAND) and $\downarrow$ (NOR) are dual of each other.

3. If any formula of proposition is valid, then its dual is also a valid formula.

Example 13. *Determine the dual of each of the following :*

(*a*) $p \wedge (q \wedge r)$ (*b*) $\sim p \vee \sim q$ (*c*) $(p \wedge \sim q) \vee (\sim p \wedge q)$

(*d*) $(p \uparrow q) \uparrow (p \uparrow q)$ (*e*) $((\sim p \vee q) \wedge (q \wedge \sim s)) \vee (p \vee F)$, *here F means false.*

Sol. To obtain the dual of all the above formulas, replace $\wedge$ by $\vee$ and $\vee$ by $\wedge$, and also replace T by F and F by T.

Also replace $\uparrow$ by $\downarrow$ and vice-versa.

(*a*) $p \wedge (q \wedge r) = p \vee (q \vee r)$ (*b*) $\sim p \vee \sim q = \sim p \wedge \sim q$

(*c*) $(p \wedge \sim q) \vee (\sim p \wedge q) = (p \vee \sim q) \wedge (\sim p \vee q)$

(*d*) $(p \downarrow q) \downarrow (p \downarrow q)$

(*e*) $((\sim p \vee q) \wedge (q \wedge \sim s)) \vee (p \vee \text{F}) = ((\sim p \wedge q) \vee (q \vee \sim s)) \wedge (p \wedge \text{T})$

EXERCISES

1. Write the following statements in symbolic form :
   (*a*) The sun is bright and the humidity is not high.
   (*b*) If you do not see me tomorrow, it means I have gone to DELHI.
2. Let p denotes the statement "The wheather is nice" and q denotes the statement "we have a picnic". Translate the following in English.
   (*a*) $\overline{p} \wedge q$ (*b*) $p \to q$ (*c*) $\overline{p} \to \overline{q}$
   (*d*) $p \leftrightarrow q$ (*e*) $\overline{(\overline{p} \vee q)} \vee (p \wedge \overline{q})$.
3. Construct the truth tables for the following statements :
   (*i*) $(\sim q \to \sim p) \to (p \to q)$ (*ii*) $(p \wedge \sim r) \leftrightarrow (q \vee r)$
   (*iii*) $(p \wedge \sim q) \to r$ (*iv*) $(p \to q) \vee \sim (p \leftrightarrow \sim q)$.
4. State converse, inverse and contrapositive of the following :
   (*i*) If p is a prime number, then p is odd or p is 2.
   (*ii*) If p is a square, then p is a rectangle.
5. Determine the dual of each of the following propositions :
   (*i*) $(p \vee q \vee r) \wedge s$ (*ii*) $p \vee \sim q \vee \sim r$
6. Construct a truth table for each of the following compound propositions :
   (*i*) $\sim p \oplus \sim q$ (*ii*) $(p \oplus q) \wedge (p \oplus \sim q)$
   (*iii*) $(p \oplus q) \vee (\sim p \oplus q)$ (*iv*) $(\sim p \leftrightarrow \sim q) \leftrightarrow (p \leftrightarrow q)$.
7. Construct the truth table for the following formula $(\sim p \vee \sim q) \leftrightarrow (p \to q)$.
   (MCA, IGNOU, June 2002)

EQUIVALENCE OF PROPOSITIONS

Two propositions are said to be logically equivalent if they have exactly the same truth values under all circumstances. The table 1 contains the fundamental logical equivalent expressions :

Table 1.

1. **De Morgan's Laws**
   $\sim (p \wedge q) \equiv \sim p \vee \sim q$
   $\sim (p \vee q) \equiv \sim p \wedge \sim q$
2. **Commutative Properties**
   $p \vee q \equiv q \vee p$; $p \wedge q \equiv q \wedge p$.
3. **Associative Properties**
   $(p \vee q) \vee r \equiv p \vee (q \vee r)$
   $(p \wedge q) \wedge r \equiv p \wedge (q \wedge r)$
4. **Distributive Properties**
   $p \wedge (q \vee r) \equiv (p \wedge q) \vee (p \wedge r)$
   $p \vee (q \wedge r) \equiv (p \vee q) \wedge (p \vee r)$
5. **Impotent Laws**
   $p \vee p \equiv p$ and $p \wedge p \equiv p$
6. **Complement Properties**
   $p \equiv \sim \sim p$
7. **Transposition**
   $(p \to q) \equiv (\sim q \to \sim p)$
8. **Material Implication**
   $(p \to q) \equiv (\sim p \vee q)$
9. **Material Equivalence**
   $(p \leftrightarrow q) \equiv [(p \to q) \wedge (q \to p)]$
   $(p \leftrightarrow q) \equiv [(p \to q) \vee (\sim p \wedge \sim q)]$
10. **Exportation**
    $[(p \wedge q) \to r] \equiv [p \to (q \to r)]$

Example 14. *Consider the following propositions*

$$\sim p \vee \sim q \text{ and } \sim (p \wedge q).$$

Are they equivalent ?

Sol. Construct the truth table for both (as in Fig. 19).

p	q	$\sim p$	$\sim q$	$\sim p \vee \sim q$	$p \wedge q$	$\sim (p \wedge q)$
T	T	F	F	F	T	F
T	F	F	T	T	F	T
F	T	T	F	T	F	T
F	F	T	T	T	F	T

Fig. 19

Since, the final values of both the propositions are same, hence the two propositions are equivalent.

TAUTOLOGIES

A proposition P is a tautology if it is true under all circumstances. It means it contains only T in the final column of its truth table.

Example 15. *Prove that the statement $(p \to q) \leftrightarrow (\sim q \to \sim p)$ is a tautology.*

Sol. Make the truth table of above statement :

p	q	$p \to q$	$\sim q$	$\sim p$	$\sim q \to \sim p$	$(p \to q) \leftrightarrow (\sim q \to \sim p)$
T	T	T	F	F	T	T
T	F	F	T	F	F	T
F	T	T	F	T	T	T
F	F	T	T	T	T	T

Fig. 20

As the final column contains all T's, so it is a tautology. (Fig. 20)

Example 16. *Prove that $p \vee p \leftrightarrow p$ is a tautology.*

Sol. Construct the truth table of the given statement :

p	$p \vee p$	$p \vee p \leftrightarrow p$
T	T	T
F	F	T

Fig. 21

As the last column contains all T's, so it is a tautology. (Fig. 21)

CONTRADICTION

A statement that is always false is called a contradiction.

Example 17. *Show that the statement $p \wedge \sim p$ is a contradiction.*

p	$\sim p$	$p \wedge \sim p$
T	*F*	*F*
F	*T*	*F*

Fig. 22

Sol. Construct the truth table of above statement.

Since, the last column contains all F's, so it is a contradiction. (Fig. 22)

CONTINGENCY

A statement that can be either true or false depending on the truth values of its variables, is called a contingency.

p	q	$p \to q$	$p \wedge q$	$(p \to q) \to (p \wedge q)$
T	T	T	T	T
T	F	F	F	T
F	T	T	F	F
F	F	T	F	F

Fig. 23

Example 18. *Prove that the statement $(p \to q) \to (p \wedge q)$ is a contingency.*

Sol. Construct the truth table of above statement.

As, the value of final column depends on the truth value of the variables, so it is a contingency. (Fig. 23)

Example 19. *From the following formulae, find out tautology, contingency and contradiction.*

(*i*) $A \equiv A \wedge (A \vee B)$ (*ii*) $(p \wedge \sim q) \vee (\sim p \wedge q)$ (*iii*) $\sim (p \vee q) \vee (\sim p \vee \sim q)$.

Sol. (*i*) Construct the truth table for $A \to A \wedge (A \vee B)$.

A	B	$A \vee B$	$A \wedge (A \vee B)$	$A \to A \wedge (A \vee B)$
T	T	T	T	T
T	F	T	T	T
F	T	T	F	T
F	F	F	F	T

Fig. 24

Since, the last column of the table contains all T's, hence it is a tautology. (Fig. 24)

(*ii*) Construct the truth table for $(p \wedge \sim q) \vee (\sim p \wedge q)$ as in Fig. 25.

p	q	$\sim p$	$\sim q$	$p \wedge \sim q$	$\sim p \wedge q$	$(p \wedge \sim q) \vee (\sim p \wedge q)$
T	T	F	F	F	F	F
T	F	F	T	T	F	T
F	T	T	F	F	T	T
F	F	T	T	F	F	F

Fig. 25

Since, the value of the final column depends on the value of the different variables, hence it is a contingency.

(*iii*) Construct the truth table of the proposition $\sim (p \vee q) \vee (\sim p \vee \sim q)$ as in Fig. 26.

p	q	$\sim p$	$\sim q$	$p \wedge q$	$\sim (p \wedge q)$	$\sim p \vee \sim q$	$\sim (p \vee q) \vee (\sim p \vee \sim q)$
T	T	F	F	T	F	F	F
T	F	F	T	F	T	T	T
F	T	T	F	F	T	T	T
F	F	T	T	F	T	T	T

Fig. 26

Since, the value of final column depends upon the value of different variables, hence it is a contingency.

Example 20. *Verify that proposition $p \vee \sim (p \wedge q)$ is tautology.*

Sol. Construct the truth table for above proposition. (Fig. 27)

p	q	$p \wedge q$	$\sim(p \wedge q)$	$p \vee \sim (p \wedge q)$
T	T	T	F	T
T	F	F	T	T
F	T	F	T	T
F	F	F	T	T

Fig. 27

Since, the last column contains all T' s, hence it is a tautology.

Example 21. *Determine whether the following is a tautology, contingency and a contradiction :*

(*i*) $p \to (p \to q)$ (*ii*) $p \to (q \to p)$ (*iii*) $p \wedge \sim p$.

Sol. (*i*) Construct truth table for $p \to (p \to q)$ as in Fig. 28.

p	q	$p \to q$	$p \to (p \to q)$
T	T	T	T
T	F	F	F
F	T	T	T
F	F	T	T

Fig. 28

Since, the value of last column depends on the value of different variables, hence it is a contingency.

(*ii*) Construct truth table for $p \to (q \to p)$ as in Fig. 29.

p	q	$q \to p$	$p \to (q \to p)$
T	T	T	T
T	F	T	T
F	T	F	T
F	F	T	T

Fig. 29

Since, the last column contains all T's, hence it is a tautology.

(*iii*) Construct truth table for $p \wedge \sim p$ as in Fig. 30.

p	$\sim p$	$p \wedge \sim p$
T	F	F
F	T	F

Fig. 30

Since, the last column contain all F's, hence it is a contradiction.

FUNCTIONALLY COMPLETE SETS OF CONNECTIVES

We have three basic and two conditional connectives *i.e.*, $\wedge, \vee, \sim \Rightarrow$ and $\Leftrightarrow$. If we have given any formula containing all these connectives, we can write an equivalent formula with certain proper subsets of these connectives.

A set of connectives is called functionally complete if every formula can be expresses on terms of an equivalent formula containing the connectives from this set.

Example 22. *Write an equivalent formula for* $p \wedge (r \Leftrightarrow s) \vee (s \Leftrightarrow p)$ *which does not involve biconditional.*

Sol. We know that $p \Leftrightarrow q \equiv (p \Rightarrow q) \wedge (q \Rightarrow p)$ ...(*i*)

So, apply eqn. (*i*) to formula to obtain the formula equivalent to

$$[p \wedge (r \Leftrightarrow s) \vee (s \Leftrightarrow p)],$$

which does not involve biconditional.

$$= [p \wedge ((r \Rightarrow s) \wedge (s \Rightarrow r)) \vee ((s \Rightarrow p) \wedge (p \Rightarrow s))].$$

Example 23. *Write an equivalent formula for* $r \vee (s \Leftrightarrow t)$*, which does not involve biconditional as well as conditional.*

Sol. We know that

$$(p \Leftrightarrow q) = (p \Rightarrow q) \wedge (q \Rightarrow p) \quad ...(i)$$

$$(p \Rightarrow q) = \sim p \wedge q \quad ...(ii)$$

So, applying the eqn. (*i*) and (*ii*) on the above formula, we can obtain an equivalent formula, which does not involve biconditional as well as conditional.

$$[r \vee (s \Leftrightarrow t)] = [r \vee ((s \Rightarrow t) \vee (t \Rightarrow s))]$$

$$= [r \vee ((\sim s \vee t) \vee (\sim t \vee s))].$$

Example 24. *Show that* $\{\sim, \wedge\}$ *is functionally complete.*

Sol. Take any formula which involve all the five connectives $\wedge, \vee, \sim, \Rightarrow$ and $\Leftrightarrow$. We can obtain an equivalence formula by first replacing biconditional and then replacing conditional and finally we can replace $\vee$.

As $$p \Leftrightarrow q = (p \Rightarrow q) \wedge (q \Rightarrow p)$$

$$= (\sim p \vee q) \wedge (\sim q \vee p) = (\sim (\sim\sim p \wedge \sim q)) \wedge (\sim (\sim \sim q \wedge \sim p))$$

Hence, $\{\sim, \wedge\}$ is functionally complete.

Similarly, we can show that $\{\sim, \vee\}$ is functionally complete.

Example 25. *Show that {~, →} is functionally complete.*

Sol. We know that $p \Rightarrow q \cong \sim p \vee q$.

So, we have $p \vee q \cong \sim p \Rightarrow q$.

Since, {~, ∨} is functionally complete. Hence from above, {~, ⇒} is also functionally complete.

i.e., given any formula which involve all the five connectives, we can obtain an equivalence formula using {~, →} by first replacing biconditional and then replacing (∧) ANDing and finally replacing ∨.

Example 26. *Express p ⇔ q in terms of {~, ∧} only.*

Sol. $(p \Leftrightarrow q) = (p \Rightarrow q) \wedge (q \Rightarrow p)$ $\quad \because (p \Leftrightarrow q) = (p \Rightarrow q) \wedge (q \Rightarrow p)$

$= (\sim p \vee q) \wedge (\sim q \vee p)$ $\quad \because p \Rightarrow q = \sim p \vee q$

$= (\sim(\sim\sim p \wedge \sim q)) \wedge (\sim(\sim\sim q \wedge \sim p))$ $\quad \because p \vee q = \sim(\sim p \wedge \sim q)$.

Example 27. *Express (p ∧ ~ q) ∨ (~ p ∧ ~ q) in terms of (~, ∨) only.*

Sol. $(p \wedge \sim q) \vee (\sim p \wedge \sim q) = (\sim(\sim p \vee \sim\sim q)) \vee (\sim(\sim\sim p \vee \sim\sim q))$ $\because p \wedge q = \sim(\sim p \vee \sim q)$

EXERCISES

1. Show that $(p \to q) \to r$ and $p \to (q \to r)$ are not equivalent.
2. Show that $\sim (p \oplus q)$ and $p \leftrightarrow q$ are logically equivalent.
3. Show that $\sim (p \vee (\sim p \wedge q))$ and $\sim p \wedge \sim q$ are logically equivalent.
4. Show that each of the following implications is a tautology by using truth tables.

   (i) $p \to (p \vee q)$ (ii) $(p \wedge q) \to (p \to q)$

   (iii) $(p \wedge (p \to q)) \to q$ (iv) $(\sim p \wedge (p \vee q)) \to q$

   (v) $\sim (p \to q) \to \sim q$ (vi) $((p \vee q) \wedge (p \to q) \wedge (q \to r)) \to r$.
5. Show that each of the following implications is a contradiction by using the truth table :

   (i) $p \wedge \sim q$ (ii) $(p \vee q) \wedge (\sim p) \wedge (\sim q)$
6. From the following formulae find which ones are tautologies or contradictions or contingencies :

   (i) $\sim (p \wedge q) \vee (\sim p \vee \sim q)$ (ii) $((\sim q \vee p) \wedge q)$

   (iii) $((\sim p \to q) \to (q \to p))$.
7. Show that ~ and ∧ forms a functionally complete collection of logical operators.
8. Express the propositional formula $G \leftrightarrow H$ only in terms of logical operations in the set {~ , ∧}.

   (IGNOU, MCA, June 2002)
9. Express $\sim (p \wedge (q \to (r \vee p)))$ in an equivalent formula which contains the connectives ~, ∧ only.

   (MCA, VTU, Karnataka, 1998)

ARGUMENT

An argument is an assertion ; that a group of propositions called premises, yields another proposition, called the conclusion. Let $P_1, P_2, P_3, \ldots, P_n$ is the group of propositions that yields the conclusion Q. Then, it is denoted as $P_1, P_2, P_n, \ldots, P_n \vdash Q$.

Conclusion. The conclusion of an argument is the proposition that is asserted on the basis of other proposition of the argument.

Premises. The propositions, which are assumed for accepting the conclusion, are called the premises of that argument.

(*a*) **Valid Argument**

An argument is called valid argument if the conclusion is true whenever all the premises are true.

The argument is also valid if and only if the ANDing of the group of propositions implies conclusion is a tautology *i.e.*, $P(p_1, p_2, p_3, \ldots, p_n) \rightarrow Q$ is a tautology, where $P(p_1, p_2, p_3, \ldots, p_n)$ is the group of propositions and Q is the conclusion.

Some common valid argument forms are given in Table 2.

(*b*) **Falacy Argument**

An argument is called falacy or an invalid argument if it is not a valid argument.

Elementary Valid Argument Forms

Table 2.

1. Modus ponens

$p \rightarrow q$

p

$\therefore\ q$

2. Modus tollens

$p \rightarrow q$

$\sim q$

$\therefore\ \sim p$

3. Hypothetical syllogism

$p \rightarrow q$

$q \rightarrow r$

$\therefore\ p \rightarrow r$

4. Disjunctive syllogism

$p \vee q$

$\sim p$

$\therefore\ q$

5. Constructive dilemma

$(p \rightarrow q) \wedge (r \rightarrow s)$

$p \vee r$

$\therefore\ q \vee s$

6. Absorption

$p \rightarrow q$

$\therefore\ p \rightarrow (p \wedge q)$

7. Simplification

$p \wedge q$

$\therefore\ p$

8. Conjunction

p

q

$\therefore\ p \wedge q$

9. Addition

p

$\therefore\ p \vee q$.

Example 28. *Show that the following rule is valid :*

$$p \vdash p \vee q \quad or \quad p \quad \therefore \quad p \vee q.$$

Sol. We can prove this rule from the truth table (Fig. 31)

p	q	$p \vee q$
T	T	T
T	F	T
F	T	T
F	F	F

Fig. 31

p is true in line 1 and 2 and $p \vee q$ is also true in line 1 and 2. Hence, argument is valid.

Example 29. *Show that the rule modus ponens is valid.*

$$\begin{array}{ll} & p \to q \\ & p \\ \therefore & q \end{array}$$

Sol. The truth table of this rule is as follows : (Fig. 32)

p	q	$p \to q$
T	T	T
T	F	F
F	T	T
F	F	T

Fig. 32

The p is true in line 1 and 2 and $p \to q$ and p both are true in line 1 and $p, p \to q$ and q all are true in line 1. Hence, argument is valid.

Example 30. *Show that the rule of hypothetical syllogism is valid*

$$\begin{array}{ll} & p \to q \\ & q \to r \\ \therefore & p \to r. \end{array}$$

Sol. The truth table of above rule is as in Fig. 33.

p	q	r	$p \to q$	$q \to r$	$p \to r$
T	T	T	T	T	T
T	T	F	T	F	F
T	F	F	F	T	F
T	F	T	F	T	T
F	T	T	T	T	T
F	T	F	T	F	T
F	F	T	T	T	T
F	F	F	T	T	T

Fig. 33

$p \to q$ is true in lines 1, 2, 5, 6, 7, 8.

$q \to r$ is true in lines 1, 3, 4, 5, 7, 8.

Both $p \to q$ and $q \to r$ is true in lines 1, 5, 7, 8, $p \to r$ is also true in lines 1, 5, 7, 8. Hence, argument is valid.

Example 31. *Show that the rule of Modus Tollens is valid*

$$\begin{array}{ll} & p \to q \\ & \sim q \\ \therefore & \sim p. \end{array}$$

Sol. The truth table of above propositions are (Fig. 34)

p	q	$\sim p$	$\sim q$	$p \to q$
T	T	F	F	T
T	F	F	T	F
F	T	T	F	T
F	F	T	T	T

Fig. 34

$p \to q$ is true in line 1, 3 and 4. $\sim q$ is true in line 2 and 4. Both $p \to q$ and $\sim q$ are true in line 4. $\sim p$ is also true in line 4. Hence, the argument is valid.

Example 32. *Show that the rule of disjunctive syllogism is valid*

$$p \vee q$$
$$\sim p$$
$$\therefore \quad q.$$

Sol. The truth table of above rule is as follows : (Fig. 35)

p	q	$\sim p$	$p \vee q$
T	T	F	T
T	F	F	T
F	T	T	T
F	F	T	F

Fig. 35

$p \vee q$ is true in line 1, 2 and 3. $\sim p$ is true in line 3. Both $p \vee q$ and $\sim p$ is true in line 3. As q is also true in line 3. Hence, argument is valid.

Example 33. *Show that the rule of simplification is valid*

$$p \wedge q$$
$$\therefore \quad p.$$

Sol. The truth table of above argument is as follows : (Fig. 36)

p	q	$p \wedge q$
T	T	T
T	F	F
F	T	F
F	F	F

Fig. 36

$p \wedge q$ is true in line 1 and p is also true in line 1. Hence, the argument is valid.

Example 34. *Show that the rule of conjunction is valid.*

$$p$$
$$q$$
$$\therefore \quad p \wedge q.$$

Sol. The argument is valid if $p \wedge q \to p \wedge q$ is a tautology. The truth table for the above proposition is as follows : (Fig. 37)

p	q	$p \wedge q$	$p \wedge q \to p \wedge q$
T	T	T	T
T	F	F	T
F	T	F	T
F	F	F	T

Fig. 37

As the proposition is a tautology. Hence, the argument is valid.

Example 35. *Show that the rule of absorption is valid*

$$p \to q$$
$$\therefore \quad p \to (p \wedge q).$$

Sol. We have to show that $(p \to q) \to [p \to (p \wedge q)]$ is tautology. The truth table of the above argu. ıent is as follows : (Fig. 38)

p	q	$p \wedge q$	$p \to q$	$p \to (p \wedge q)$	$(p \to q) \to [p \to (p \wedge q)]$
T	T	T	T	T	T
T	F	F	F	F	T
F	T	F	T	T	T
F	F	F	T	T	T

Fig. 38

Since, the argument is a tautology. Hence, it is a valid argument.

PROOF OF VALIDITY

We can test the validity of any argument by constructing the truth tables. But as the no. of variable statements increases, the truth tables grow unwieldly. So, a more efficient method to test the validity of the argument is to deduce its conclusion from its premises by a sequence of elementary arguments each of which is known to be valid.

Example 36. *Prove that the argument* $p \to \sim q, r \to q, r \vdash \sim p$ *is valid without using truth tables.*

Sol. (*i*)	$p \to \sim q$	(Given)
(*ii*)	$r \to q$	(Given)
(*iii*)	$\sim q \to \sim r$	Contrapositive of (*ii*)
(*iv*)	$p \to \sim r$	Hypothetical syllogism using (*i*) and (*iii*)
(*v*)	$r \to \sim p$	Contrapositive of (*iv*)

(*vi*)	r is true	(Given)
(*vii*)	$\sim p$ is true	Modus ponens using (*v*) and (*vi*).

Example 37. *Prove that the argument* $p \to q, q \to r, r \to s, \sim s, p \vee t \vdash t$ *is valid without using truth tables.*

Sol. (*i*)	$p \to q$	(Given)
(*ii*)	$q \to r$	(Given)
(*iii*)	$r \to s$	(Given)
(*iv*)	$\sim s$	(Given)
(*v*)	$p \vee t$	(Given)
(*vi*)	$p \to r$	Hypothetical syllogism using (*i*) and (*ii*)
(*vii*)	$p \to s$	Hypothetical syllogism using (*vi*) and (*iii*)
(*viii*)	$\sim p$	Modus tollens using (*vii*) and (*iv*)
(*ix*)	t	Disjunctive syllogism using (*v*) and (*viii*).

Example 38. *Prove that the argument* $p, q \vdash (p \vee r) \wedge q$ *is valid without using truth tables.*

Sol. (*i*)	p	(Given)
(*ii*)	$p \vee r$	Rule of addition using (*i*)
(*iii*)	q	(Given)
(*iv*)	$(p \vee r) \wedge q$	Rule of conjunction using (*ii*) and (*iii*).

Example 39. *Prove that the argument* $p \to q, p \wedge r \vdash q$ *is valid without using truth table.*

Sol. (*i*)	$p \to q$	(Given)
(*ii*)	$p \wedge r$	(Given)
(*iii*)	p	Rule of simplification using (*i*)
(*iv*)	q	Modus ponens using (*i*) and (*iii*).

Example 40. *Prove that the argument* $(p \to q) \wedge (r \to s), (p \vee r) \wedge (q \vee r) \vdash q \vee s$

Sol. (*i*)	$(p \to q) \wedge (r \to s)$	(Given)
(*ii*)	$(p \vee r) \wedge (q \vee r)$	(Given)
(*iii*)	$(p \vee r)$	Simplification using (*iii*)
(*iv*)	$q \vee s$	Constructive dilemma using (*i*) and (*iii*).

Example 41. *Prove that the argument* $(p \wedge q) \vee (r \to s), t \to r, \sim (p \wedge q) \vdash t \to s$ *is valid without using truth tables.*

Sol. (*i*)	$(p \wedge q) \vee (r \to s)$	(Given)
(*ii*)	$t \to r$	(Given)
(*iii*)	$\sim (p \wedge q)$	(Given)
(*iv*)	$r \to s$	Disjunctive syllogism using (*i*) and (*iii*)
(*v*)	$t \to s$	Hypothetical syllogism using (*ii*) and (*iii*).

Example 42. *Prove that the argument* $(p \to q) \wedge (r \to s), q \to s, (q \to s) \to (p \vee r) \vdash q \vee s$ *is valid using deduction method.*

Sol. (*i*) $(p \to q) \wedge (r \to s)$ (Given)

(*ii*) $q \to s$ (Given)

(*iii*) $(q \to s) \to (p \vee r)$ (Given)

(*iv*) $p \vee r$ Modus ponens using (*iii*) and (*ii*)

(*v*) $q \vee s$ Constructive dilemma using (*i*) and (*iv*).

Example 43. *Prove that the argument* $p \to (q \vee r), (s \wedge t) \to q, (q \vee r) \to (s \wedge t) \vdash p \to q$ *is valid without using truth table.*

Sol. (*i*) $p \to (q \vee r)$ (Given)

(*ii*) $(s \wedge t) \to q$ (Given)

(*iii*) $(q \vee r) \to (s \wedge t)$ (Given)

(*iv*) $p \to (s \wedge t)$ Hypothetical syllogism using (*i*) and (*iii*)

(*v*) $p \to q$ Hypothetical syllogism using (*ii*) and (*iv*).

Example 44. *Prove that the argument* $p \vee (q \to p), \sim p \wedge r \vdash \sim q$ *is valid without using truth tables.*

Sol. (*i*) $p \vee (q \to p)$ (Given)

(*ii*) $\sim p \wedge r$ (Given)

(*iii*) $\sim p$ Rule of simplification using (*ii*)

(*iv*) $q \to p$ Disjunctive syllogism using (*i*) and (*iii*)

(*v*) $\sim q$ Modus tollens using (*iv*) and (*iii*).

Example 45. *Test the validity of following argument. If I will select in IAS examination, then I will not be able to go to London. Since, I am going to London, I will not select in IAS examination.*

Sol. Let p be "I will select in IAS examination" and q be "I am going to London". Then the above argument can be written in symbolic form as follows :

$$p \to \sim q$$
$$q$$
$$\therefore \quad \sim p$$

Construct the truth table for above argument (Fig. 39)

p	q	$\sim p$	$\sim q$	$p \to \sim q$
T	T	F	F	F
T	F	F	T	T
F	T	T	F	T
F	F	T	T	T

Fig. 39

$p \to \sim q$ is true in line 2, 3 and 4. q is true in line 1 and 4 and $\sim p$ is true in line 3 and 4. Hence, all three are true in line 4. So it is a valid statement.

Example 46. *Consider the following argument and determine whether it is valid.*

Either I will get good marks or I will not graduate. If I did not graduate I will go to Canada. I get good marks. Thus, I would not go to Canada.

Sol. Let p be "I will get good marks" and q be "I will graduate" and r be "I will go to Canada". Thus, the above argument can be written in symbolic form as follows :

$$p \vee \sim q$$
$$\sim q \to r$$
$$p$$
$$\therefore \quad \sim r$$

The truth table of above proposition is (Fig. 40)

p	q	r	$\sim q$	$\sim r$	$p \vee \sim q$	$\sim q \to r$
T	T	T	F	F	T	T
T	**T**	**F**	**F**	**T**	**T**	**T**
T	F	F	T	T	T	F
T	F	T	T	F	T	T
F	T	T	F	F	F	T
F	T	F	F	T	F	T
F	F	T	T	F	T	T
F	F	F	T	T	T	F

Fig. 40

$p \vee \sim q$ is true in line 1, 2, 3, 4, 7, 8 and $\sim q \to r$ is true in line 1, 2, 4, 5, 6, 7 and p is true in line 1, 2, 3 and 4. $\sim r$ is true in 2, 3, 6 and 8. All the above are true in line 2. Hence, the argument is valid.

Example 47. *Determine the validity of the following argument without using truth tables.*

Either I will pass the examination, or, I will not graduate. If I do not graduate, I will go to Canada. I failed : Thus, I will go to Canada.

Sol. Let p be "I will pass the examination" and q be "I will graduate" and t be "I will go to Canada". Thus the above argument, in symbolic form can be written as

$$p \vee \sim q$$
$$\sim q \to t$$
$$\sim p$$
$$\therefore \quad t$$

Thus to prove the validity of the argument, use the standard results as follows :

(*i*)	$p \vee \sim q$	(Given)
(*ii*)	$\sim q \to t$	(Given)
(*iii*)	$\sim p$	(Given)
(*iv*)	$\sim q$	Disjunctive syllogism using (*i*) and (*iii*)
(*v*)	t	Modus ponens using (*ii*) and (*iv*)

Hence proved.

Example 48. *Determine the validity of the following argument using deduction method.*

If I study, then I will pass examination. If I do not go to picnic, then I will study. But I failed examination. Therefore, I went to picnic.

Sol. Let p be "I study" and q be "I will pass examination" and t be "I go to picnic". Then the above argument is written in symbolic form as follows :

$$p \to q$$
$$\sim t \to p$$
$$\sim p$$
$$\therefore \quad t$$

Thus to prove the validity of the argument use the rules of inference.

(*i*)	$p \to q$	(Given)
(*ii*)	$\sim t \to p$	(Given)
(*iii*)	$\sim p$	(Given)
(*iv*)	$\sim \sim t$	Modus tollens using (*ii*) and (*iii*)
(*v*)	t	Complement property using (*iv*)

Hence proved.

Example 49. *Prove the validity of the following argument using truth tables as well as deduction method.*

"If the market is free then there is no inflation. If there is no inflation then there are price controls. Since there are price controls, therefore, the market is free".

Sol. Let p be "The market is free" and q be "There is inflation" and r be "There are price controls". Then the above argument can be written in symbolic form as follows :

$$p \to \sim q$$
$$\sim q \to r$$
$$r$$
$$\therefore \quad p$$

Ist Method. By using truth tables

Construct the truth table of above argument (Fig. 41)

p	q	r	$\sim q$	$p \to \sim q$	$\sim q \to r$
T	T	T	F	F	T
T	T	F	F	F	T
T	F	F	T	T	F
T	F	T	T	T	T
F	T	T	F	T	T
F	T	F	F	T	T
F	F	T	T	T	T
F	F	F	T	T	F

Fig. 41

$p \to \sim q$ is true in line 3, 4, 5, 6, 7 and 8, $\sim q \to r$ is true in line 1, 2, 4, 5, 6, 7, r is true in line 1, 4, 5, 7. All the above three are true in line 4 and 5. Also p is true in line 4. Hence the argument is valid.

IInd Method. Using deduction method

(*i*)	$p \to \sim q$	(Given)
(*ii*)	$\sim q \to r$	(Given)
(*iii*)	$p \to r$	Hypothetical syllogism using (*i*) and (*ii*)
(*iv*)	$\sim p \to \sim r$	Transposition using (*iii*)
(*v*)	r	(Given)
(*vi*)	$\sim \sim p$	Modus tollens using (*iv*) and (*v*)
(*vii*)	p	Complement of (*vi*).

EXERCISES

1. Show that $R \wedge (P \vee Q)$ is a valid conclusion from the premises $P \vee Q$, $Q \to R$, $P \to M$ and $\sim M$. *(MCA, VTU, Karnataka, 1999)*
2. Prove the validity of the following arguments using truth tables as well as deduction method :
   (*i*) $p \to \sim q \;\; r \to q, r \vdash \sim p$ (*ii*) $p \to r, q \to r, p \vee q \vdash r$
   (*iii*) $p \to q, r \vee \sim q, \sim r \vdash \sim p$ (*iv*) $p \to q, \sim q \vdash \sim p$
3. From the given premises
   (*i*) $p \leftrightarrow Q$ (*ii*) $R \vee \sim S$
   (*iii*) $Q \to S$ (*iv*) $\sim P \to R$
   $\therefore$ R
   Check the validity. *(MCA, IGNOU, June 2002)*
4. Formulate the following arguments symbolically and determine whether each is valid.
   (*i*) If I study hard or I get rich, then I get A's
   I get A's
   ―――――――――――
   $\therefore$ If I don't study hard, then I get rich.
   (*ii*) If I study hard, then I get A's
   If I don't get rich, then I don't get A's
   ―――――――――――
   $\therefore$ I get rich.
5. Determine whether each argument is valid.
   (*i*) $p \to (q \to r)$
   $q \to (p \to r)$
   ―――――――――――
   $\therefore \; (p \vee q) \to r$
   (*ii*) $(p \to q) \wedge (r \to s)$
   $p \vee q$
   ―――――――――――
   $\therefore \; q \vee s$

EXISTENTIAL QUANTIFIER

If $p(x)$ is a proposition over the universe U. Then it is denoted as $\exists x\, p(x)$ and read as "There exists at least one value in the universe of variable x such that $p(x)$ is true. The quantifier $\exists$ is called the existential quantifier.

There are several ways to write a proposition, with an existential quantifier *i.e.*,

$(\exists x \in A)\, p(x)$ or $\exists x \in A$ such that $p(x)$ or $(\exists x)\, p(x)$ or $p(x)$ is true for some $x \in A$.

UNIVERSAL QUANTIFIER

If $p(x)$ is a proposition over the universe U. Then it is denoted as $\forall x, p(x)$ and read as "For every $x \in$ U, $p(x)$ is true". The quantifier $\forall$ is called the universal quantifier.

There are several ways to write a proposition, with a universal quantifier.

$\forall x \in A, p(x)$ or $p(x), \forall x \in A$

or $\forall x, p(x)$ or $p(x)$ is true for all $x \in A$.

Example 50. *Let A(x) : x has a white colour, B(x) : x is a polar bear, C(x) : x is found in cold regions, over the universe of animals. Translate the following into simple sentences :*

(i) $\exists x\ (B(x) \wedge \sim A(x))$ *(ii)* $(\exists x)\ (\sim C(x))$

(iii) $(\forall x)\ (B(x) \wedge C(x) \rightarrow A(x))$.

Sol. *(i)* There exists a polar bear whose colour is not white.

(ii) There exists an animal that is not found in cold regions.

(iii) Every polar bear that is found in cold regions has a white colour.

Example 51. *Let K(x) : x is a two-wheeler, L(x) : x is a scooter. M(x) : x is manufactured by Bajaj. Express the following using quantifiers.*

(i) Every two wheeler is a scooter.

(ii) There is a two wheeler that is not manufactured by Bajaj.

(iii) There is a two wheeler manufactured by Bajaj that is not a scooter.

(iv) Every two wheeler that is a scooter is manufactured by Bajaj.

Sol. *(i)* $(\forall x)\ (K(x) \rightarrow L(x))$

(ii) $(\exists x)\ (K(x) \wedge M(x))$

(iii) $(\exists x)\ (K(x) \wedge M(x) \rightarrow \sim L(x))$

(iv) $(\forall x)\ (K(x) \wedge L(x) \rightarrow M(x))$

NEGATION OF QUANTIFIED PROPOSITIONS

When we negate a quantified proposition *i.e.,* when a universally quantified proposition is negated, we obtain an existentially quantified proposition and when an existentially quantified proposition is negated, we obtain a universally quantified proposition.

The two rules for negation of quantified proposition are as follows. These are also called De Morgan's law.

(i) $\sim \exists x\ p(x) \equiv \forall x \sim p(x)$ *(ii)* $\sim \forall x\ p(x) \equiv \exists x \sim p(x)$.

Example 52. *Negate each of the following propositions :*

(i) All boys can run faster than all girls.

(ii) Some girls are more intelligent than all boys.

(iii) Some students do not live in hostel.

(iv) All students pass the semester exams.

(v) Some of the students are absent and the classroom's empty.

Sol. *(i)* Some boys can run faster than some girls.

(ii) All girls are more intelligent than some boys.

(iii) All students live in hostel.

(*iv*) Some students do not pass the semester exams.

(*v*) All students are present and the class-room is full.

Example 53. *Negate each of the following propositions :*

(*i*) $\forall x\, P(x) \wedge \exists y\, q(y)$ (*ii*) $\forall x\, p(x) \wedge \forall y\, q(y)$

(*iii*) $\exists x\, p(x) \vee \forall y\, q(y)$ (*iv*) $\exists x\, p(x) \vee \exists y\, q(y)$

(*v*) $(\exists x \in U)\,(x + 6 = 25)$ (*vi*) $(\forall x \in U)\,(x < 25)$.

Sol. (*i*)
$$\sim (\forall x\, p(x) \wedge \exists y\, q(y))$$
$$\cong \sim \forall x\, p(x) \vee \sim \exists y\, q(y) \qquad (\because\ \sim (p \wedge q) = \sim p \vee \sim q)$$
$$\cong \exists x \sim p(x) \vee \forall y \sim q(y)$$

(*ii*)
$$\sim (\forall x\, p(x) \wedge \forall y\, q(y))$$
$$\cong \sim \forall x\, p(x) \vee \sim \forall y\, q(y) \qquad (\because\ \sim (p \wedge q) = \sim p \sim q)$$
$$\cong \exists x \sim p(x) \vee \exists y \sim q(y)$$

(*iii*)
$$\sim (\exists x\, p(x) \vee \forall y\, q(y))$$
$$\cong \sim \exists x\, p(x) \wedge \sim \forall y\, q(y) \qquad (\because\ \sim (p \vee q) = \sim p \wedge \sim q)$$
$$\cong \forall x \sim p(x) \wedge \exists y \sim q(y)$$

(*iv*)
$$\sim (\exists x\, p(x) \vee \exists y\, q(y))$$
$$\cong \sim \exists x\, p(x) \wedge \sim \exists y\, q(y) \qquad (\because\ \sim (p \vee q) = \sim p \wedge \sim q)$$
$$\cong \forall x \sim p(x) \wedge \forall y \sim q(y)$$

(*v*)
$$\sim (\exists x \in \text{U})\,(x + 6) = 25$$
$$\cong \forall x \in \text{U} \sim (x + 6) = 25$$
$$\cong (\forall x \in \text{U})\,(x + 6) \neq 25$$

(*vi*)
$$\sim (\forall x \in \text{U})\,(x < 25)$$
$$\cong \exists x \in \text{U} \sim (x < 25)$$
$$\cong (\exists x \in \text{U})\,(x \geq 25)$$

PROPOSITIONS WITH MULTIPLE QUANTIFIERS

The proposition having more than one variable can be quantified with multiple quantifiers. The multiple universal quantifiers can be arranged in any order without altering the meaning of the resulting proposition. Also the multiple existential quantifiers can be arranged in any order without altering the meaning of proposition.

The proposition which contains both universal and existential quantifiers, the order of these quantifiers can't be exchanged without altering the meaning of proposition *e.g.*, the proposition $\exists x\, \forall y\, p(x, y)$ means " There exists some x such that $p(x, y)$ is true for every y".

Example 54. *Write the negation for each of the following. Determine whether the resulting statement is true or false. Assume* $U = R$.

(*i*) $\forall x\, \exists m(x^2 < m)$ (*ii*) $\exists m\, \forall x(x^2 < m)$.

Sol. (*i*) Negation of $\forall x\, \exists m(x^2 < m)$ is $\exists x\, \forall m\, (x^2 \geq m)$. The meaning of $\exists x\, \forall m(x^2 \geq m)$ is that there exists some x such that $x^2 \geq m$, for every m. The statement is true as there is some greatest x such that $x^2 \geq m$, for every m.

(*ii*) Negation of $\exists\, m\ \forall\, x\ (x^2 < m)$ is $\forall\, m\ \exists\, x\ (x^2 \geq m)$. The meaning of $\forall\, m\ \exists\, x\ (x^2 \geq m)$ is that for every m, there exists some x such that $(x^2 \geq m)$. The statement is true as for every m, there exists some greatest x such that $x^2 \geq m$.

Example 55. *Check the validity of following formula under given interpretation.*

(*a*) $\forall\, x\ \exists\, y\ P(x, y)$ *under interpretation domain* = {1, 2} *and* $P(1, 1) = T$, $P(1, 2) = F$, $P(2, 1) = T$, $P(2, 2) = T$ *where T and F refer to true and false respectively.*

(*b*) $\forall\, x\ (P(x) \to Q(a, f(x))$ *under interpretation domain* = {1, 2} *and* $a = 1$, $f(1) = 2$, $f(2) = 1$, $P(1) = F$, $P(2) = T$, $Q(1, 1) = T$, $Q(1, 2) = T$, $Q(2, 1) = F$ *and* $Q(2, 2) = T$. *Here T and F refer to true and false respectively.*

Sol. (*a*) $\forall\, x\ \exists\, y\ P(x, y)$ means for every x, there exists some y such that $P(x, y)$ is true.

Since $P(1, 1) = T$ and $P(2, 1) = T$. Hence, the formula is valid under the domain = {1, 2}.

(*b*) $\forall\, x\ (P(x) \to Q(a, F(x))$ means for every x, whenever $P(x)$ is true implies $Q(a, f(x))$ is also true. So, under the domain {1, 2}, the different values of $\forall\, x\ (P(x) \to Q(a, f(x))$ are as follows :

x	$P(x)$	$Q(a, f(x))$
1	F	T
2	T	T

So, when $P(x)$ is true, $Q(a, f(x))$ is also true. Hence, the formula is valid.

EXERCISES

1. Represent the following statements symbolically :
   (*i*) All cars run faster than some dogs
   (*ii*) Some dogs can run faster than some scooters.
2. Let $P(x, y)$: x is younger than y
   $Q(x, y)$: x is wiser than y
   $R(x)$: x is handsome.
   Then translate the following symbolic statements into English :
   (*i*) $(x)\ (y)\ [Q(x, y) \to R(y)]$ (*ii*) $(x)\ (\exists y)\ [P(x, y) \to (Q(x, y) \wedge R(y))]$
   (*iii*) $(x)\ [(y) \sim P(x, y) \to R(x)]$
3. Let $P(x, y)$: x is slower than y
   $Q(x, y)$: y is taller than x
   $R(x)$: x is less than 100 kg
   Then translate the following into English :
   (*i*) P(x, MONU) (*ii*) Q (MONU, SONU) $\wedge$ R(MONU)
   (*iii*) $Q(x, y) \to \sim R(x)$.
4. Negate the following :
   (*i*) $(\exists x)\ (\forall y)\ [P(x) \vee Q(y)]$ (*ii*) $(\forall\, x)\ [P(x) \to Q(x)]$
5. Use the universal and existential quantifiers to symbolize the following statements :
   (*i*) All dogs are black (*ii*) All cats are white
   (*iii*) Some cats mew sharply (*iv*) All men are equal.

SOLVED PROBLEMS

Problem 1. *Translate the following statements in proposition logic :*

(i) If you study, you will get good marks. If you do not study, you will enjoy. Therefore, either you will get good marks or you will enjoy.

(ii) If the catalogue is correct, then if the seeds are planted on April, flowers will bloom in July.

(iii) If John is elected class representative, then either Mary is elected treasurer or Alice is elected vice-treasurer.

(iv) Either taxes are increased or if expenditures rise, then the debt ceiling is raised.

Sol. (*i*) Let p be "You study" and q be "You will get good marks" and r be "You will enjoy". Then the proposition logic is

$$p \to q$$
$$\sim p \to r$$
$$\therefore \quad q \vee r$$

(*ii*) Let p be "The catalogue is correct" and q be "Seeds are planted in April" and r be "The flowers bloom in July". Then, proposition logic is

$$p \to (q \to r)$$

(*iii*) Let p be "John is elected class representative" and q be "Mary is elected treasurer" and r be "Alice is elected vice-treasurer". Then, proposition logic is

$$p \to q \vee r$$

(*iv*) Let p be "Taxes are increased" and q be "Expenditures rise" and r be "The debt ceiling is raised". Then proposition logic is

$$p \vee (q \to r).$$

Problem 2. *The meaning of proposition $p \to q$ is "If p then q" or "q is a necessary condition for p" or "p only if q" or "p is a sufficient condition for q".*

Write the following statements in terms of above.

1. p : *New Delhi is capital of India* $\quad q$: *India is in Asia.*
2. $p : x^2 = 4 \quad q : x = 2$
3. $p : x^2 = y^2 \quad q : x = y.$
4. p = *he works hard* $\quad q$: *he is a Gold Medalist.*

Sol. 1. (*i*) If New Delhi is capital of India, then India is in Asia.

(*ii*) India is in Asia is a necessary condition for New Delhi to be capital of India.

(*iii*) New Delhi is capital of India only if India is in Asia.

(*iv*) The fact that New Delhi is capital of India is a sufficient condition that India is in Asia.

2. (*i*) If $x^2 = 4$, then $x = 2$

(*ii*) $x = 2$ is necessary for $x^2 = 4$

(*iii*) $x^2 = 4$ only if $x = 2$

(*iv*) $x^2 = 4$ is sufficient for $x = 2$.

3. (*i*) If $x^2 = y^2$ then $x = y$

(*ii*) $x = y$ is necessary for $x^2 = y^2$

(*iii*) $x^2 = y^2$ only if $x = y$

(*iv*) $x^2 = y^2$ is sufficient for $x = y$.

4. (*i*) If he works hard then he is a Gold Medalist.

(*ii*) Gold medal is necessary for hard work.

(*iii*) He works hard only if he is a Gold Medalist.

(*iv*) Hard work is sufficient condition for Gold Medalist.

Problem 3. *Construct the truth table for the following statements*

(*i*) $(p \to p) \to (p \to \sim p)$ (*ii*) $(\sim q \to \sim p) \to (p \to q)$.

Sol. (*i*) The truth table for $(p \to p) \to (p \to \sim p)$ is as follows. (Fig. 42)

p	$\sim p$	$p \to p$	$p \to \sim p$	$(p \to p) \to (p \to \sim p)$
T	F	T	F	F
F	T	T	T	T

Fig. 42

(*ii*) The truth table of statement $(\sim q \to \sim p) \to (p \to q)$ is as follows. (Fig. 43)

p	q	$\sim p$	$\sim q$	$\sim q \to \sim p$	$p \to q$	$(\sim q \to \sim p) \to (p \to q)$
T	T	F	F	T	T	T
T	F	F	T	F	F	T
F	T	T	F	T	T	T
F	F	T	T	T	T	T

Fig. 43

Problem 4. *Construct the truth table for following statements*

(*i*) $(p \to (q \to r)) \to ((p \to q) \to (p \to r))$

(*ii*) $p \leftrightarrow (\sim p \vee \sim q)$

(*iii*) $(p \to p) \vee (p \to \sim p)$.

Sol. (*i*) Truth table for $(p \to (q \to r)) \to ((p \to q) \to (p \to r))$ is as follows. (Fig. 44)

p	q	r	$q \to r$	$p \to q$	$p \to r$	$\sim(p \to (q \to r))$	$((p \to q) \to (p \to r))$	$K \to L$
T	T	T	T	T	T	T	T	T
T	T	F	F	T	F	F	F	T
T	F	F	T	F	F	T	T	T
T	F	T	T	F	T	T	T	T
F	T	T	T	T	T	T	T	T
F	T	F	F	T	T	T	T	T
F	F	T	T	T	T	T	T	T
F	F	F	T	T	T	T	T	T

Fig. 44

(*ii*) Truth table for $p \leftrightarrow (\sim p \vee \sim q)$ is as following. (Fig. 45)

p	q	$\sim p$	$\sim q$	$\sim p \vee \sim q$	$p \leftrightarrow (\sim p \vee \sim q)$
T	T	F	F	F	F
T	F	F	T	T	T
F	T	T	F	T	F
F	F	T	T	T	F

Fig. 45

(*iii*) The truth table for $(p \to p) \vee (p \to \sim p)$ is as follows. (Fig. 46)

p	$\sim p$	$p \to p$	$p \to \sim p$	$(p \to p) \vee (p \to \sim p)$
T	F	T	F	T
F	T	T	T	T

Fig. 46

Problem 5. *Assume the value of $p \to q$ is false. Determine the value of $(\sim p \vee \sim q) \to q$.*

Sol. Construct the truth table for both the statements and determine the value of $(\sim p \vee \sim q) \to q$ against the false values of $p \to q$. (Fig. 47)

p	q	$\sim p$	$\sim q$	$p \to q$	$\sim p \vee \sim q$	$(\sim p \vee \sim q) \to q$
T	T	F	F	T	F	T
T	F	F	T	F	T	F
F	T	T	F	T	T	T
F	F	T	T	T	T	F

Fig. 47

So, when $p \to q$ is false, the value of $(\sim p \vee \sim q) \to q$ is also false.

Problem 6. *Given the value of $p \to q$ is true. Determine the value of $\sim p \vee (p \leftrightarrow q)$.*

Sol. Construct truth table for both statements. (Fig. 48)

p	q	$\sim p$	$p \to q$	$p \leftrightarrow q$	$\sim p \vee (p \leftrightarrow q)$
T	T	F	T	T	T
T	F	F	F	F	F
F	T	T	T	F	T
F	F	T	T	T	T

Fig. 48

So, when the value of $p \to q$ is true *i.e.*, in line 1, 3 and 4. The value of $\sim p \vee (p \leftrightarrow q)$ is also true.

Problem 7. *Prove that the negation of conditional statement $\sim (p \to q)$ is equivalent of $p \wedge \sim q$.*

Sol. The truth table of the above propositions are as follows. (Fig. 49)

p	q	$\sim q$	$p \to q$	$\sim (p \to q)$	$p \wedge \sim q$
T	T	F	T	F	F
T	F	T	F	T	T
F	T	F	T	F	F
F	F	T	T	F	F

Fig. 49

As the values of both the propositions are same, hence $\sim (p \to q) \equiv p \wedge \sim q$.

Problem 8. *Prove that the negation of biconditional statement $\sim (p \leftrightarrow q)$ is equivalent to $p \leftrightarrow \sim q$ or $\sim p \leftrightarrow q$.*

Sol. The truth table of all the above propositions is as follows. (Fig. 50)

p	q	$\sim p$	$\sim q$	$p \leftrightarrow q$	$\sim (p \leftrightarrow q)$	$p \leftrightarrow \sim q$	$\sim p \leftrightarrow q$
T	T	F	F	T	F	F	F
T	F	F	T	F	T	T	T
F	T	T	F	F	T	T	T
F	F	T	T	T	F	F	F

Fig. 50

As the values of $\sim (p \leftrightarrow q)$ and $p \leftrightarrow \sim q$ and $\sim p \leftrightarrow q$ are same, hence they all are equivalent.

Problem 9. *From the following formulae find out tautology, contingency and contradiction.*

(*i*) $\sim (A \to B) \vee (\sim A \vee (A \wedge B))$ (*ii*) $(H \to (I \wedge J)) \to \sim (H \to I)$

(*iii*) $(p \leftrightarrow Q) \equiv (p \wedge Q) \vee (\sim p \wedge Q)$.

Sol. (*i*) Construct the truth table for $\sim (A \to B) \vee (\sim A \vee (A \wedge B))$ as in Fig. 51.

A	B	$\sim$ A	$\sim A \vee (A \wedge B)$	$A \to B$	$\sim (A \to B)$	$\sim (A \to B) \vee (\sim A \vee (A \wedge B))$
T	T	F	T	T	F	T
T	F	F	F	F	T	T
F	T	T	T	T	F	T
F	F	T	T	T	F	T

Fig. 51

As the last column of the table contains all T's, hence it is a tautology.

(*ii*) Construct truth table for $(H \to I \wedge J)) \to \sim (H \to I))$ as in (Fig. 52)

H	I	J	$(I \wedge J)$	$H \to (I \wedge J)$	$H \to I$	$\sim (H \to I)$	$(H \to I \wedge J) \to \sim (H \to I)$
T	T	T	T	T	T	F	F
T	T	F	F	F	T	F	T
T	F	F	F	F	F	T	T
T	F	T	F	F	F	T	T
F	T	T	T	T	T	F	F
F	F	T	F	T	T	F	F
F	T	F	F	T	T	F	F
F	F	F	F	T	T	F	F

Fig. 52

As the value of last column depends upon the value of the variables, hence it is a contingency.

(*iii*) Construct the truth table for $(p \leftrightarrow q) \rightarrow (p \wedge q) \vee (\sim p \wedge q)$ as in (Fig. 53)

p	q	$\sim p$	$p \leftrightarrow q$	$p \wedge q$	$\sim p \wedge q$	$(p \wedge q) \vee (\sim p \wedge q)$	$(p \leftrightarrow q) \rightarrow (p \wedge q) \vee (\sim p \wedge q)$
T	T	F	T	T	F	T	T
T	F	F	F	F	T	T	T
F	T	T	F	F	T	T	T
F	F	T	T	F	F	F	T

Fig. 53

As the last column contains all T's, hence the above formulae is a tautology.

Problem 10. *Prove that following is a tautology*

$$A \vee (\overline{B \wedge C}) \equiv (A \vee \overline{B}) \vee \overline{C}.$$

Sol. Construct the truth table for $A \vee (\overline{B \wedge C}) \rightarrow (A \vee \overline{B}) \vee \overline{C}$ as in (Fig. 54)

A	B	C	$\overline{B}$	$\overline{C}$	$\overline{B \wedge C}$	$A \vee (\overline{B \wedge C})$	$(A \vee \overline{B})$	$(A \vee \overline{B}) \vee \overline{C}$	$A \vee (\overline{B \wedge C}) \rightarrow (A \vee \overline{B}) \vee \overline{C}$
T	T	T	F	F	F	T	T	T	T
T	T	F	F	T	T	T	T	T	T
T	F	F	T	T	T	T	T	T	T
T	F	T	T	F	T	T	T	T	T
F	T	T	F	F	F	F	F	F	T
F	T	F	F	T	T	T	F	T	T
F	F	T	T	F	T	T	T	T	T
F	F	F	T	T	T	T	T	T	T

Fig. 54

As the last column of the table contains all T's, hence it is a tautology.

Problem 11. *Determine whether the following are equivalent, using biconditional statement.*

(*i*) $p \leftrightarrow q \equiv (p \wedge q) \vee (\sim p \wedge \sim q)$

(*ii*) $(p \rightarrow q) \rightarrow t \equiv (p \wedge \sim q) \rightarrow t$.

Sol. To prove that the above pairs are equivalent. Prove that $(p \leftrightarrow q) \leftrightarrow ((p \wedge q) \vee (\sim p \wedge \sim q))$ and $((p \rightarrow q) \rightarrow t) \leftrightarrow ((p \wedge \sim q) \rightarrow t)$ are tautologies.

(*i*) Construct truth table for $(p \leftrightarrow q) \leftrightarrow ((p \wedge q) \vee (\sim p \wedge \sim q))$ as in (Fig. 55)

p	q	$\sim p$	$\sim q$	$p \leftrightarrow q$	$p \wedge q$	$\sim p \wedge \sim q$	$(p \wedge q) \vee (\sim p \wedge \sim q)$	$(p \leftrightarrow q) \leftrightarrow (p \wedge q) \vee (\sim p \wedge \sim q)$
T	T	F	F	T	T	F	T	T
T	F	F	T	F	F	F	F	T
F	T	T	F	F	F	F	F	T
F	F	T	T	T	F	T	T	T

Fig. 55

As the proposition $(p \leftrightarrow q) \leftrightarrow ((p \wedge q) \vee (\sim p \wedge \sim q))$ is a tautology, hence they are equivalent.

(*ii*) Construct the truth table for $((p \to q) \to t) \leftrightarrow ((p \wedge \sim q) \to t)$ as in Fig. 56.

p	q	t	$\sim q$	$p \to q$	$(p \to q) \to t$	$p \wedge \sim q$	$p \wedge \sim q \to t$	$((p \to q) \to t) \leftrightarrow ((p \wedge \sim q) \to t))$
T	T	T	F	T	T	F	T	T
T	T	F	F	T	F	F	T	F
T	F	F	T	F	T	T	F	F
T	F	T	T	F	T	T	T	T
F	T	T	T	T	T	F	T	T
F	T	F	F	T	F	F	T	F
F	F	T	T	T	T	F	T	T
F	F	F	F	T	F	F	T	F

Fig. 56

Since, the proposition is not a tautology, hence they are not equivalent.

Problem 12. *Determine the dual of each of the following :*

(*a*) $(p \wedge q) \vee (\sim p \vee (\sim p \vee q))$

(*b*) $\sim p \wedge (\sim q \wedge r) \vee (q \wedge r) \vee (p \wedge r)$

(*c*) $(p \downarrow q) \uparrow r$.

Sol. The dual of above propositions are as follows :

(*a*) $(p \wedge q) \vee (\sim p \vee (\sim p \vee q)) = (p \vee q) \wedge (\sim p \wedge (\sim p \wedge q))$

(*b*) $\sim p \wedge (\sim q \wedge r) \vee (q \wedge r) \vee (p \wedge r) = \sim p \vee (\sim q \vee r) \wedge (q \vee r) \wedge (p \vee r)$

(*c*) $(p \downarrow q) \uparrow r = (p \uparrow q) \downarrow r$.

Problem 13. *Express $p \Rightarrow q$ using NAND $\uparrow$ only.*

Sol.
$$p \Rightarrow q \cong \sim p \vee q \cong \sim (\sim \sim p \wedge \sim q) \qquad \because \; p \vee q \cong \sim (\sim p \wedge \sim q)$$
$$\cong \sim (p \wedge \sim q) \qquad \because \; \sim \sim p \equiv p$$
$$\cong p \uparrow \sim q \qquad \because \; p \uparrow q \cong \sim (p \wedge q)$$
$$\cong p \uparrow q \uparrow q \qquad \because \; \sim p = p \uparrow p.$$

Problem 14. *Express $p \downarrow q$ using $\uparrow$ only.*

Sol.
$$p \downarrow q \cong \sim (p \vee q) \cong (p \vee q) \uparrow (p \vee q)$$
$$\cong [(p \uparrow p) \uparrow (q \uparrow q)] \uparrow [(p \uparrow p) \uparrow (q \uparrow q)] \qquad \because \; p \vee q \cong (p \uparrow q) \uparrow (p \uparrow q)$$

Problem 15. *Express the following formula using only $\sim$ and $\wedge$.*

$$(p \downarrow q) \uparrow r$$

where $\downarrow$ denotes NOR and $\uparrow$ denotes NAND.

Sol.
$$(p \downarrow q) \uparrow r \cong \sim (p \vee q) \uparrow r \qquad \because \; p \downarrow q \cong \sim (p \vee q)$$
$$\cong \sim ((\sim p \vee q) \wedge r) \qquad \because \; p \uparrow q \cong \sim (p \wedge q)$$

$$\equiv \sim ((\sim \sim \sim (p \wedge \sim q)) \wedge r) \qquad \because \quad p \vee q \equiv \sim (\sim p \wedge \sim q)$$

$$\equiv \sim ((\sim (p \wedge \sim q)) \wedge r) \qquad \because \quad \sim \sim p = p$$

Note. ↑ (NAND) and ↓ (NOR) are dual of each other.

Problem 16. *Show that the connective ↑ (NAND) is functionally complete.*

Sol. To show that the connective ↑ is functionally complete. We have to show that the set of connectives ($\wedge$, ~) and ($\vee$, ~) can be expressed in terms of ↑ alone which is shown as follows :

$$\sim p \equiv \sim p \vee \sim p \equiv \sim (p \wedge p) \equiv p \uparrow p$$

$$p \vee q \equiv \sim (\sim p \wedge \sim q) \equiv \sim p \uparrow \sim q \equiv (p \uparrow p) \uparrow (q \uparrow q)$$

and

$$p \wedge q \equiv \sim (p \uparrow q) \equiv (p \uparrow q) \uparrow (p \uparrow q)$$

Since the above expresses $\wedge$, $\vee$ and ~ in terms of ↑ alone. Hence ↑ is functionally complete.

Problem 17. *Prove the validity of following arguments without using truth tables.*

1. $p \vee q, \sim p \vdash q$ *2.* $p, p \to q, q \to r \vdash r$

3. $p, q, (p \wedge q) \to r \vdash r$ *4.* $p, (p \wedge \sim q) \to \sim p \vdash p \to q$.

Sol. 1. (*i*) $p \vee q$ (Given)

(*ii*) $\sim p$ (Given)

(*iii*) q Disjunctive syllogism

2. (*i*) p (Given)

(*ii*) $p \to q$ (Given)

(*iii*) $q \to r$ (Given)

(*iv*) $p \to r$ Hypothetical syllogism using (*ii*) and (*iii*)

(*v*) r Modus ponens using (*iv*) and (*i*)

3. (*i*) p (Given)

(*ii*) q (Given)

(*iii*) $(p \wedge q) \to r$ (Given)

(*iv*) $p \wedge q$ Rule of conjunction using (*i*) and (*ii*)

(*v*) r Modus ponens using (*iii*) and (*iv*)

4. (*i*) p (Given)

(*ii*) $(p \wedge \sim q) \to \sim p$ (Given)

(*iii*) $\sim (p \wedge \sim q)$ Modus tollens using (*ii*) and (*i*)

(*iv*) $\sim \sim (p \to q)$ As $\sim (p \to q) \equiv (p \wedge \sim q)$

(*v*) $p \to q$ Complement property using (*iv*).

Problem 18. *Prove that the argument* $(p \wedge q) \to r, p \to q \vdash p \to ((p \wedge q) \wedge r)$ *is valid without using truth table.*

Sol. (*i*) $(p \wedge q) \to r$ (Given)

(*ii*) $p \to q$ (Given)

(*iii*) $p \to (p \wedge q)$ Rule of absorption using (*ii*)

(*iv*)	$(p \wedge q) \to ((p \wedge q) \wedge r)$	Rule of absorption using (*i*)
(*v*)	$p \to ((p \wedge q) \wedge r)$	Hypothetical syllogism using (*iii*) and (*iv*).

Problem 19. *Prove that the following arguments are valid without using truth tables.*

1. $(p \vee q) \to \sim r, r \vee t, p \vdash t$

2. $(p \wedge q) \to r, (r \to q), (r \to q) \to (q \wedge r) \vdash (p \wedge q) \to (q \wedge r)$

Sol. 1. (*i*)	$(p \vee q) \to \sim r$	(Given)
(*ii*)	$r \vee t$	(Given)
(*iii*)	p	(Given)
(*iv*)	$p \vee q$	Rule of addition using (*iii*)
(*v*)	$\sim r$	Modus ponens using (*i*) and (*iv*)
(*vi*)	t	Disjunctive syllogism using (*ii*) and (*v*)
2. (*i*)	$p \wedge q \to r$	(Given)
(*ii*)	$r \to q$	(Given)
(*iii*)	$(r \wedge q) \to (q \wedge r)$	(Given)
(*iv*)	$r \to (r \wedge q)$	Rule of absorption using (*ii*)
(*v*)	$(p \wedge q) \to (r \wedge q)$	Hypothetical syllogism using (*i*) and (*iv*)
(*vi*)	$(p \wedge q) \to (q \wedge r)$	Hypothetical syllogism using (*v*) and (*iii*)

Problem 20. *Prove the validity of following argument using deduction system.*

(1) $A \to B$ *(2)* $B \to \sim C$

(3) $\sim C \wedge D$ *(4)* $A \to D$

$\therefore B.$

Sol. (*i*)	$A \to B$	(Given)
(*ii*)	$B \to \sim C$	(Given)
(*iii*)	$\sim C \wedge D$	(Given)
(*iv*)	$A \to D$	(Given)
(*v*)	$A \to \sim C$	Hypothetical syllogism using (*i*) and (*ii*)
(*vi*)	$\sim C$	Simplification of (*iii*)
(*vii*)	$\sim A \to C$	Transposition of (*iv*)
(*viii*)	A	Modus tollens using (*vii*) and (*vi*)
(*ix*)	B	Modus ponens using (*i*) and (*viii*)

Problem 21. *Prove the validity of following by deduction method.*

(A)	(B)	(C)
(*i*) $p \to q$	(*i*) $p \vee q$	(*i*) $(q \to r) \wedge (s \to t)$
(*ii*) $\sim q \vee r$	(*ii*) $q \to r$	(*ii*) $(u \to v) \wedge (w \to x)$
(*iii*) $\sim (r \wedge \sim s)$	(*iii*) $r \wedge s$	(*iii*) $q \vee u$
(*iv*) p	(*iv*) $p \to s$	$\therefore r \vee v$
$\therefore s$	(*v*) p	
	$\therefore s$	

Sol. (A) (*i*)	$p \to q$	(Given)
(*ii*)	$\sim q \vee r$	(Given)
(*iii*)	$\sim (r \wedge \sim s)$	(Given)
(*iv*)	p	(Given)
(*v*)	q	Modus ponens using (*i*) and (*iv*)
(*vi*)	$r \vee \sim q$	Commutative property using (*ii*)
(*vii*)	$\sim r$	Rule of Modus tollens using (*vi*) and (*v*)
(*viii*)	$\sim r \vee s$	De Morgan's Law using (*iii*)
(*ix*)	s	Modus ponens using (*viii*) and (*vii*)
(B) (*i*)	$p \vee q$	(Given)
(*ii*)	$q \to r$	(Given)
(*iii*)	$r \wedge s$	(Given)
(*iv*)	$p \to s$	(Given)
(*v*)	p	(Given)
(*vi*)	s	Modus ponens using (*iv*) and (*v*)
(C) (*i*)	$(q \to r) \wedge (s \to t)$	(Given)
(*ii*)	$(u \to v) \wedge (w \to x)$	(Given)
(*iii*)	$q \vee u$	(Given)
(*iv*)	$(q \to r)$	Simplification using (*i*)
(*v*)	$(u \to v)$	Simplification using (*ii*)
(*vi*)	$(q \to r) \wedge (u \to v)$	Conjunction using (*iv*) and (*v*)
(*vii*)	$r \vee v$	Constructive dilemma using (*vi*) and (*iii*)

Problem 22. *Prove the validity of following argument using truth tables.*

"If it rains then it will be cold. If it is cold then I shall stay at home. Since it rains therefore, I shall stay at home".

Sol. Let p be "It rains" and q be "It will be cold" and r be "I shall stay at home". Then the above argument in symbolic form is

$$p \to q$$
$$q \to r$$
$$p$$
$$\therefore \quad r$$

Construct the truth table of $(p \to q) \wedge (q \to r) \wedge p \to r$ as in (Fig. 57).

p	q	r	$p \to q$	$q \to r$	$(p \to q) \wedge (q \to r)$	$(p \to q) \wedge (q \to r) \wedge p$	$(p \to q) \wedge (q \to r) \wedge p \to r$
T	T	T	T	T	T	T	T
T	T	F	T	F	F	F	T
T	F	F	F	T	F	F	T
T	F	T	F	T	F	F	T
F	T	T	T	T	T	F	T
F	T	F	T	F	F	F	T
F	F	T	T	T	T	F	T
F	F	F	T	T	T	F	T

Fig. 57

As the last column contain all T's, hence the argument is valid.

Problem 23. *Translate the following into symbolic form and test the validity of the argument.*

If 6 is even then 2 does not divide 7. Either 5 is not prime or 2 divides 7. But 5 is prime, therefore, 6 is odd.

Sol. Let p be "6 is even" and q be "2 divide 7" and r be "5 is primed". Thus, the above argument in symbolic form can be written as

$$p \to \sim q$$
$$\sim r \vee q$$
$$r$$
$$\therefore \quad \sim p$$

Construct the truth table to test the validity. (Fig. 58)

p	q	r	$\sim p$	$\sim q$	$\sim r$	$p \to \sim q$	$\sim r \vee q$
T	T	T	F	F	F	F	T
T	T	F	F	F	T	F	T
T	F	F	F	T	T	T	F
T	F	T	F	T	F	T	T
F	T	T	T	F	F	T	T
F	T	F	T	F	T	T	T
F	F	T	T	T	F	T	F
F	F	F	T	T	T	T	F

Fig. 58

$p \to \sim q$ is true in line 3, 4, 5, 6, 7 and 8, $\sim r \vee q$ is true in line 1, 2, 3, 4, 5 and r is true in line 1, 4, 5 and 7. All the three are true in line 4 and 5 and $\sim p$ is true in line 5. Hence, argument is valid.

Problem 24. *Determine the negation of the following statements*

(*i*) $\forall x \, \forall y \, \forall z, p(x, y, z)$ (*ii*) $\forall x \, \exists_y, p(x, y)$

(*iii*) $\exists_y \, \forall_x \, \forall_z, p(x, y, z)$.

Sol. (*i*) $\sim (\forall_x \forall_y \forall_z, p(x, y, z)) \cong \exists_x \exists_y \exists_z \sim p(x, y, z)$

(*ii*) $\sim (\forall_x \exists_y, p(x, y)) \cong \exists_x \forall_y, \quad p(x, y)$

(*iii*) $\sim (\exists_y \forall_x \forall_z, p(x, y, z)) \cong \forall_y \exists_x \exists_z, \sim p(x, y, z)$.

Problem 25. *Determine the negation of the following statements :*

(*i*) $\exists_x \forall_y (p(x) \vee q(y))$ (*ii*) $\forall_x \exists_y (p(x, y) \to q(x, y))$

(*iii*) $\forall_x \forall_y (p(x) \wedge q(y))$.

Sol. (*i*) $\sim (\exists x \forall_y, (p(x) \vee q(y)) \cong \forall_x \exists_y \sim (p(x) \vee q(y))$

$\cong \forall_x \exists_y (\sim p(x) \wedge \sim q(y))$

(*ii*) $\sim \forall_x \exists_y (p(x, y) \to q(x, y)) \cong \exists_x \forall_y \sim (p(x, y) \to q(x, y))$

(*iii*) $\cong \forall_x \forall_y (p(x) \wedge p(y)) \cong \exists_x \exists_y \sim (p(x) \wedge p(y)) \cong \exists_x \exists_y (\sim p(x) \vee \sim p(y))$.

Problem 26. *Let $U = Q$. Use quantifiers to express the following statements :*

(*i*) $\sqrt{5}$ *is not rational.*

(*ii*) *Subtraction of any two rational numbers is rational.*

Sol. (*i*) $\sim (\exists x) (x^2 = 5)$

(*ii*) $(\forall x) (\forall y)$ ($x - y$ is a rational).

SECTION III

Chapter 8

Lattices

LATTICES

Lattice

A lattice L is a poset in which every pair of elements has a least upper bound (LUB) *i.e.,* supremum and a greatest lower bound (GLB) or infimum.

Join

Consider a poset L under the ordering $\leq$. Let $a, b \in$ L. Then LUB (a, b) or SUP (a, b) is denoted by $a \vee b$ or $a \cup b$ and is called the join of a and b *i.e.,* $a \vee b =$ SUP (a, b).

Meet

Consider a poset L under the ordering $\leq$. Let $a, b \in$ L. Then GLB (a, b) or inf (a, b) is denoted by $a \wedge b$ or $a \cap b$ and is called the meet of a and b *i.e.,* $a \wedge b =$ inf (a, b).

From the above, it follows that a lattice L is a mathematical structure with two binary operations $\vee$ (Join) and $\wedge$ (meet). It is denoted by $\{L, \vee, \wedge\}$. The lattice L for any elements a, b and c satisfies the following properties :

(*a*) **Commutative Property**

(*i*) $a \wedge b = b \wedge a$

(*ii*) $a \vee b = b \vee a$.

(*b*) **Associative Property**

(*i*) $(a \wedge b) \wedge c = a \wedge (b \wedge c)$

(*ii*) $(a \vee b) \vee c = a \vee (b \vee c)$.

(*c*) **Absorption Property**

(*i*) $a \wedge (a \vee b) = a$ (*ii*) $a \vee (a \wedge b) = a$.

Theorem I. *Prove that if L be a lattice then $a \wedge b = a$ if and only if $a \vee b = b$.*

Proof. Let us first assume that $a \wedge b = a$.

Using absorption property, we have

$$b = b \vee (b \wedge a) = b \vee (a \wedge b) = b \vee a = a \vee b \quad ...(i)$$

Conversely, let us assume $a \vee b = b$.

Again using the absorption property, we have

$$a = a \wedge (a \vee b) = a \vee b \quad ...(ii)$$

From eqn. (*i*) and (*ii*), we have

$$a \wedge b = a \text{ if and only if } a \vee b = b.$$

Theorem II. *Prove that for elements of lattice*

(*i*) $a \wedge a = a$ (*ii*) $a \vee a = a$ Idempotent property

Proof. (*i*) $a \wedge a = a \wedge (a \vee (a \wedge b))$ {Using absorption property $c(ii)$}

$= a$ {Using absorption property $c(i)$}

(*ii*) $a \vee a = a \vee (a \wedge (a \vee b))$ {Using absorption property $c(i)$}

$= a$ {Using absorption property $c(ii)$}

Theorem III. *Consider a lattice L. Prove that the relation $a \leq b$ defined by either $a \wedge b = a$ or $a \vee b = b$ is a partial ordering on lattice L.*

Proof. For any element $a \in$ L, we have $a \wedge a = a$

i.e., $a \leq a$. Therefore, the relation $\leq$ is reflexive. Now assume $a \leq b$ and $b \leq a$. Then we have

$$a \wedge b = a \quad \text{and} \quad b \wedge a = b$$

Thus, $a = a \wedge b = b \wedge a = b$, therefore the relation $\leq$ is antisymmetric.

At last, we assume $a \leq b$ and $b \leq c$. So, we have

$$a \wedge b = a \text{ and } b \wedge c = b$$

i.e., $$a \wedge c = (a \wedge b) \wedge c = a \wedge (b \wedge c)$$

$$= a \wedge b = a$$

Therefore, $a \leq c$. So, the relation $\leq$ is transitive. From above, we can say, $\leq$ is a partial order on L.

Example 1. *Let P(S) be the power set of the set S = {1, 2, 3}. Construct the Hasse diagram of the partial order induced on P(S) by the lattice (P(S), $\wedge$, $\vee$).*

Sol. The Hasse diagram obtained by a lattice is same as obtained under the partial ordering of set inclusion. In the lattice, $a \leq b$ whenever $a \wedge b = a$. Thus, in the above case $a \leq b$, whenever $a \wedge b = a$

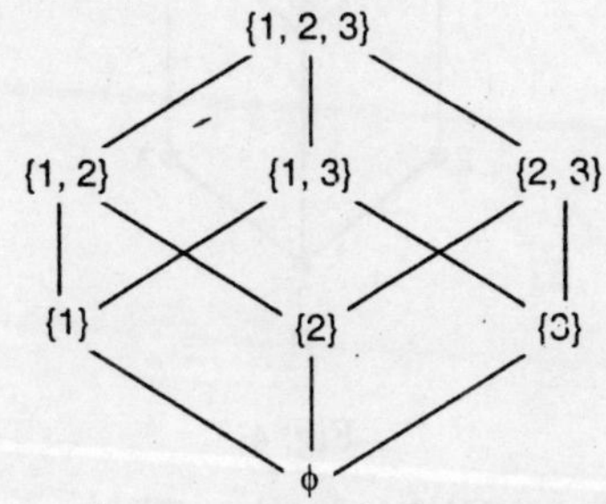

Fig. 1

Example 2. *Determine which of the posets shown in Fig. 2 are lattices.*

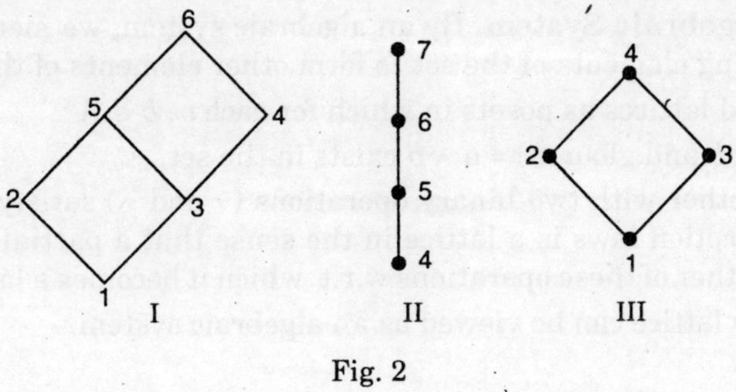

Fig. 2

Sol. All the posets shown in Fig. 2 are lattices

Example 3. *Determine whether the posets shown in Fig. 3 are lattices or not.*

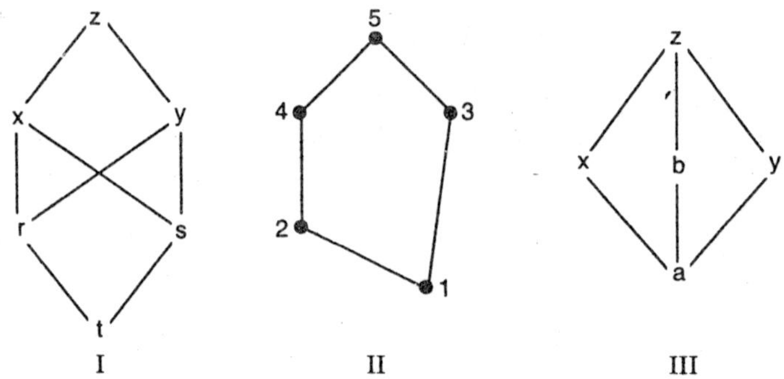

Fig. 3

Sol. The posets shown in Figs. 3(II) and 3(III) are lattices. The posets shown in Fig. 3(I) is not lattice as (x, y) has three lower bounds *i.e.*, r, s and t but the inf (x, y) does not exist. Also, sup (r, s) does not exist.

Example 4. *Determine whether the posets shown in Fig. 4 are lattices or not.*

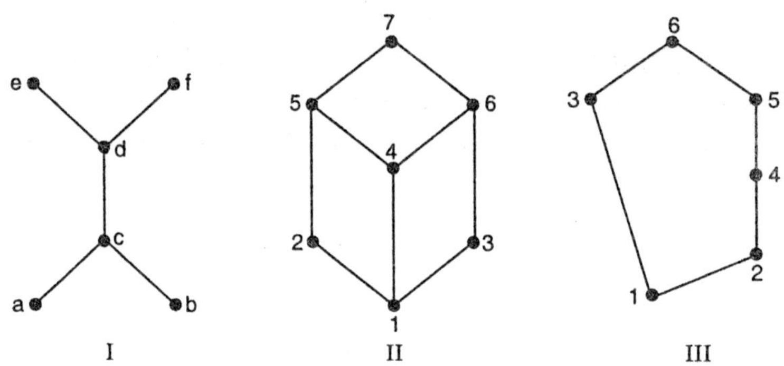

Fig. 4

Sol. The poset shown in Fig. 4(II) is a lattice. The poset shown in Fig. 4(I) is not lattice since the elements e and f have no upper bound, hence sup (e, f) does not exist. Similarly, the elements a and b have no lower bound, hence inf (a, b) does not exist.

The poset shown in Fig. 4(III) is lattice.

Lattice As Algebraic System. By an algebraic system, we mean a set together with few rules for combining elements of the set to form other elements of the set.

We had defined lattices as posets in which for each $a, b \in A$

lub $(a, b) = a \vee b$ and glb$(a, b) = a \wedge b$ exists in the set.

Any set L together with two binary operations ($\vee$ and $\wedge$) satisfying the commutative, associative and absorption laws is a lattice in the sense that a partial order relation can be defined in terms of either of these operations w.r.t. which it becomes a lattice with sup and inf.

Therefore, any lattice can be viewed as an algebraic system.

BOUNDED LATTICES

A lattice L is called a bounded lattice if it has a greatest element 1 and a least element 0.

Example. *(i)* The power set P(S) of the set S under the operations of intersection and union is a bounded lattice since ϕ is the least element of P(S) and the set S is the greatest element of P(S).

(ii) The set of +ve integers I_+ under the usual order of $\leq$ is not a bounded lattice since it has a least element 1 but the greatest element does not exist.

Properties of Bounded Lattices

It L is a bounded lattice, then for any element $a \in$ L, we have the following identities :

(i) $a \vee 1 = 1$ *(ii)* $a \wedge 1 = a$ *(iii)* $a \vee 0 = a$ *(iv)* $a \wedge 0 = 0$.

Theorem IV. *Prove that every finite lattice* $L = \{a_1, a_2, a_3, \ldots\ldots, a_n\}$ is bounded.

Proof. We have given the finite lattice :

$$L = \{a_1, a_2, a_3, \ldots\ldots, a_n\}$$

Thus, the greatest element of lattice L is $a_1 \vee a_2 \vee a_3 \vee \ldots\ldots \vee a_n$.

Also, the least element of lattice L is $a_1 \wedge a_2 \wedge a_3 \wedge \ldots\ldots \wedge a_n$.

Since, the greatest and least elements exist for every finite lattice. Hence L is bounded.

SUBLATTICES

Consider a non empty subset L_1 of a lattice L. Then L_1 is called a sublattice of L if L_1 itself is a lattice w.r.t. the operations of L *i.e.*, if $a \vee b \in L_1$ and $a \wedge b \in L_1$ whenever $a \in L_1$ and $b \in L_1$.

Example 5. *Consider the lattice of all +ve integers* I_+ *under the operation of divisibility. The lattice* D_n *of all divisors of* $n > 1$ *is a sublattice of* I_+.

Determine all the sublattices of D_{30} *that contain at least four elements,* $D_{30} = \{1, 2, 3, 5, 6, 10, 15, 30\}$.

Sol. The sublattices of D_{30} that contain at least four elements are as follows :

(i) {1, 2, 6, 30} *(ii)* {1, 2, 3, 30} *(iii)* {1, 5, 15, 30}
(iv) {1, 3, 6, 30} *(v)* {1, 5, 10, 30} *(vi)* {1, 3, 15, 30}
(vii) {2, 6, 10, 30}.

Example 6. *Consider the lattice L = {1, 2, 3, 4, 5} as shown in Fig. 5. Determine all the sublattices with three or more elements.*

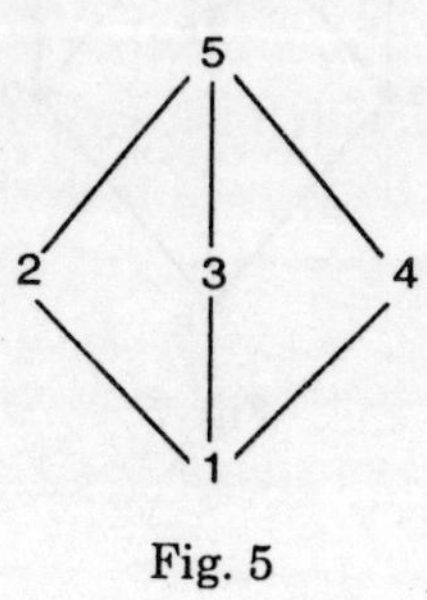

Fig. 5

Sol. All the sublattices with three or more elements are those whose supremum and infimum exists for every pair of elements which are as follows .

(*i*) {1, 2, 5} (*ii*) {1, 3, 5}
(*iii*) {1, 4, 5} (*iv*) {1, 2, 3, 5}
(*v*) {1, 3, 4, 5} (*vi*) {1, 2, 3, 4, 5}
(*vii*) {1, 2, 4, 5}.

Example 7. *Consider the lattice L as shown in Fig. 6. Determine whether or not each of the following is a sublattice of L.*

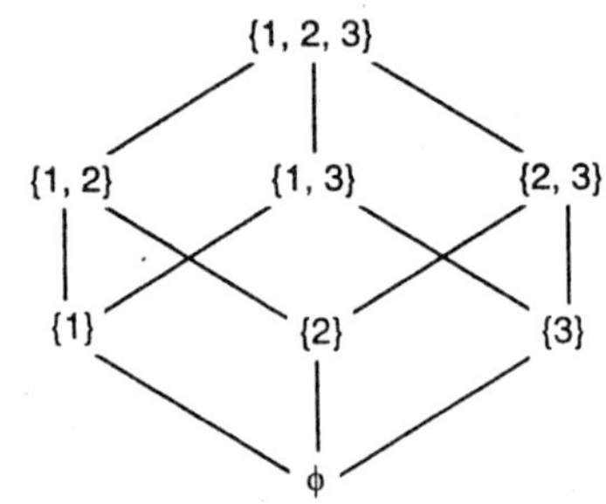

Fig. 6

$A = \{\phi, \{1, 2\}, \{2, 3\}, \{1, 2, 3\}\}$
$B = \{\phi, \{1\}, \{1, 2\}, \{1, 2, 3\}\}$
$C = \{\phi, \{3\}, \{1, 3\}, \{1, 2, 3\}\}$
$D = \{\{1\}, \{3\}, \{1, 3\}, \{1, 2, 3\}\}$
$E = \{\phi, \{3\}, \{1, 2\}, \{1, 2, 3\}\}$.

Sol. A is not a sublattice since $\{1, 2\} \wedge \{2, 3\} = \{2\}$ which does not exist in A.

B is a sublattice since sup and inf of every pair of elements exist.

C is a sublattice since sup and inf of every pair of elements exist.

D is not a sublattice since $\{1\} \wedge \{3\} = \phi$ which does not exist in D.

E is a sublattice since sup and inf of every pair of elements exist.

EXERCISES

1. Determine whether a subset of a linearly ordered poset is a lattice.
2. Determine all sublattices D_{30} which contain at least three elements.
3. Let L be a lattice in which L = {*a*, *b*, *c*, *d*, *e*, *f*, *g*, *h*} as shown in Fig. 7.
   If $S_1 = \{a, b, d, f\}$, $S_2 = \{c, e, g, h\}$ and $S_3 = \{a, b, d, h\}$ be three subsets of L, then determine whether S_1, S_2 and S_3 are the sublattices of L.

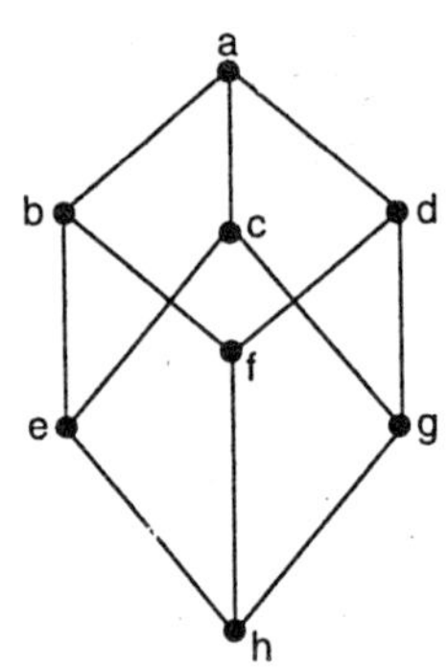

Fig. 7

4. Determine whether the following Hasse diagrams represent a lattice.

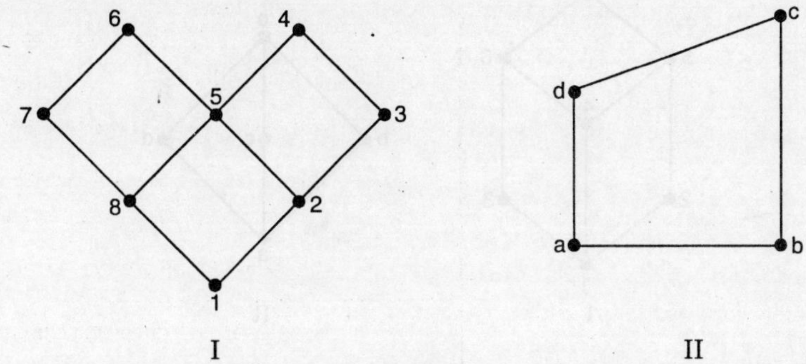

Fig. 8

5. Draw the Hasse diagrams of the lattices $\{D_n, 1\}$; $n = 4, 6, 10, 12, 15$.
6. For any lattice, prove that
   (*i*) $a \vee (b \vee c) = (a \vee b) \vee c$ (*ii*) $a \vee (a \wedge b) = a$.
7. Draw Hasse diagrams of all lattices with upto five elements.
8. Find all sub lattices of D_{24} that contain five or more elements.
9. Show that every finite lattice has a least element and a greatest element.

ISOMORPHIC LATTICES

Two lattices L_1 and L_2 are called isomorphic lattices if there is a bijection from L_1 to L_2 *i.e.*, $f : L_1 \rightarrow L_2$, such that

$$f(a \wedge b) = f(a) \wedge f(b) \text{ and } f(a \vee b) = f(a) \vee f(b)$$

for every element a, b belongs to L_1.

Example 8. *Determine whether the lattices shown in Fig. 9 are isomorphic.*

Sol. The lattices shown in Fig. 9 are isomorphic. Consider the mapping $f = \{(a, 1), (b, 2), (c, 3), (d, 4)\}$. For example $f(b \wedge c) = f(a) = 1$. Also we have $f(b) \wedge f(c) = 2 \wedge 3 = 1$.

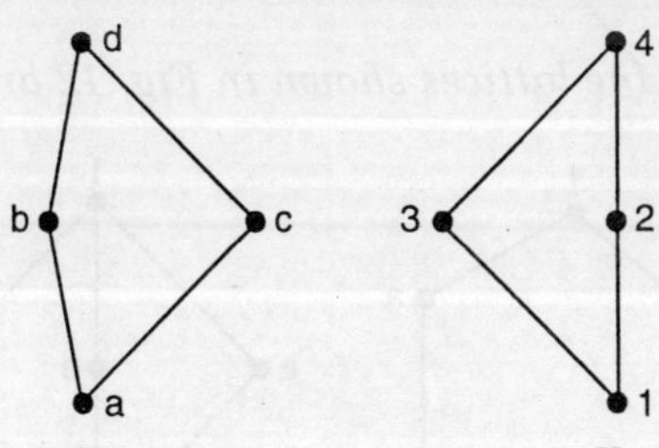

Fig. 9

Example 9. *Determine whether the lattices shown in Fig. 10 are isomorphic.*

Sol. The lattices shown in Fig. 10 are not isomorphic since bijection is not possible as the element of the two lattices are not same.

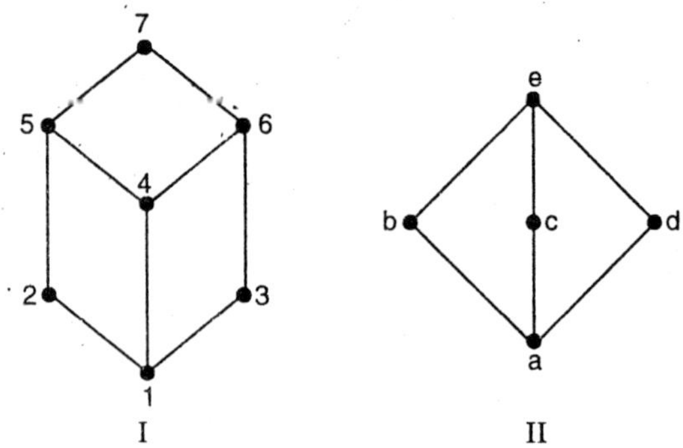

Fig. 10

DISTRIBUTIVE LATTICE

A lattice L is called distributive lattice if for any elements a, b and c of L, it satisfies following distributive properties :

(*i*) $a \wedge (b \vee c) = (a \wedge b) \vee (a \wedge c)$

(*ii*) $a \vee (b \wedge c) = (a \vee b) \wedge (a \vee c)$.

If the lattice L does not satisfies the above properties, it is called a non-distributive lattice.

For example :

1. The power set P(S) of the set S under the operations of intersection and union is a distributive function. Since,

$$a \cap (b \cup c) = (a \cap b) \cup (a \cap c)$$

and also $a \cup (b \cap c) = (a \cup b) \cap (a \cup c)$ for any sets a, b and c of P(S).

2. The lattice shown in Fig. 11 is distributive. Since it satisfies the distributive properties for all ordered triples which are taken from 1, 2, 3 and 4.

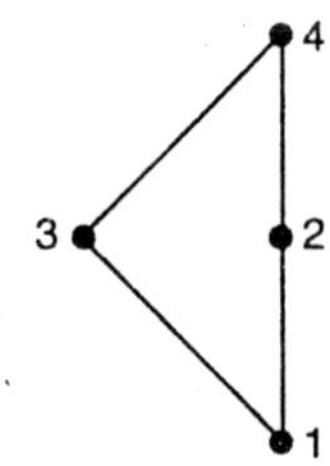

Fig. 11

Example 10. *Show that the lattices shown in Fig. 12 are non-distributive.*

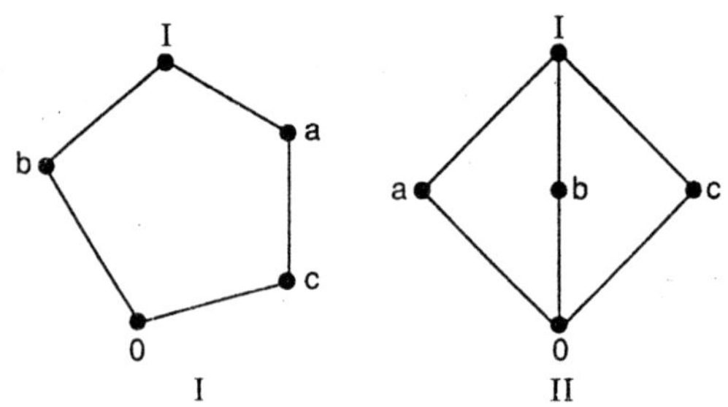

Fig. 12

Sol.

I. From the (i) property of distributivity, we have

$$a \wedge (b \vee c) = a \wedge \text{I} = a$$

But, $(a \wedge b) \vee (a \wedge c) = 0 \vee c = c$

Since $a \wedge (b \vee c) \neq (a \wedge b) \vee (a \wedge c)$

Hence, the lattice is not distributive.

II. Again, from the (i) property of distributivity, we have

$$a \wedge (b \vee c) = a \wedge \text{I} = a$$

But $(a \wedge b) \vee (a \wedge c) = 0 \vee 0 = 0$

Since $a \wedge (b \vee c) \neq (a \wedge b) \vee (a \wedge c)$

Hence, the lattice is not distributive.

Note. A lattice L is non-distributive if and only if it contains a sublattice that is isomorphic to one of the two lattices shown in Figs. 12(I) and (II).

Theorem V. *Prove that in a distributive lattice* $(L, \wedge, \vee)$, $(a \wedge b) \vee (b \wedge c) \vee (c \wedge a) = (a \vee b) \wedge (b \vee c) \wedge (c \vee a)$ *holds for all* $a, b, c \in L$.

Proof. We have given that L is distributive, so using distributive property, we have

$$\begin{aligned}
(a \wedge b) \vee (b \wedge c) \vee (c \wedge a) &= \{[(a \vee b) \vee b] \wedge [(a \wedge b) \vee c)]\} \vee (c \wedge a) \\
&= [b \wedge \{(a \vee c) \wedge (b \vee c)\}] \vee (c \wedge a) \\
&= [(a \vee c) \wedge \{b \wedge (b \vee c)\}] \vee (c \wedge a) \\
&= [(a \vee c) \wedge b] \vee (c \wedge a) \\
&= [(a \vee c) \vee (c \wedge a)] \wedge [b \vee (c \wedge a)] \\
&= [\{(a \vee c) \vee c\} \wedge \{(a \vee c) \vee a\}] \wedge [(b \vee c) \wedge (b \vee a)] \\
&= (a \vee c) \wedge (a \vee c) \wedge (b \vee c) \wedge (a \vee b) \\
&= (a \vee c) \wedge (b \vee c) \wedge (a \vee b) = \text{R.H.S.}
\end{aligned}$$

Theorem VI. *In a distributive lattice* $(L, \wedge, \vee)$,

$$a \wedge b = a \wedge c \textit{ and } a \vee b = a \vee c \textit{ imply that } b = c.$$

Proof. We have

$$\begin{aligned}
b &= b \wedge (a \vee b) && \text{(by absorption law)} \\
&= b \wedge (a \vee c) && (\because \ a \vee b = a \vee c) \\
&= (b \wedge a) \vee (b \wedge c) && \text{(distributive law)} \\
&= (a \wedge b) \vee (b \wedge c) && \text{(commutative law)} \\
&= (a \wedge c) \vee (b \wedge c) && (\because \ a \wedge b = a \wedge c) \\
&= (c \wedge a) \vee (c \wedge b) && \text{(commutative law)} \\
&= c \wedge (a \vee c) && \text{(distributive law)} \\
&= c.
\end{aligned}$$

Hence proved.

JOIN-IRREDUCIBLE

Consider a lattice $(L, \wedge, \vee)$. An element $a \in$ L is called join-irreducible if it can be expressed as the join of two distinct elements of L *i.e.,* $a \in$ L is join-irreducible if

$$a = x \vee y \Rightarrow a = x \text{ or } a = y, \text{ where } x, y \in \text{L}.$$

Example 11. *Determine the join-irreducible elements of the lattices as shown in Fig. 13.*

Sol. 1. The join irreducible elements of Fig. 13(I) are a, b and d.

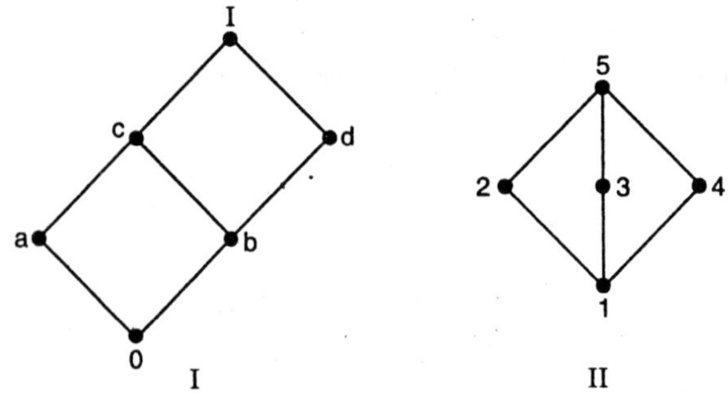

Fig. 13

2. The join-irreducible elements of Fig. 13(II) are 2, 3 and 4.

MEET-IRREDUCIBLE

Consider a lattice (L, ∧, ∨). An element $a \in$ L is called meet-irreducible if it can be expressed as the meet of two distinct elements of L *i.e.,* $a \in$ L is meet-irreducible if

$$a = x \wedge y \quad \Rightarrow \quad a = x \text{ or } a = y,$$

where $x, y \in$ L.

Example 12. *Determine the meet-irreducible elements of the lattices as shown in Fig. 14.*

Sol. 1. The meet-irreducible elements of Fig. 14(I) are a, b and c.

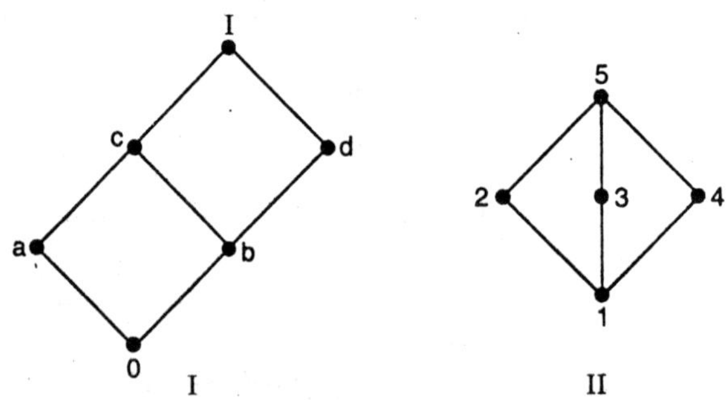

Fig. 14

2. The meet-irreducible elements of Fig. 14(II) are 2, 3 and 4.

ATOM

Consider a lattice (L, ∧, ∨) with a lower bound (1). An element a is called an atom if it is an immediate successor of 0 *i.e.,* $a \neq 0$ is an atom if $0 \leq b \leq a \Rightarrow b = 0$ or $b = a$.

ANTIATOM

Consider a lattice (L, ∧, ∨) with an upper bound 1. An element a is called an antiatom if it is an immediate predecessor of 1 *i.e.*, $a \neq 1$ is an antiatom if

$$a \leq b \leq 1 \;\Rightarrow\; b = a \quad \text{or} \quad b = 1.$$

Example 13. *Determine the atoms and antiatoms of the following lattices shown in Fig. 15.*

Sol. Atoms. 1. Atoms of Fig. 15(I) are a and b.

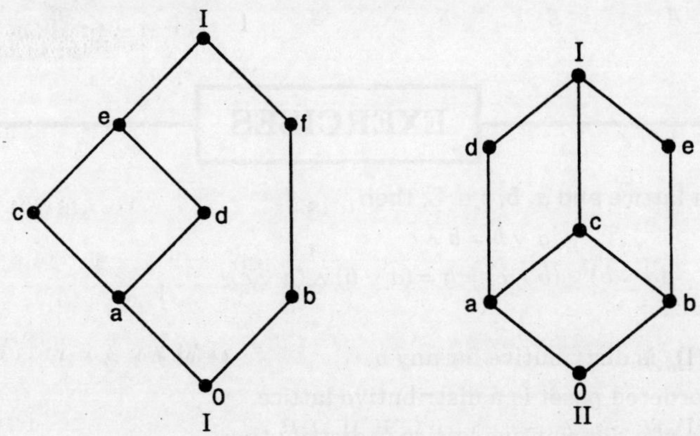

Fig. 15

2. Atoms of Fig. 15(II) are a and b.

Antiatoms. 1. Antiatoms of Fig. 15(I) are e and f.

2. Antiatoms of Fig. 15(II) are d and e.

Example 14. *Construct the meet and join table of the lattice (L, ∨, ∧) as shown in Fig. 16.*

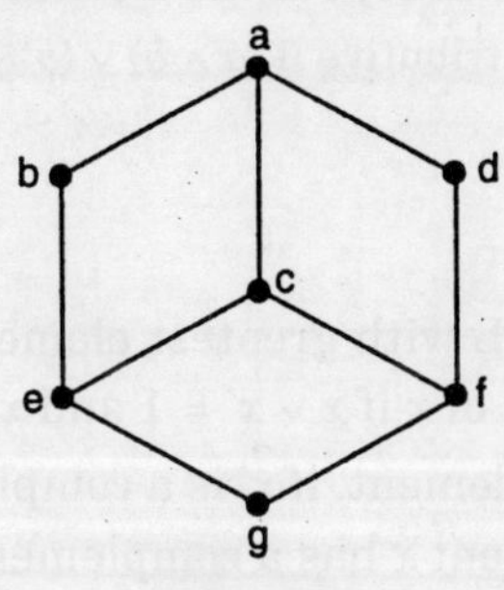

Fig. 16

Sol. The following tables show the meet and join table of the lattice (L, ∧, ∨).

∨	*a*	*b*	*c*	*d*	*e*	*f*	*g*
a	*a*	*a*	*a*	*a*	*a*	*a*	*a*
b	*a*	*b*	*a*	*a*	*b*	*a*	*b*
c	*a*	*a*	*c*	*a*	*c*	*c*	*c*
d	*a*	*a*	*a*	*d*	*a*	*d*	*d*
e	*a*	*b*	*c*	*a*	*e*	*c*	*e*
f	*a*	*a*	*c*	*d*	*c*	*f*	*f*
g	*a*	*b*	*c*	*d*	*e*	*f*	*g*

$\wedge$	a	b	c	d	e	f	g
a	a	b	c	d	e	f	g
b	b	b	e	g	e	g	g
c	c	e	c	f	e	f	g
d	d	g	f	d	g	f	g
e	e	e	e	g	e	g	g
f	f	g	f	f	g	f	g
g	g	g	g	g	g	g	g

EXERCISES

1. Let (L, ≤) be a lattice and $a, b, c \in$ L, then

$$a \vee b = b \wedge c$$
$$(a \wedge b) \vee (b \wedge c) = b = (a \vee b) \wedge (a \vee c).$$

2. Show whether
   (*i*) the lattice D_n is distributive for any n.
   (*ii*) a linearly ordered poset is a distributive lattice.
   (*iii*) a sublattice of a distributive lattice is distributive.
3. Show whether the lattice L = I = set of positive integers under the partial order relation ≤, is distributive.
4. Show that the lattice $\{S_n, R\}$; $n = 216$ is isomorphic to the direct product of lattices $n = 8$ and $n = 27$.
5. Show whether the dual of a distributive lattice is distributive ?
6. Let S = $\{a, b, c\}$ and Let L = power set P(S). Prove that (L, ≤) is isomorphic to D_{42}.
7. Show that a lattice (L, ≤) is distributive if $(a \wedge b) \vee (b \wedge c) \vee (c \wedge a) = (a \vee b) \wedge (b \vee c) \wedge (c \vee a)$.

COMPLEMENTED LATTICES

Consider a bounded lattice L with greatest element I and the least element 0. An element $x \in$ L is called a complement of x if $x \vee x' = 1$ and $x \wedge x' = 0$.

From the definition of complement, if x' is a complement of x, then x is a complement of x'. It is not necessary that an element x has a complement. Also the complements need not be unique *i.e.*, an element have more than one complement.

Note. That $1' = 0$ and $0' = 1$.

Definition. A lattice L is called a complemented lattice if L is bounded and every element in L has a complement.

Example 15. *Determine the complement of a and c in Fig. 17.*

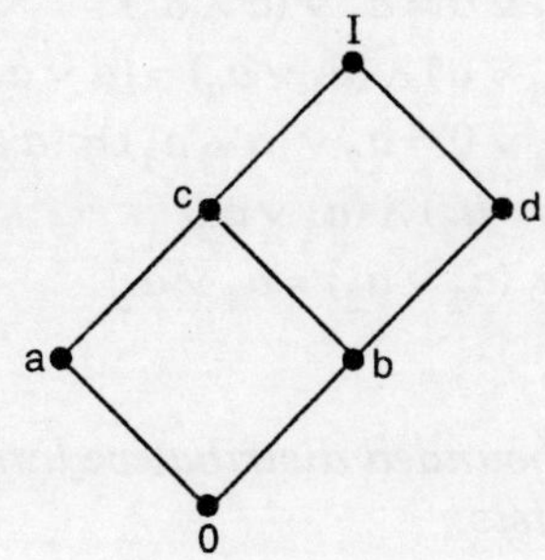

Fig. 17

Sol. The complement of a is d. Since, $a \vee d = 1$ and $a \wedge d = 0$.

The complement of c does not exist. Since, there does not exist any element c such that $c \vee c' = 1$ and $c \wedge c' = 0$.

Theorem VII. *Prove that 0 and 1 are complement of each other.*

Proof. To show that 1 is the only complement of 0, consider that $c \neq 1$ is a complement of 0 and $c \in$ L.

Then, $\quad 0 \wedge c = 0 \quad$ and $\quad 0 \vee c = 1$

But $\quad 0 \vee c = c \qquad$ {Property of bounded lattice}

and $c \neq 1$ leads to a contradiction.

Similarly, we can show that 0 is the only complement of 1.

Example. The power set P(S) of the set S under the operations of intersection and union is a complemented lattice L since each element of L has a unique complement.

Example. The lattices shown below are complemented lattices Fig. 18.

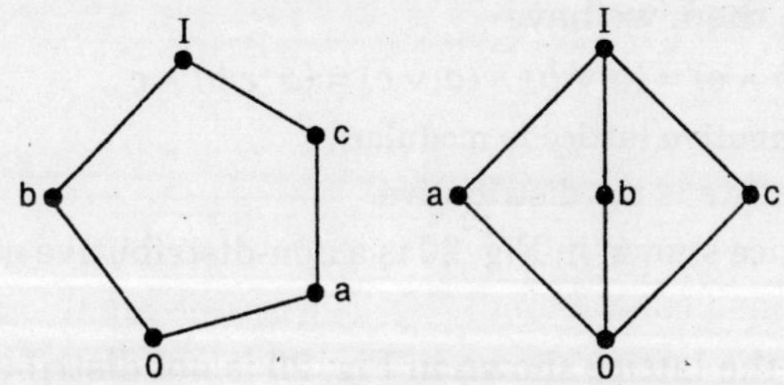

Fig. 18

But the complements of some of the elements are not unique *e.g.*, b has two complements a and c in both the cases.

Theorem VIII. *Prove that for a bounded distributive lattice L, the complements are unique if they exist.*

Proof. Consider a_1 and a_2 be complements of some elements $a \in$ L. Then, we have

$$a \vee a_1 = \text{I} \quad \text{and} \quad a \vee a_2 = \text{I}$$

Also $\quad a \wedge a_1 = 0 \quad$ and $\quad a \wedge a_2 = 0$

Now using the distributive property, we have

$$a_1 = a_1 \vee 0 = a_1 \vee (a \wedge a_2)$$
$$= (a_1 \vee a) \wedge (a_1 \vee a_2) = (a \vee a_1) \wedge (a_1 \vee a_2) = \mathrm{I} \wedge (a_1 \vee a_2) = a_1 \vee a_2$$

Similarly, $$a_2 = a_2 \vee 0 = a_2 \vee (a \wedge a_1) = (a_2 \vee a) \wedge (a_2 \vee a_1)$$
$$= (a \vee a_2) \wedge (a_1 \vee a_2)$$
$$= \mathrm{I} \wedge (a_1 \vee a_2) = a_1 \vee a_2$$

Thus, $a_1 = a_2$. Hence proved.

Example 16. *Consider the bounded distributive lattice as shown in Fig. 19. Show that every complement is unique if it exists.*

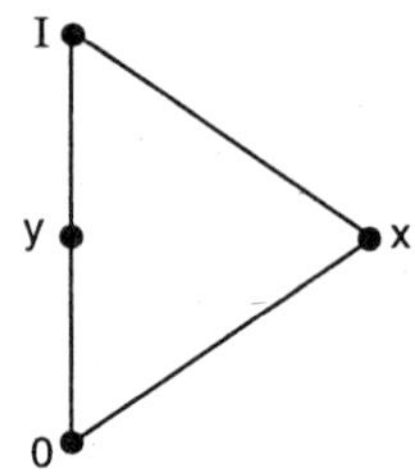

Fig. 19

Sol. The complement of x is y and vice-versa. Similarly, the complement of 0 is I and vice-versa. Hence, all the complements are unique.

MODULAR LATTICE

A lattice (L, $\wedge$, $\vee$) is called a modular lattice if $a \vee (b \wedge c) = (a \vee b) \wedge c$ whenever $a \leq c$.

Theorem IX. *Show that* $a \vee (b \wedge c) = (a \vee b) \wedge c$ *whenever* $a \leq c$.

Proof. If $a \leq c$, then $a \vee c = c$

If L is distributive, then, we have

$$a \vee (b \wedge c) = (a \vee b) \wedge (a \vee c) = (a \vee b) \wedge c \qquad | \because a \vee c = c$$

Note. (*i*) Every distributive lattice is modular.

(*ii*) Every modular lattice is not distributive.

Example. The lattice shown in Fig. 20 is a non-distributive modular lattice.

We can check that the lattice shown in Fig. 20 is not distributive.

$$a \wedge (b \vee c) = a \wedge 1 = a$$

but $$(a \wedge b) \vee (a \wedge c) = 0 \vee 0 = 0$$

To, check the modularity holds for $a = c$.

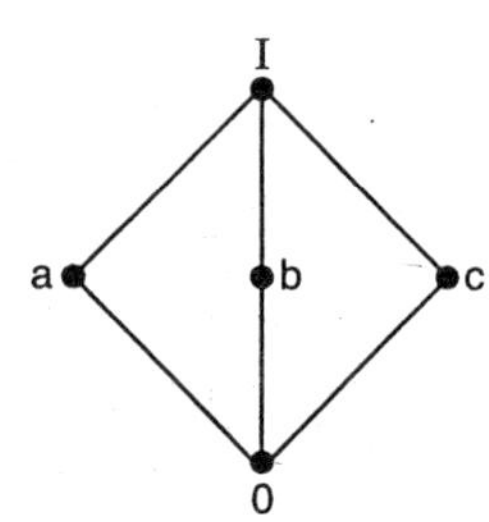

Fig. 20

Theorem X. *Prove that every distributive lattice is Modular.*

Proof. Let us assume that (L, $\leq$) is a distributive lattice and $a, b, c \in$ L such that $a \leq c$. Thus if $a \leq c$ then we have $a \vee c = c$

Now $$a \vee (b \wedge c) = (a \vee b) \wedge (a \vee c)$$
$$= (a \vee b) \wedge c$$

Hence, every distributive lattice is modular.

Example 17. *Show that the pentagonal lattice is not modular.*

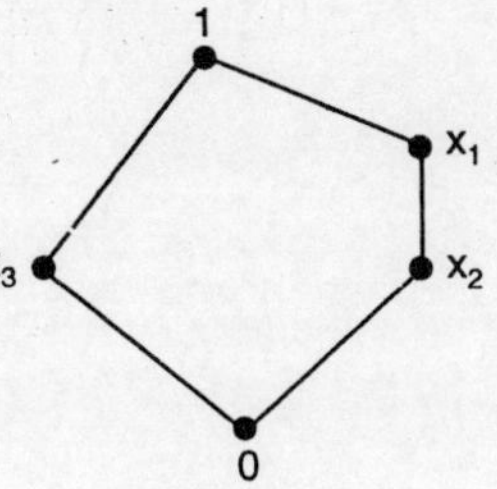

Fig. 21

Sol. Let us assume that $a = x_2$, $b = x_3$ and $c = x_1$ then $a < c$.

But $$a \vee (b \wedge c) = x_2 \vee (x_3 \wedge x_1) = x_3 \vee 0 = x_3$$

whereas $$(a \vee b) \wedge c = (x_2 \vee x_3) \wedge x_1 = 1 \wedge a_1 = a_1$$

which are unequal.

Hence proved.

Note. A lattice $(L, \wedge, \vee)$ is modular iff it does not contain a sublattice isomorphic with the pentagonal lattice.

DIRECT PRODUCT OF LATTICES

Let $(L_1, \vee_1, \wedge_1)$ and $(L_2, \vee_2, \wedge_2)$ be two lattices. Then $(L, \vee, \wedge)$ is the direct product of lattices, where $L = L_1 \times L_2$ in which the binary operations $\vee$(join) and $\wedge$(meet) on L are such that for any (a_1, b_1) and (a_2, b_2) in L.

$$(a_1, b_1) \vee (a_2, b_2) = (a_1 \vee_1 a_2, b_1 \vee_2 b_2)$$

and $$(a_1, b_1) \wedge (a_2, b_2) = (a_1 \wedge_1 a_2, b_1 \wedge_2 b_2).$$

Example 18. *Consider a lattice $(L, \leq)$ as shown in Fig. 22, where L = {1, 2}. Determine the lattice $(L^2, \leq)$, where $L^2 = L \times L$.*

(L, ≤)

Fig. 22

Sol. The lattice $(L^2, \leq)$ is as shown in Fig. 23.

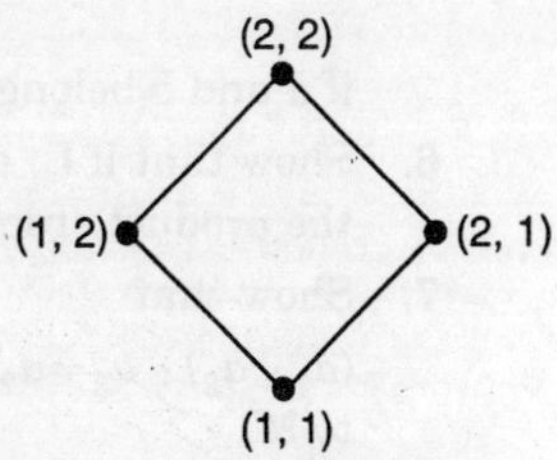

$(L^2, \leq)$

Fig. 23

Example 19. *Let $(L_1, \leq)$ and $(L_2, \leq)$ be two lattices as shown in Figs. 24 and 25. Determine the lattice $(L, \leq)$, where $L = L_1 \times L_2$.*

Fig. 24

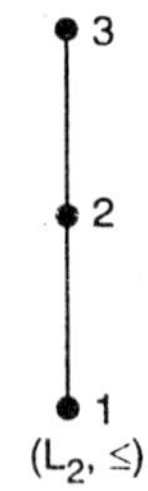

Fig. 25

Sol. The lattice $(L, \leq)$ is a direct product of the lattices $(L_1, \leq)$ and $(L_2, \leq)$ as shown in Fig. 26.

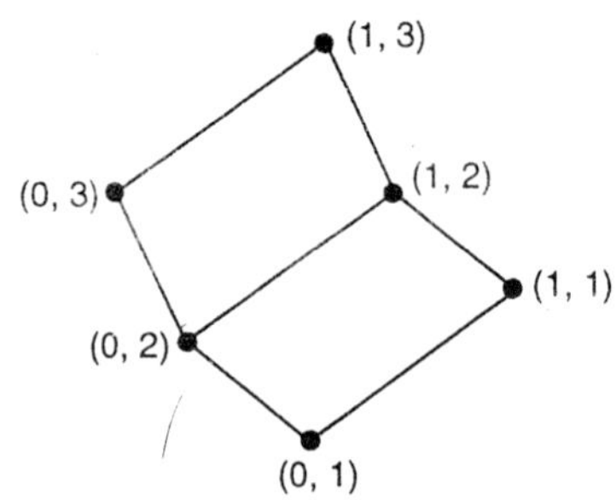

Fig. 26

EXERCISES

1. Show whether the following holds in a complemented, distributive lattice :
   $a \leq b \Leftrightarrow a \wedge b' = 0 \Leftrightarrow a' \vee b = 1 \Leftrightarrow b' \leq a'$
2. If a chain has three or more elements, then show that it will not be complemented.
3. Show whether the lattices $(S_n, \leq)$; $n = 30$ and 42 are complemented or distributive.
4. Determine the complement of each element in L_{75} and L_{105} .
5. Show that

$$a_1 \vee (a_1' \wedge a_2) = a_1 \vee a_2$$
$$a_1 \wedge (a_1' \vee a_2) = a_1 \wedge a_2$$

   if a and b belong to a bounded, distributive lattice. Also a' is the complement of a.
6. Show that if L_1 and L_2 are distributive lattices, then $L = L_1 \times L_2$ is also distributive, where L is the product order of the orders in L_1 and L_2.
7. Show that
   $(a_1 \vee a_2) \wedge a_3 = a_2 \wedge a_3 \quad \Rightarrow \quad (a_3 \vee a_2) \wedge a_1 = a_2 \vee a_1$ for a modular lattice M and a_1 , a_2 and a_3 belong to M.

SOLVED PROBLEMS

Problem 1. *Determine whether the posets shown in Fig. 27 are lattices or not ?*

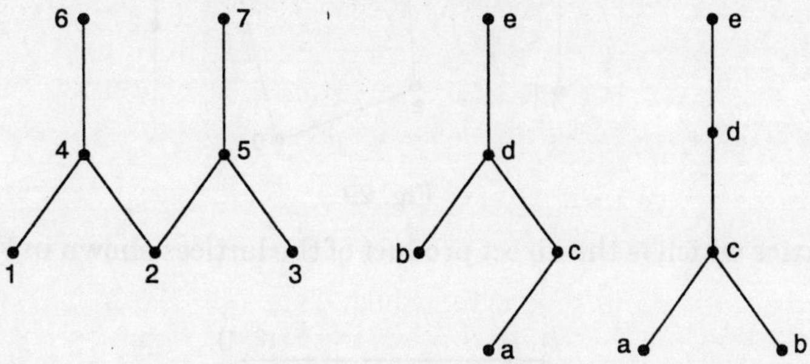

Fig. 27

Sol. (*i*) The poset shown in Fig. 27(I) is not a lattice since the elements 6 and 7 have no upper bound. Similarly, the elements 1 and 2, 1 and 3, 2 and 3 have no lower bound.

(*ii*) The poset shown in Fig. 27(II) is not a lattice since the elements b and c have no lower bound *i.e.,* inf (b, c) does not exist.

(*iii*) The poset shown in Fig. 27(III) is not a lattice since the elements a and b have no lower bound *i.e.,* inf (a, b) does not exist.

Problem 2. *Consider the lattice as shown in Fig. 28.*

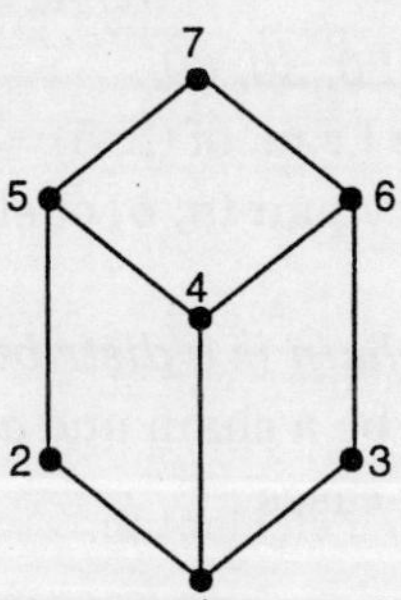

Fig. 28

(*i*) *Determine the join-irreducible elements of the lattice.*

(*ii*) *Determine the meet-irreducible elements of the lattice.*

(*iii*) *Determine the atoms of the lattice.*

(*iv*) *Determine the antiatoms of the lattice.*

Sol. (*i*) The join-irreducible elements of the lattice are 2, 3 and 4.

(*ii*) The meet-irreducible elements of the lattice are 5, 6, 2 and 3.

(*iii*) The atoms of the lattice are 2, 3 and 4.

(*iv*) The antiatoms of the lattice are 5 and 6.

Problem 3. *Draw the diagram of a lattice which is the direct product of the following lattices shown in Fig. 29.*

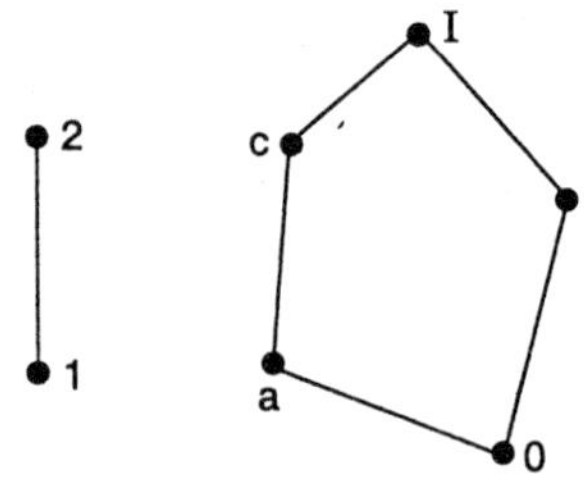

Fig. 29

Sol. The lattice which is the direct product of the lattices shown in Fig. 29 is as shown in Fig. 30.

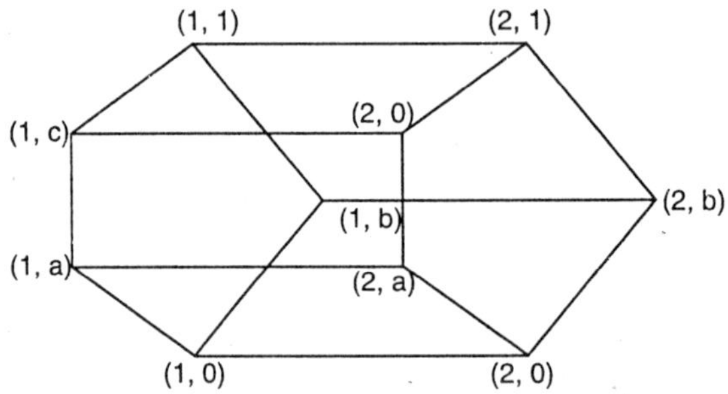

Fig. 30

Problem 4. *Determine whether or not each of the following sets is a lattice w.r.t. divisibility :*

(*i*) $A_1 = \{1, 2, 4, 5, 20\}$ (*ii*) $A_2 = \{2, 4, 8, 10, 20, 40\}$

(*iii*) $A_3 = \{1, 2, 3, 4, 5, 6, 10, 15, 20, 30, 60\}$.

Sol. (*i*) A_1 is not a lattice since l.c.m. of (2, 5) = 10, which does not belong to set A_1.

(*ii*) A_2 is a lattice since for every pair (a, b) of elements $\inf(a, b)$ and $\sup(a, b)$ exists.

(*iii*) A_3 is a lattice.

Problem 5. *Show that every chain is a distributive lattice.*

Sol. Let us assume that $(L, \leq)$ be a chain and $a, b, c \in L$.

We will consider the following cases :

(*i*) $a \leq b$ or $a \leq c$ and

(*ii*) $a \geq b$ or $a \geq c$

Now, we have to show that distributive law is satisfied by a, b, c.

Case I. We have $a \wedge (b \vee c) = a$

and $(a \wedge b) \vee (a \wedge c) = a$ ...(1)

Case II. We have $a \wedge (b \vee c) = b \vee c$

and $(a \wedge b) \vee (a \wedge c) = b \vee c$ ...(2)

From (1) and (2), we can say that

$$a \wedge (b \vee c) = (a \wedge b) \vee (a \wedge c)$$

This shows that every chain is a distributive lattice.

Problem 6. *Show that the posets given below are lattices. Also draw the Hasse diagrams.*

(i) $(S_8, /)$ *(ii)* $(S_{15}, /)$ *(iii)* $(S_{24}, /)$

Where S_n *is the set of all divisors of n and / denotes the division operation.*

Sol. We know that $S_8 = \{1, 2, 4, 8\}$

If we take any two elements of S_8 then their GLB and LUB will also be in S_8. Therefore $(S_8, /)$ is a lattice.

By giving similar arguments, we can show that

$$S_{15} = \{1, 3, 5, 15\}$$

and $$S_{24} = \{1, 2, 3, 4, 6, 8, 12, 24\}$$

are also lattices.

The Hasse diagram of these lattices are shown in Fig. 31.

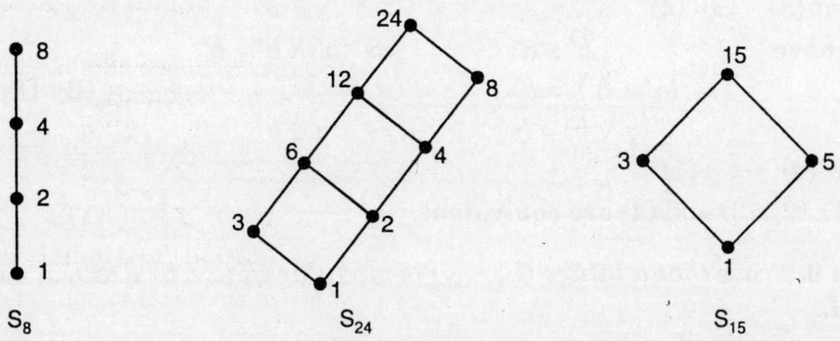

Fig. 31

Problem 7. *Give an example of an infinite lattice L with finite length.*

Sol. Let $L = \{0, 1, a_1, a_2, a_3, ..., a_n, ...\}$ be ordered as such that $0 < a_n < 1 \quad \forall\, n \in N$

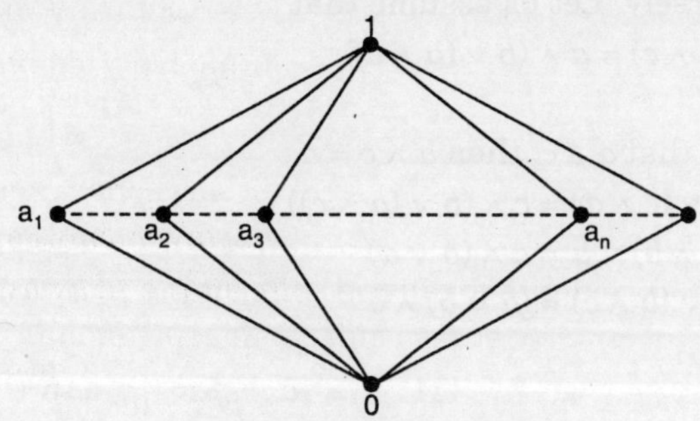

Fig. 32

Thus, L is an infinite lattice having infinite length.

Problem 8. *Show that in a complemented distributive lattice (L, ≤), the following are equivalent*

(1) $a \le b$ *(2)* $a \wedge b' = 0$ *(3)* $a' \vee b = 1$ *(4)* $b' \le a'$

Sol. We know, $a \le b \Rightarrow a \vee b = b$

$\Rightarrow \quad (a \vee b) \wedge b' = 0 \quad \{\because\ b \wedge b' = 0\}$

$\Rightarrow \quad (a \wedge b') \vee (b \wedge b') = 0$ {By distributive law}

$\Rightarrow \quad a \wedge b' = 0$

which shows that (1) $\Rightarrow$ (2)

Now, we know that

$a \wedge b' = 0 \quad \Rightarrow \quad (a \wedge b')' = 1$

$\Rightarrow \quad a' \vee (b')' = 1 \quad \Rightarrow \quad a' \vee b = 1$

which shows that (2) $\Rightarrow$ (3)

Now, we know that

$a' \vee b = 1 \quad \Rightarrow \quad (a' \vee b) \wedge b' = b'$ $\{\because \; 1 \wedge b' = b'$

$\Rightarrow \quad (a' \wedge b') \vee (b \wedge b') = b'$ {By distributive law}

$\Rightarrow \quad a' \wedge b' = b'$ $\{\because \; b \wedge b' = 0\}$

$\Rightarrow \quad b' \leq a'$

which shows that (3) $\Rightarrow$ (4)

Now, we have $\quad b' \leq a' \quad \Rightarrow \quad a' \wedge b' = b'$

$\Rightarrow \quad (a' \wedge b')' = b$ {By De Morgan's law}

$\Rightarrow \quad (a \vee b) = b \quad \Rightarrow \quad a \leq b$

which show that (4) $\Rightarrow$ (1)

Hence, (1), (2), (3) and (4) are equivalent.

Problem 9. *Prove that a lattice (L, $\wedge$, $\vee$) is modular iff $(a \wedge b) \vee (a \wedge c) = a \wedge (b \vee (a \wedge c))$ for all $a, b, c \in L$.*

Sol. Let us assume that (L, $\wedge$, $\vee$) is a modular lattice.

Then $\quad (a \wedge b) \vee (a \wedge c) = (a \wedge c) \vee (a \wedge b)$

$= (a \wedge c) \vee (b \wedge a)$

$= ((a \wedge c) \vee b) \wedge a$ (by modularity)

$= a \wedge (b \vee (a \wedge c))$

Now to show conversely, Let us assume that

$$(a \wedge b) \vee (a \wedge c) = a \wedge (b \vee (a \wedge c))$$

for all $a, b, c \in$ L.

Let us also assume that $a \leq c$, then $a \wedge c = a$.

Now $\quad (c \wedge b) \vee (c \wedge a) = c \wedge (b \vee (a \wedge c))$

Therefore $\quad (c \wedge b) \vee a = c \wedge (b \vee a)$

i.e., $\quad a \vee (b \wedge c) = (a \vee b) \wedge c$

Hence proved.

Chapter 9

Boolean Algebra

BOOLEAN ALGEBRA

Definition

A complemented distributive lattice is called a Boolean algebra. It is denoted by $(B, \wedge, \vee, ', 0, 1)$, where B is a set on which two binary operations $\wedge$ (*) and $\vee$ (+) and a unary operation $'$ (complement) are defined. Here 0 and 1 are two distinct elements of B.

Since $(B, \wedge, \vee)$ is a complemented distributive lattice, therefore each element of B has a unique complement.

ALTERNATE DEFINITION

Consider a set B on which two binary operations * and + and a unary operation $'$ (complement) are defined. Also let 0 and 1 are two distinct elements of B. Then it is called a Boolean algebra, if the following properties are satisfied for any elements a, b and c of the set B by it.

1. Commutative Properties

(i) $a + b = b + a$

(ii) $a * b = b * a$

2. Distributive Properties

(i) $a + (b * c) = (a + b) * (a + c)$

(ii) $a * (b + c) = (a * b) + (a * c)$

3. Identity Properties

(i) $a + 0 = a$

(ii) $a * 1 = a$

4. Complement Laws

(i) $a + a' = 1$

(ii) $a * a' = 0$

The Boolean algebra is denoted by $(B, +, *, ', 0, 1)$.

Example. Consider a set B = {0, 1}. Also consider the operations *, + and $'$ on the set B as shown in Fig. 1. Since the algebra $(B, *, +, ', 0, 1)$ satisfies all the properties. Hence, it is a Boolean algebra.

*	0	1
0	0	0
1	0	1

+	0	1
0	0	1
1	1	1

x	x'
0	1
1	0

Fig. 1

Example. Consider a non-empty set S and power set of S is P(S). Since P(S) is a collection of set S closed under the operations $\cup$ (union), $\cap$ (intersection) and ~ (complement). Hence

P(S) is a Boolean algebra, with the empty set ϕ as the zero element and unit element is U (universal set). The complement of any subset $A \subseteq U$ is $\sim A = U - A$, called the relative complement of the set A.

Note 1. If a finite lattice L does not contain 2^n element for some positive integer n, then L cannot be a Boolean algebra.

2. If $|L| = 2^n$, then L may or may not be a Boolean algebra.

Example 1. *Determine whether the poset given in Fig. 2 is a Boolean algebra or not. Give reason.*

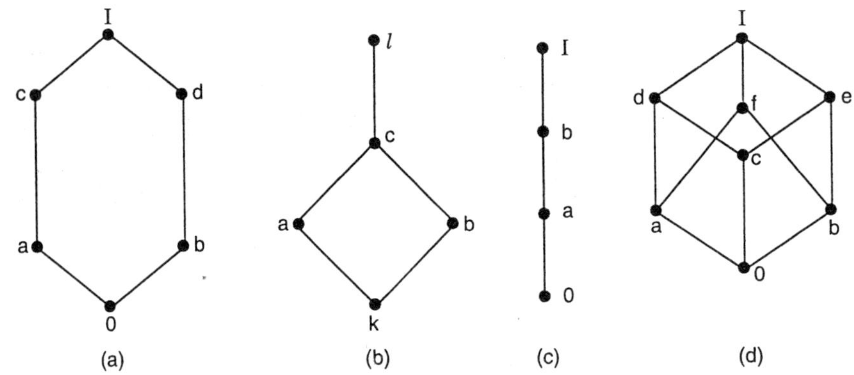

Fig. 2

Sol. 1. The poset shown in Fig. 2(a) is not a Boolean algebra since it contains 6 elements and $6 \neq 2^n$ for any integer $n \geq 0$.

2. The poset shown in Fig. 2(b) is not a Boolean algebra since it contains 5 elements and $5 \neq 2^n$ for any integer $n \geq 0$.

3. The poset shown in Fig. 2(c) is not a Boolean algebra since it does not satisfy the properties of Boolean algebra *i.e.*, complement of a and b does not exist.

4. The poset shown in Fig. 2(d) is a Boolean algebra since it is isomorphic to $(P(S), \subseteq)$ which is a Boolean algebra. Here P(S) is a power set of S having three elements.

Theorem I. *Prove that the complement of every element on a Boolean algebra B is unique.*

Proof. Consider an element $a \in B$, let a has two complements a_1' an a_2'' in B.

Therefore,
$$\left.\begin{aligned} a + a_1' = 1 = a_1' + a \\ a * a_1' = 0 = a_1' * a \end{aligned}\right\} \quad ...(1)$$

and also
$$\left.\begin{aligned} a + a_2' = 1 = a_2' + a \\ a * a_2' = 0 = a_2' * a \end{aligned}\right\} \quad ...(2)$$

We know that
$$\begin{aligned} a_1' &= a_1' * 1 && \{\because\ a_1' * 1 = a_1'\} \\ &= a_1' * (a + a_2') && \{\because\ a + a_2' = 1\} \\ &= (a_1' * a) + (a_1' * a_2') && \{\because\ \text{B is distributive}\} \end{aligned}$$

$$= 0 + (a_1' * a_2') \qquad \{\because \; a_1' * a = 0\}$$
$$= a * a_2' + a_1' * a_2' \qquad \{\because \; a * a_2' = 0\}$$
$$= (a + a_1') * a_2' \qquad \{\because \; \text{B is distributive}\}$$
$$= 1 * a_2' \qquad \{\because \; a + a_1' = 1\}$$
$$= a_2' \qquad \{\because \; 1 * a_2' = a_2'\}$$

Hence, the complement is unique.

SUB-ALGEBRA

Consider a Boolean algebra (B, ∗, +, ′, 0, 1) and let A ⊆ B. Then (A, ∗, +, ′, 0, 1) is called a sub-algebra or sub-Boolean algebra of B if A itself is a Boolean algebra *i.e.,* A contains the elements 0 and 1 and is closed under the operations ∗, + and ′.

Example. Consider the Boolean algebra D_{70} whose Hasse diagram is shown in Fig. 3.

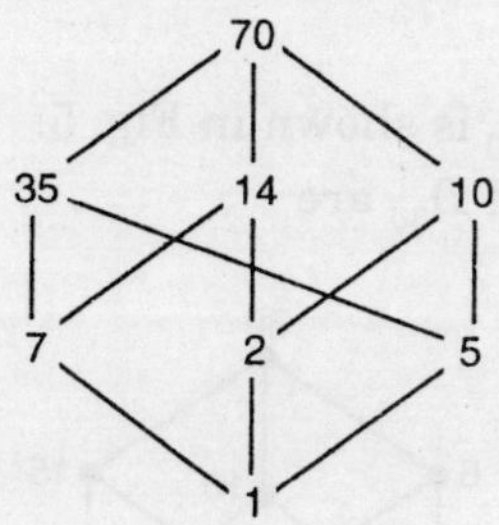

Fig. 3

Clearly A = {1, 7, 10, 70} and B = {1, 2, 35, 70} is sub-algebra of D_{70}. Since, both A and B are closed under operation ∧, ∨ and ′.

Note. A subset of a Boolean algebra can be a Boolean algebra but it may or may not be a sub-algebra as it may not closed w.r.t. the operations in B.

Example 2. *Consider the Boolean algebra (B, +, ∗, ′, 0, 1) as shown in Fig. 4. Let the subsets of S be as follows :*

(i) S_1 = {c, b, 0, 1} *(ii)* S_2 = {a, b, 0, 1}

(iii) S_3 = {a, b, 0, f} *(iv)* S_4 = {a, c, 0, 1}

(v) S_5 = {a, c, f, 1}.

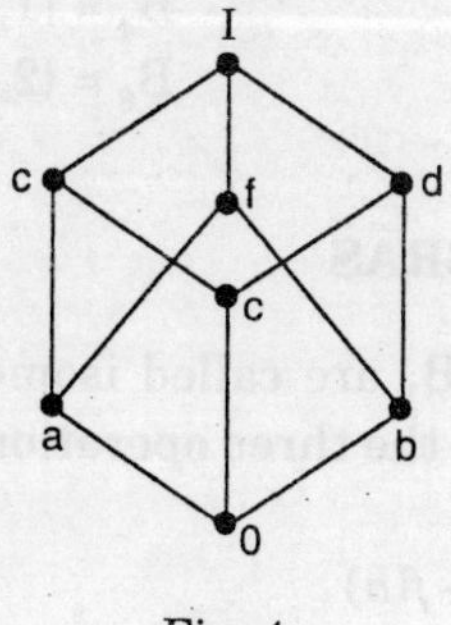

Fig. 4

Determine whether the S_1, S_2, S_3, S_4, S_5 are Boolean algebra, sub-algebra, either both or none.

Sol. (*i*) The subset S_1 is a sub-algebra, since it is closed under the operations $*$, $+$, $'$. So, it is also a Boolean algebra.

(*ii*) The subset S_2 is neither a sub-algebra nor Boolean algebra.

(*iii*) The subset S_3 is a Boolean algebra, but not a sub-algebra of B.

(*iv*) The subset S_4 is neither a Boolean algebra nor a sub-algebra of B.

(*v*) The subset S_5 is not a sub-algebra, but it is a Boolean algebra.

Note 1. Every Boolean sub-algebra of a Boolean algebra B is a Boolean algebra. But there may be Boolean algebras which are not Boolean sub-algebras of B.

2. The Boolean algebra B and {0, 1} are always sub-algebras of B.

Example 3. *Consider the Boolean algebra D_{30}. Determine the following :*

(*a*) *All the Boolean sub-algebra of D_{30}.*

(*b*) *All Boolean algebras which are not Boolean sub-algebras of D_{30} having at least four elements.*

Sol. The Hasse diagram of D_{30} is shown in Fig. 5.

(*a*) All Boolean sub-algebras of D_{30} are

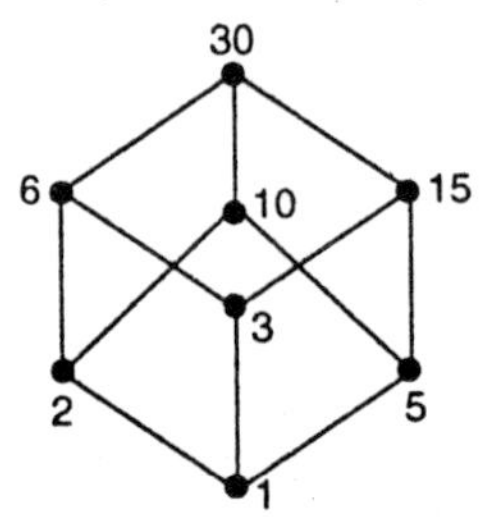

Fig. 5

$B_1 = \{1, 2, 15, 30\}$ $B_4 = \{1, 30\}$

$B_2 = \{1, 5, 6, 30\}$ $B_5 = \{1, 2, 3, 5, 6, 10, 15, 30\}$.

$B_3 = \{1, 3, 10, 30\}$

(*b*) All Boolean algebras which are not Boolean sub-algebras of D_{30} are :

$B_1 = \{1, 2, 3, 6\}$ $B_2 = \{1, 3, 5, 15\}$

$B_3 = \{3, 6, 5, 30\}$ $B_4 = \{1, 2, 5, 10\}$

$B_5 = \{5, 10, 15, 30\}$ $B_6 = \{2, 6, 10, 30\}$.

ISOMORPHIC BOOLEAN ALGEBRAS

Two Boolean algebras B and B_1 are called isomorphic if there is a one-to-one correspondence $f : B \to B_1$ which preserves the three operations $+$, $*$ and $'$ for any elements a, b in B *i.e.*,

$$f(a + b) = f(a) + f(b)$$
$$f(a * b) = f(a) * f(b) \quad \text{and} \quad f(a') = f(a)'.$$

Example. The following are two distinct Boolean algebras with two elements, which are isomorphic.

1. The first one is a Boolean algebra that is derived from a power set P(S) under $\subseteq$ (set inclusion) *i.e.,* let S = {*a*}, then B = {P(S), $\cup$, $\cap$, ′} is a boolean algebra with two elements P(S) = {ϕ, {*a*}}.

2. The second one is a Boolean algebra {B, $\vee$, $\wedge$, ′} with two elements 1 and *p* {here *p* is a prime number} under operation divides *i.e.,* let B = {1, *p*}. So, we have $1 \wedge p = 1$ and $1 \vee p = p$ also $1' = p$ and $p' = 1$.

The table shows all the basic properties of a Boolean algebra (B, $*$, +, ′, 0, 1) for any elements *a, b, c* belongs to B. The greatest and least elements of B are denoted by 1 and 0 respectively.

Table I

1. $a \le b$ iff $a + b = b$	**2.** $a \le b$ iff $a * b = a$
3. Idempotent Laws	**4. Commutative Property**
(*i*) $a + b = a$	(*i*) $a + b = b + a$
(*ii*) $a * a = a$	(*ii*) $a * b = b * a$
5. Associative Property	**6. Absorption Laws**
(*i*) $a + (b + c) = (a + b) + c$	(*i*) $a + (a * b) = a$
(*ii*) $a * (b * c) = (a * b) * c$	(*ii*) $a * (a + b) = a$
7. Identity Laws	**8. Null Laws**
(*i*) $a + 0 = a$	(*i*) $a * 0 = 0$
(*ii*) $a * 1 = a$	(*ii*) $a + 1 = 1$
9. Distributive Laws	**10. Complement Laws**
(*i*) $a * (b + c) = (a * b) + (a * c)$	(*i*) $0' = 1$
(*ii*) $a + (b * c) = (a + b) * (a + c)$	(*ii*) $1' = 0$
	(*iii*) $a + a' = 1$
	(*iv*) $a * a' = 0$
11. Involution Law	**12. De Morgan's Laws**
$(a')' = a$	(*i*) $(a * b)' = (a' + b')$
	(*ii*) $(a + b)' = (a' * b')$.

Note. (*i*) $0 \le a \le 1$ for every $a \in$ B.

(*ii*) Every element *b* has a unique complement *b*′.

Theorem II. *Prove the idempotent laws i.e.,*

(*i*) $a * a = a$ (*ii*) $a + a = a$.

Proof. (*i*)
$$a = a * 1 = a * (a + a')$$
$$= (a * a) + (a * a') \quad \text{...\{Distributive Law\}}$$
$$= (a * a) + 0 = a * a.$$

(*ii*) It follows from the duality of (*i*).

Theorem III. *Prove the Boundedness (NULL) laws i.e.,*

(*i*) $a * 0 = 0$ (*ii*) $a + 1 = 1$.

Proof. (*i*)
$$a * 0 = (a * 0) + 0$$
$$= (a * 0) + (a * a') \quad \text{...\{Complement law\}}$$

$= a * (0 + a')$ ...{Distributive law}

$= a * (a' + 0)$

$= a * a'$

$= 0.$

(*ii*) It is directly followed from the duality of proof (*i*).

Theorem IV. *Prove the absorption laws i.e.,*

(*i*) $a + (a * b) = a$ (*ii*) $a * (a + b) = a.$

Proof. (*ii*) $a * (a + b) = (a + 0) * (a + b)$

$= a + (0 * b)$ ...{Distributive law}

$= a + (b * 0)$

$= a + 0$ ...{Null law}

$= a$

(*i*) It is directly followed from duality of proof (*ii*).

Theorem V. *Prove De Morgan's laws i.e.,*

(*i*) $(a + b)' = a' * b'$ (*ii*) $(a * b)' = a' + b'.$

Proof. (*i*) We have to show that

$$(a + b) + (a' * b') = 1 \quad \text{and} \quad (a + b) * (a' * b') = 0$$

Then, we have $a' * b' = (a + b)'$

Now, $(a + b) + (a' * b') = b + a + (a' * b')$

$= b + (a + a') * (a + b')$ ...{Distributive law}

$= b + [1 * (a + b')] \Rightarrow b + (1 * a) + (1 * b')$

$= b + a + b' = b + b' + a$

$= 1 + a = 1$

Also, $(a + b) * (a' * b') = ((a + b) * a') * b'$

$= ((a * a') + (b * a')) * b'$

$= (0 + (b * a')) * b'$

$= (b * a') * b'$

$= (b * b') * a'$

$= 0 + a'$

$= 0$

Thus, $a' * b' = (a + b)'.$

(*ii*) It follows directly from duality of proof (*i*).

Example 4. *If $a = b$ then show that $ab' + a'b = 0$.*

Sol. Let us assume that $a = b$, then

$ab' + a'b = bb' + a'b$

$= bb' + a'a$ $\{\because \ a = b\}$

$= 0 + 0$

$= 0$

Thus $ab' + a'b = 0$ iff $a = b$.

Example 5. *For a Boolean algebra B, show that*

(i) $ab + (c(a' + b')) = ab + c$ *(ii)* $(a + a'b)(a' + ab) = b$.

Sol. *(i)* Take the L.H.S., then we have

$= ab + (c(a' + b'))$

$= ab + c(ab)'$ {Demorgan's law}

$= (ab + c)(ab + (ab)')$

$= (ab + c) . 1$

$= (ab + c) =$ R.H.S.

(ii) Take the L.H.S., then we have

$= (a + a'b)(a' + ab)$

$= (a + a')(a + b)(a' + a)(a' + b)$

$= 1 . (a + b) . 1. (a' + b)$

$= (a + b)(a' + b)$

$= (b + a)(b + a')$

$= b + aa'$

$= b + 0 = b =$ R.H.S.

Example 6. *If a, b, c are elements of any Boolean algebra, then show that*

$$ba = ca \quad and \quad ba' = ca' \Rightarrow b = c.$$

Sol. Let us assume $ba = ca$ and $ba' = ca'$

Then we have $b = b . 1$

$= b(a + a')$

$= ba + ba'$

$= ca + ca'$ $\{\because \; b = c\}$

$= c(a + a')$

$= c . 1 = c$

Hence proved.

Example 7. *Show that for a Boolean algebra B,*

$$(a + b)' + (a + b')' = a' \quad \forall\, a, b \in B.$$

Sol. From De Morgan's law, we know that

$(ab)' = a' + b'$

Then $(a + b)' + (a + b') = ((a + b)(a + b'))'$

$= (a + bb')'$ {By distributive law}

$= (a + 0)'$ $\{\because \; bb' = 0\}$

$= a'$

Thus $(a + b)' + (a + b')' = a'$

BOOLEAN FUNCTIONS

Consider the Boolean algebra (B, ∨, ∧, ′, 0, 1). A function from A^n to A is called a Boolean function if it can be specified by a Boolean expression of *n* variables.

For the two valued Boolean algebra any function from $\{0, 1\}^n$ to $\{0, 1\}$ is a Boolean function.

e.g., 1. The table shows a function *f* from $\{0, 1\}^3$ to $\{0, 1\}$

Table II

(x, y, z)	f
(0, 0, 0)	0
(0, 0, 1)	0
(0, 1, 0)	1
(0, 1, 1)	0
(1, 0, 0)	1
(1, 0, 1)	1
(1, 1, 0)	0
(1, 1, 1)	1

2. The table shows a function f from $\{0, 1, 2, 3\}^2$ to $\{0, 1, 2, 3\}$.

Table III

(x, y)	f
(0, 0)	1
(0, 1)	0
(0, 2)	0
(0, 3)	3
(1, 0)	1
(1, 1)	1
(1, 2)	0
(1, 3)	3
(2, 0)	2
(2, 1)	0
(2, 2)	1
(2, 3)	1
(3, 0)	3
(3, 1)	0
(3, 2)	0
(3, 3)	2

Note. A function can always be described in tabular form. An alternative way of describing the functions is specifying the function by an expression.

BOOLEAN EXPRESSION

Consider a Boolean algebra (B, $\vee$, $\wedge$, $'$, 0, 1). A Boolean expression over Boolean algebra B is defined as

(*i*) Every element of B is a Boolean expression

(*ii*) Every variable name is a Boolean expression

(*iii*) If a_1 and a_2 are Boolean expressions, then a_1', $a_1 \vee a_2$ and $a_1 \wedge a_2$ are Boolean expressions.

e.g., Consider the Boolean algebra ({0, 1, 2, 3}, $\vee$, $\wedge$, $'$, 0, 1)

(*i*) $0 \vee x$ (*ii*) $(\overline{2 \wedge 3})$ (*iii*) $(x_1 \vee \overline{x_2}) \wedge (\overline{x_1 \wedge x_3})$

are Boolean expressions over the Boolean algebra.

A Boolean expression that contains n distinct variables is usually referred to as a Boolean expression of n variables.

EVALUATION OF BOOLEAN EXPRESSION

Let $E(x_1, x_2, \ldots\ldots, x_n)$ be a Boolean expression of n variables over a Boolean algebra B. By an assignment of values to the variables $x_1, x_2, \ldots\ldots, x_n$, means an assignment of elements of A to be the values of the variables.

We can evaluate the expression $E(x_1, x_2, \ldots\ldots, x_n)$ by substituting the variables in the expression by their values.

Example. Consider the Boolean expression

$$E(x_1, x_2, x_3) = (x_1 \vee x_2) \wedge (\overline{\overline{x_1} \vee x_2}) \wedge (\overline{x_2 \vee x_3})$$

over the Boolean algebra $(\{0, 1\}, \vee, \wedge, ')$.

By assigning the values $x_1 = 0, x_2 = 1, x_3 = 0$ yields

$$E(0, 1, 0) = (0 \vee 1) \wedge (\overline{\overline{0} \vee 1}) \wedge (\overline{1 \vee 0}) = 1 \wedge 1 \wedge 0 = 0.$$

EQUIVALENT BOOLEAN EXPRESSIONS

Two Boolean expressions of n variables are said to be equivalent if they assume the same value for every assignment of values to the n variables.

e.g., The following two Boolean algebras

$$(x_1 \wedge x_2) \vee (x_1 \wedge \overline{x_3}) \quad \text{and} \quad x_1 \wedge (x_2 \vee \overline{x_3})$$

are equivalent.

We may write $E_1(x_1, x_2, \ldots\ldots, x_n) = E_2(x_1, x_2, \ldots\ldots, x_n)$ to mean the two expressions $E_1(x_1, x_2, \ldots\ldots, x_n)$ and $E_2(x_1, x_2, \ldots\ldots, x_n)$ are equivalent.

e.g., The Boolean expression

$$(\overline{x_1} \wedge x_2 \wedge \overline{x_3}) \vee (x_1 \wedge \overline{x_2}) \vee (x_1 \wedge x_3)$$

over the Boolean algebra $(\{0, 1\}, \vee, \wedge, ')$ defines the function f in figure.

Minterm. A Boolean expression of n variables $x_1, x_2, \ldots\ldots, x_n$ is said to be a minterm if it is of the form

$$\tilde{x}_1 \wedge \tilde{x}_2 \wedge \tilde{x}_3 \wedge \ldots\ldots \wedge \tilde{x}_n$$

where $\tilde{x}_i$ is used to denote x_i or x_i'.

DISJUNCTIVE NORMAL FORM OR SUM OF PRODUCTS OR SOP

A Boolean expression over $(\{0, 1\}, \vee, \wedge, ')$ is said to be in disjunctive normal form if it is a join of minterms *e.g.*,

$$(x_1' \wedge x_2' \wedge x_3') \vee (x_1' \wedge x_2 \wedge x_3') \vee (x_1 \wedge x_2 \wedge x_3)$$

is a Boolean expression in disjunctive normal form.

Since there are three minterms $x_1' \wedge x_2' \wedge x_3'$, $x_1' \wedge x_2 \wedge x_3$ and $x_1 \wedge x_2 \wedge x_3$.

Maxterm. A Boolean expression of n variables $x_1, x_2, \ldots, x_n$ is said to be a maxterm if it is of the form $\tilde{x}_1 \vee \tilde{x}_2 \vee \ldots \vee \tilde{x}_n$, where $\tilde{x}_i$ is used to denote x_i or x_i'.

CONJUNCTIVE NORMAL FORM OR PRODUCTS OF SUMS OR POS

A Boolean expression over ({0, 1}, ∨, ∧, ′) is said to be in conjunctive normal form if it is a meet of maxterms *e.g.*,

$$(x_1 \vee x_2 \vee x_3) \wedge (x_1 \vee x_2 \vee x_3) \wedge (x_1 \vee x_2 \vee x_3) \wedge (x_1' \vee x_2 \vee x_3') \wedge (x_1' \wedge x_2' \wedge x_3)$$

is a Boolean expression in conjunctive normal form consisting of five maxterms.

OBTAINING A DISJUNCTIVE NORMAL FORM

Consider a function from $\{0, 1\}^n$ to {0, 1}. A Boolean expression can be obtained in disjunctive normal form corresponding to this function by having a minterm corresponding to each ordered n-tuple of 0's and 1's for which the value of function is 1.

OBTAINING A CONJUNCTIVE NORMAL FORM

Consider a function from $\{0, 1\}^n$ to {0, 1}. A Boolean expression can be obtained in conjunctive normal form corresponding to this function by having a maxterm corresponding to each ordered n-tuple of 0's and 1's at which the value of function is 0.

Example 8. *Express the following function in*
(i) Disjunctive normal form *(ii) Conjunctive normal form*

	f		*f*
(0, 0, 0)	*1*	*(1, 0, 0)*	*0*
(0, 0, 1)	*0*	*(1, 0, 1)*	*1*
(0, 1, 0)	*1*	*(1, 1, 0)*	*0*
(0, 1, 1)	*0*	*(1, 1, 1)*	*1*

Sol. (*i*) $(x_1' \wedge x_2' \wedge x_3') \vee (x_1' \wedge x_2 \wedge x_3') \vee (x_1 \wedge x_2' \wedge x_3) \vee (x_1 \wedge x_2 \wedge x_3)$
(*ii*) $(x_1' \vee x_2' \vee x_3') \wedge (x_1' \vee x_2 \vee x_3) \wedge (x_1 \vee x_2' \vee x_3') \wedge (x_1 \vee x_2 \vee x_3')$.

Example 9. *If $f(x, y, z) = (x \vee y) \wedge (x \vee y') \wedge (x' \vee z)$ be a given Boolean function. Determine its DN form.*

Sol. First determine all values of $f(x, y, z)$ when x, y, z take values 0, 1 and then from this table, we will find the DN form

x	y	z	$f(x, y, z)$	$(x \vee y) \wedge (x \vee y') \wedge (x' \vee z)$
0	0	0	0	$= (0 \vee 0) \wedge (0 \vee 1) \wedge (1 \vee 0)$
0	0	1	0	$= (0 \vee 0) \wedge (0 \vee 1) \wedge (1 \vee 1)$
0	1	0	0	$= (0 \vee 1) \wedge (0 \vee 0) \wedge (1 \vee 0)$
0	1	1	0	$= (0 \vee 1) \wedge (0 \vee 0) \wedge (1 \vee 1)$
1	0	0	0	$= (1 \vee 0) \wedge (1 \vee 1) \wedge (0 \vee 0)$
1	0	1	1	$= (1 \vee 0) \wedge (1 \vee 1) \wedge (0 \vee 1)$
1	1	0	0	$= (1 \vee 1) \wedge (1 \vee 0) \wedge (0 \vee 0)$
1	1	1	1	$= (1 \vee 1) \wedge (1 \vee 0) \wedge (0 \vee 1)$

The disjunctive normal form of the function is

$$f(x, y, z) = (x \wedge y' \wedge z) \vee (x \wedge y \wedge z).$$

Example 10. *Determine the disjunctive normal form of the following Boolean expression :*

$$x \wedge (y \vee z).$$

Sol. First determine all values of $f(x, y, z)$ when x, y, z take values 0, 1 and then from this table we will write disjunctive form. Thus,

x	y	z	$f(x, y, z)$		$x \wedge (y \vee z)$
0	0	0	0	=	$0 \wedge (0 \vee 0)$
0	0	1	0	=	$0 \wedge (0 \vee 1)$
0	1	0	0	=	$0 \wedge (1 \vee 0)$
0	1	1	0	=	$0 \wedge (1 \vee 1)$
1	0	0	0	=	$1 \wedge (0 \vee 0)$
1	0	1	1	=	$1 \wedge (0 \vee 1)$
1	1	0	1	=	$1 \wedge (1 \vee 0)$
1	1	1	1	=	$1 \wedge (1 \vee 1)$

Thus, the DNF of the function is

$$f(x, y, z) = (x \wedge y \wedge z) \vee (x \wedge y' \wedge z) \vee (x \wedge y \wedge z').$$

Example 11. *Find the Boolean expression in conjunctive normal form that defines the function f given by*

x	y	z	f
0	0	0	1
0	0	1	0
0	1	0	1
0	1	1	0
1	0	0	0
1	0	1	0
1	1	0	0
1	1	1	1

Sol. To obtain the expression in conjunctive normal form, take the values of $f(x, y, z)$ when it is 0 *i.e.,* take the max. terms of the function. Thus, the expression in CN form is

$$(x \vee y \vee z') \wedge (x \vee y' \vee z) \wedge (x \vee y' \vee z') \wedge (x' \vee y \vee z) \wedge (x' \vee y' \vee z).$$

PRINCIPLE OF DUALITY

The dual of any expression E is obtained by interchanging the operations + and * and also interchanging the corresponding identity elements 0 and 1, in original expression E.

Example 12. *Write the dual of following Boolean expressions :*

(i) $(x_1 * x_2) + (x_1 * \overline{x_3})$ (ii) $(1 + x_2) * (x_1 + 1)$

(iii) $(a \wedge (b \wedge c))$.

Sol. (i) $(x_1 + x_2) * (x_1 + \overline{x_3})$ (ii) $(0 * x_2) + (x_1 * 0)$

(iii) $(a \vee (b \wedge c))$.

Note. The dual of any theorem in a Boolean algebra is also a theorem.

EXERCISES

1. Prove that in a Boolean algebra B, $a \le b$ iff $(a \wedge b') \vee (a' \vee b) = a$
2. In a Boolean algebra, show that

   (i) $(a + b)(a' + c) = ac + a'b = ac + a'b + bc$

   (ii) $a = b \Leftrightarrow ab' + a'b = 0$.
3. Show that for any a, b in a Boolean algebra the following holds :

   (i) $a \wedge (a' \vee b) = a \wedge b$ (ii) $b \wedge (a \vee (a' \wedge (b \vee b'))) = b$

   (iii) $(a \wedge b) \vee (a \wedge b') = a$.
4. Show whether the following Boolean polynomials are equivalent :

   (i) $x_1 \wedge (x_2 \vee (x_2' \wedge (x_2 \vee x_2')))$; x

   (ii) $(x_1 \vee x_2)$; $[(x_1 \wedge x_3) \vee (x_2' \vee x_3)'] \vee [(x_2 \wedge x_3) \vee (x_1 \wedge x_3')]$
5. Simplify the following Boolean expressions

   (i) $X = a(a + b)(a + ab)$ (ii) $X = a'bc' + abc' + ab'c' + a'b'c'$

   (iii) $A = xy + x'yz' + yz$ (iv) $X = (xy' + z)(x + y')z$.
6. Determine the dual form of each of the followings :

   (i) $x(y + y'z) + y'z$ (ii) $0 + (x_1' + x_2 + x_3)$
7. Represent each of the following Boolean functions in disjunctive normal form :

   (i) $f(x, y, z) = (yz + xz')(xy' + z)'$ (ii) $f(x, y, z) = (x + x'y + x'yz')(xy + (xz)')(y + xyz')$

   (iii) $f(x_1, x_2) = (x_1 + x_2)(x_1' + x_2')$.
8. Determine the disjunctive normal forms of the functions whose truth table are shown in figure :

(i)

x	y	z	$f(x, y, z)$
0	0	0	0
0	0	1	1
0	1	0	1
0	1	1	1
1	0	0	0
1	0	1	0
1	1	0	0
1	1	1	1

(ii)

x	y	$f(x, y)$
0	0	1
0	1	1
1	0	0
1	1	1

9. Represent each of the following Boolean expression in conjunctive normal form :

   (i) $f(x, y, z) = xy'$ (ii) $f(x, y, z) = (x + z)y$.

DESIGN OF DIGITAL CIRCUITS

Schematic or Graphical Representation of Boolean Expressions

The Boolean expressions can be graphically represented by using logic circuits. These logic circuits can be constructed using solid state devices called gates, which are capable of switching voltage levels. If x and y are variables, then the basic expressions $x \wedge y$ (AND), $x \vee y$ (OR) and $\overline{x'}$ (NOT) are shown graphically as follows :

AND Gate. An AND gate receives inputs x and y and produces output denoted $x \wedge y$, as shown in logic table.

x	y	$x \wedge y$
0	0	0
0	1	0
1	0	0
1	1	1

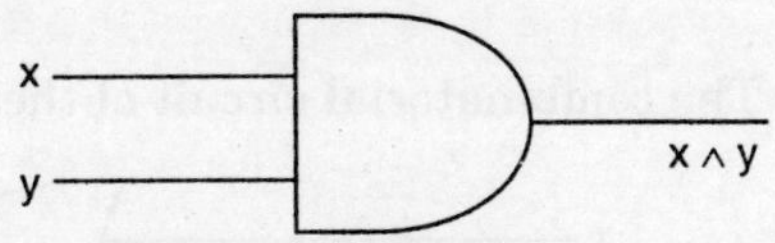

AND Gate

Fig. 6

OR Gate. An OR gate receives inputs x and y and produces output denoted $x \vee y$ as shown in logic table.

x	y	$x \vee y$
0	0	0
0	1	1
1	0	1
1	1	1

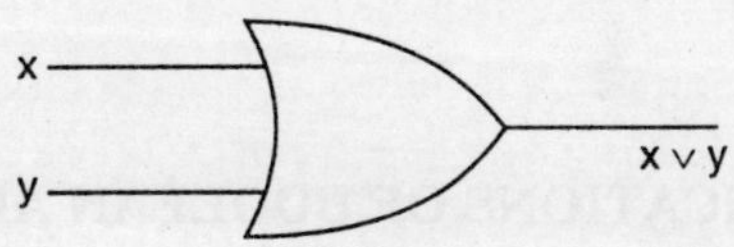

OR Gate

Fig. 7

NOT Gate. A NOT gate receives input x and produces output y denoted $\overline{x}$, as shown in logic table.

x	$\overline{x}$
0	1
1	0

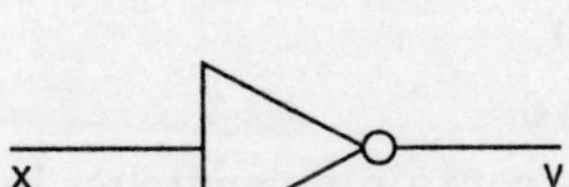

Fig. 8

We can interconnect these devices to form an electronic circuit that realizes any given Boolean expression.

Example 13. *Design a combinatorial circuit for the following Boolean expression :*

(i) $(x \wedge \overline{y}) \vee (\overline{x} \wedge y)$ *(ii)* $(x_1 \vee x_2) \wedge x_1$.

Sol. *(i)* The combinatorial circuit of the Boolean expression is shown in Fig. 9.

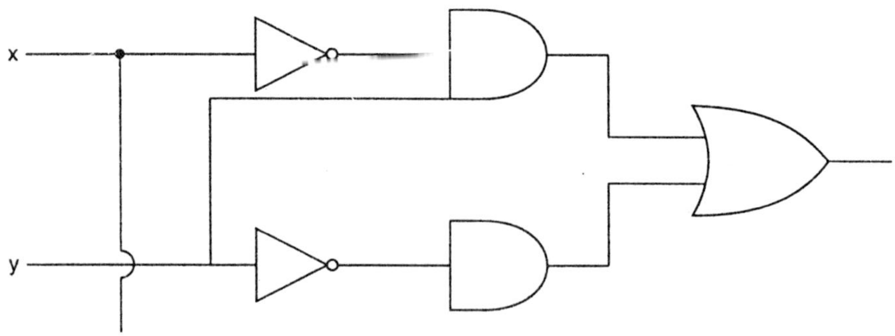

Fig. 9

(*ii*) The combinatorial circuit of the Boolean expression is shown in Fig. 10.

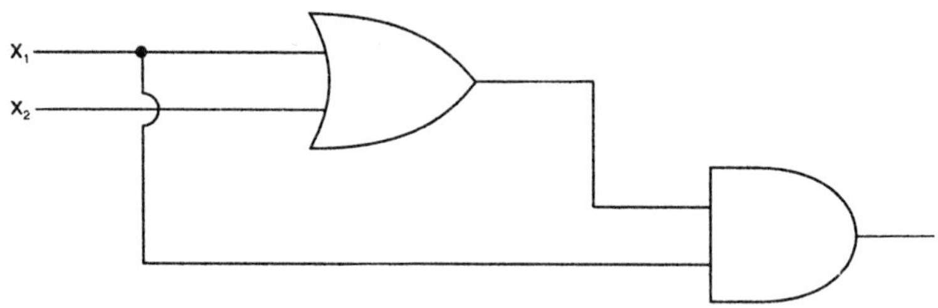

Fig. 10

APPLICATIONS OF BOOLEAN ALGEBRA IN SWITCHING THEORY

The simplest switching device is ON-OFF switch. A switch is a device in an electric circuit which lets or does not let the current to flow through the circuit.

The switch has two states ON or OFF (closed or open)

— If the switch is closed, current will pass through it.

— If the switch is open, current will not pass through it.

— ON is denoted by 1.

— OFF is denoted by 0.

The electrical circuits containing switches by Boolean expressions, if ON is denoted by true or 1 and OFF is denoted by False or 0.

SERIES AND PARALLEL CONNECTION

There are 2 ways in which switches are connected with each other.

(*i*) Series connection (*ii*) Parallel connection.

SERIES CONNECTION

Two switches x_1 and x_2 are said to be connected in series if current can pass only when both the switches are ON (closed) and the current does not flow if both the switches or any one are OFF (open). Symbolically, series connection of two switches x_1 and x_2 is denoted by $x_1 \wedge x_2$ or $x_1 * x_2$. It is represented diagrammatically as follows :

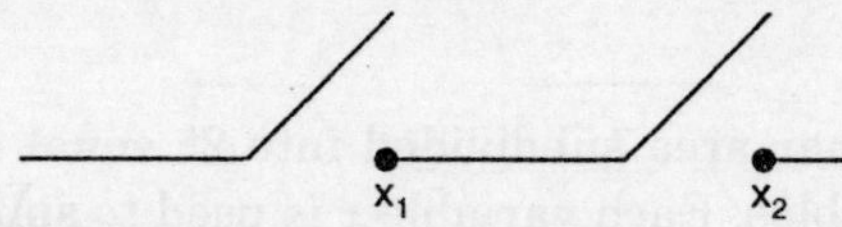

PARALLEL CONNECTION

Two switches x_1 and x_2 are said to be connected in parallel if current flows when both or any one of the switches are ON (closed) and current does not flow when both are OFF (open). Symbolically, parallel connection of two switches x_1 and x_2 is denoted by $x_1 \vee x_2$ or $x_1 + x_2$. It is represented diagrammatically as follows :

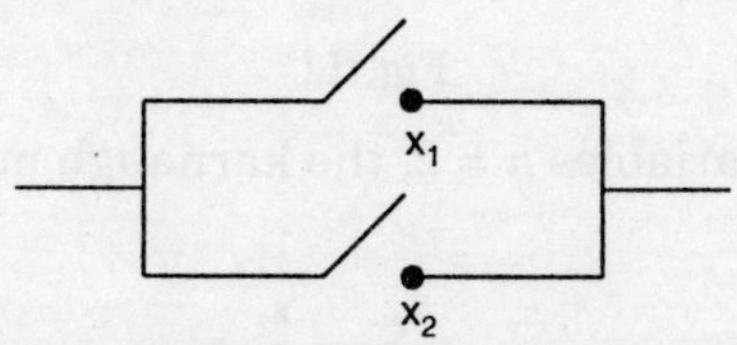

Example 14. *Construct the circuit diagram represented by Boolean expression* $a \wedge (b \vee c)$.

Sol. For the given Boolean expression, we will see that the current would flow when a and b or a and c are ON (closed). So, it is represented as follows :

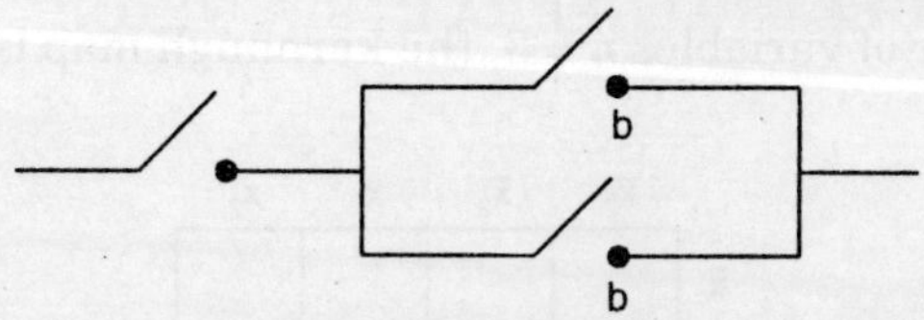

Example 15. *Draw the circuit diagram represented by following Boolean expressions :*

(*i*) $(a \wedge b \wedge c) \vee (a' \wedge b' \wedge c')$ (*ii*) $(x_1 \wedge x_2) \vee (x_1' \wedge x_2')$.

Sol. (*i*) For the given expression, we will see that the current would flow when a, b, c or a', b', c' are ON (closed). It is represented as follows :

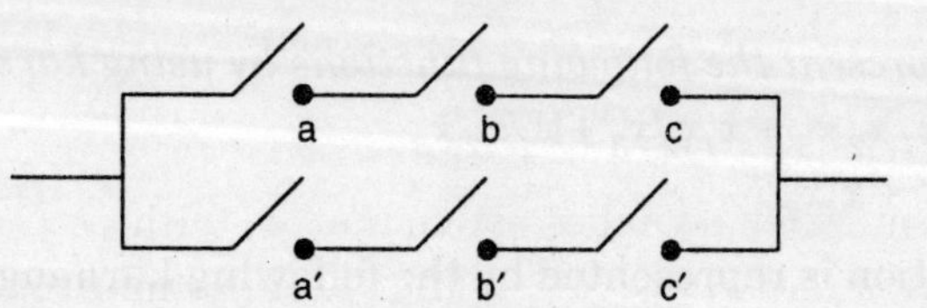

(*ii*) The circuit diagram for the expression is represented as follows :

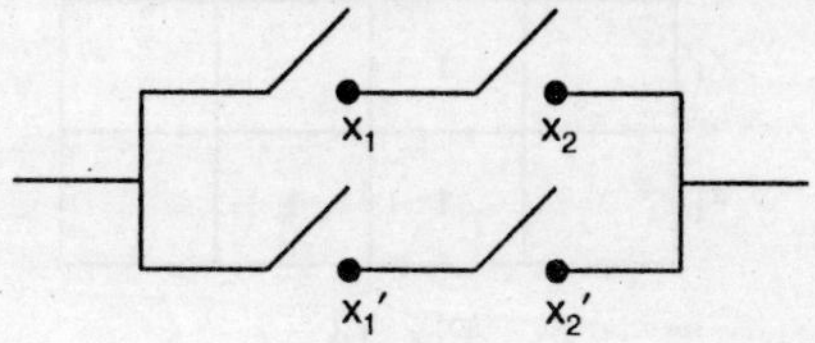

KARNAUGH MAPS

A karnaugh map is a planar area subdivided into 2^n equal cells each representing a point B_2^n for functions of n variables. Each variable x is used to split the area into two equal halves in a different ways *i.e.,* one for x and other for x'. The cells corresponding to the arguments for which the function has the value 1 contains 1.

Examples 1. When the number of variables $n = 1$, the karnaugh map is like as shown in Fig. 11.

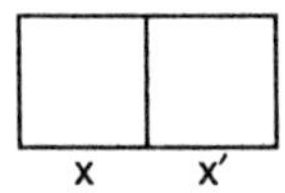

Fig. 11

2. When the number of variables $n = 2$, the karnaugh map is like as shown in Fig. 12.

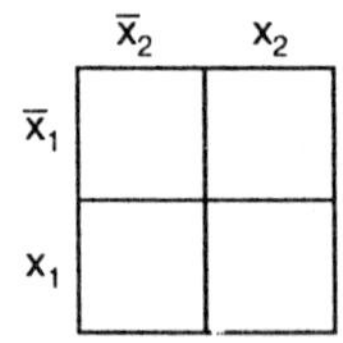

Fig. 12

3. When the number of variables $n = 3$, the karnaugh map is like as shown in Fig. 13.

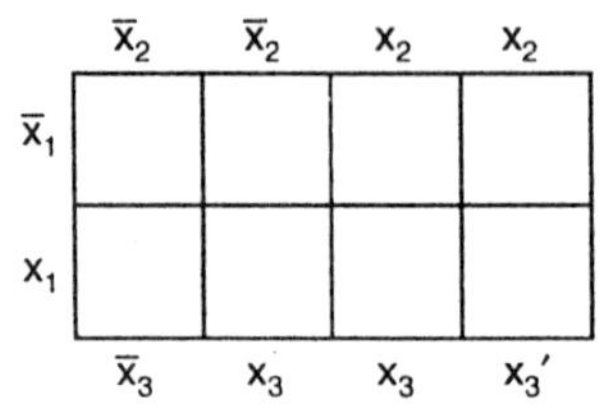

Fig. 13

Example 16. *Represent the following functions by using karnaugh maps :*

(*a*) $f(x_1, x_2, x_3) = x_1 x_2 x_3 + x_1x_2'x_3 + x_1'x_2'x_3$

(*b*) $f(x_1, x_2) = x_1x_2' + x_1x_2$.

Sol. (*a*) The function is represented by the following karnaugh map (Fig. 14).

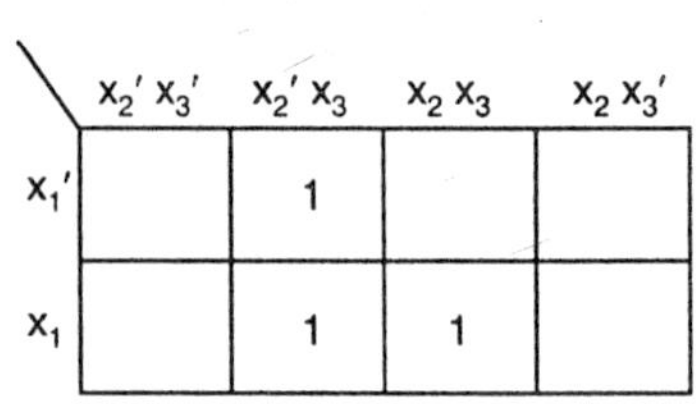

Fig. 14

(*b*) The function is represented by following karnaugh map shown in Fig. 15.

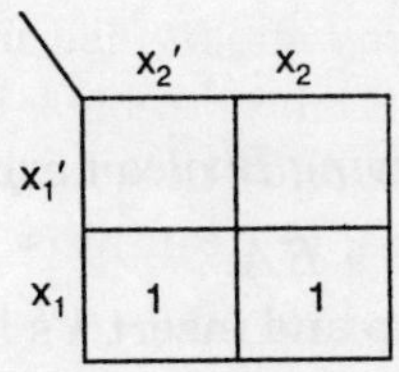

Fig. 15

Note. It is widely used for representation and simplification of Boolean functions. It is the most efficient method for simplification of functions having upto 5 variables.

SIMPLIFICATION OF BOOLEAN FUNCTIONS USING K-MAP

Boolean functions can be simplified with *k*-map. It is based on the principle of combining terms in adjacent cells. Two cells are said to be adjacent if they differ in only one variable. In adjacent cells one of the variable is same, whereas the other variable appears in uncomplemented form in one and in the complemented form in the other cell.

MINIMISATION OF SOP FORM

The following algorithm can be used by which minimized expression can be obtained :

1. Identify the ones which cannot be combined with any other ones encircle them.

2. Identify the ones that can be combined in groups of two in only one way and encircle them as groups.

3. Identify the ones that can be combined with three other ones, to make a group of four adjacent ones, in only one way and encircle them as groups.

4. Identify the ones that can be combined with seven other ones, to make a group of eight adjacent ones, in only way and encircle them as groups.

5. After identifying the essential groups of 2, 4 and 8 ones, if there still remains some ones which have not been encircled then these are to be combined with each other or with other already encircled ones *i.e.*, we should combine the left over ones in largest possible groups and in as few groupings as possible.

Example 17. *Minimize the following Boolean expression using k-map :*

$$f(A, B) = \overline{A}B + BA.$$

Sol. First of all draw the 2-variable *k*-map and insert 1's in the corresponding cells as shown in Fig. 16.

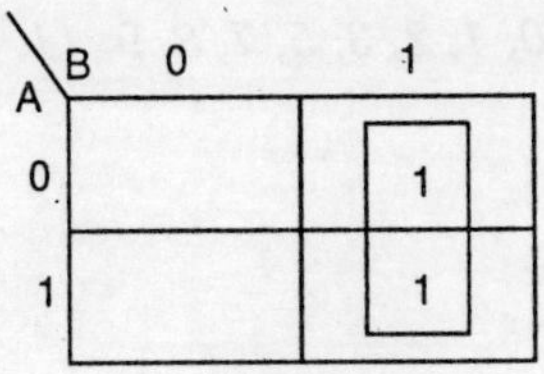

Fig. 16

Now, group the 1's as described earlier. Then the required minimized Boolean expression is

$$f = B.$$

Example 18. *Minimize the following Boolean expression using k-map :*

$$AB + \overline{A}B + B\overline{A}.$$

Sol. Draw the two variable *k*-map and insert 1's in the corresponding cells as shown in Fig. 17.

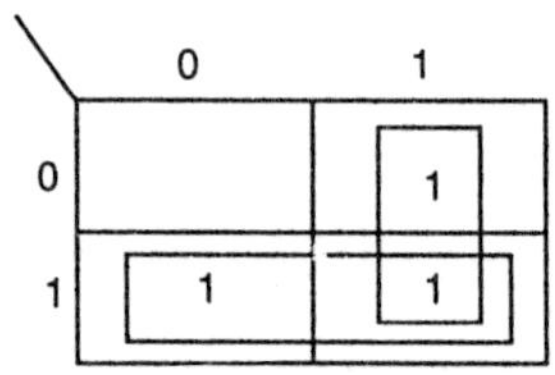

Fig. 17

Now, group the 1's as described earlier. Then the required minimized Boolean expression is

$$f = A + B.$$

Example 19. *Minimize following Boolean expression using k-map :*

$$f(A, B, C) = A\overline{B}C + \overline{A}BC + AB + \overline{A}\,\overline{B}C.$$

Sol. Draw the 3-variable *k*-map and insert 1's in the corresponding cells as shown in Fig. 18.

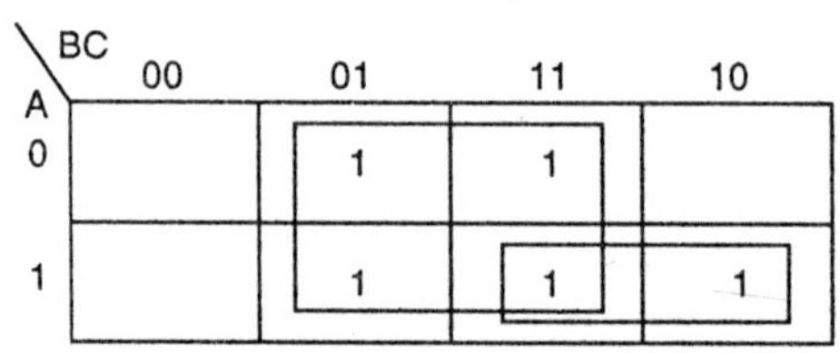

Fig. 18

Now, group the 1's as described earlier. Then the required minimized Boolean expression is

$$f = AB + \overline{C}.$$

Example 20. *Minimize the four-variable logic function using k-map :*

$$f(A, B, C, D) = \Sigma\,(0, 1, 2, 3, 5, 7, 8, 9, 11, 14).$$

Sol.

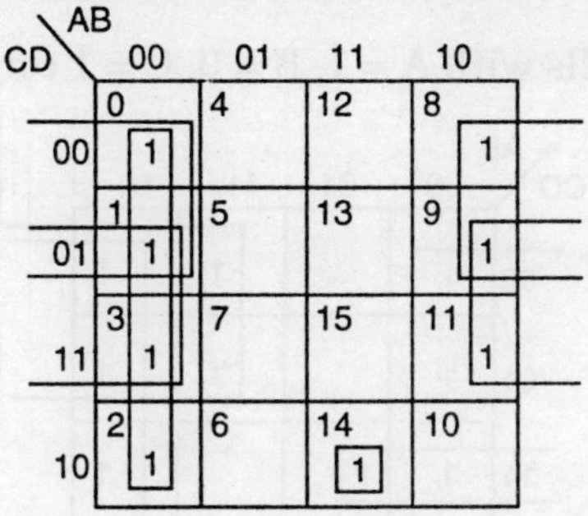

Fig. 19

(*a*) Encircle 1 in cell 14 which cannot be combined with any other 1. The term for this is $ABC\overline{D}$.

(*b*) There is only one possible group of four adjacent ones involving each of the cells 8, 11, 5 or 7 and 2, and these are (8, 9, 0, 1), (11, 9, 1, 3), (5, 7, 3, 1) and (2, 3, 10) respectively. Encircle these groups. The terms corresponding to these groups are $\overline{B}\,\overline{C}$, $\overline{B}\,\overline{D}$. $\overline{A}D$ and $\overline{A}\,\overline{B}$ respectively. Since all the ones have been encircled, therefore, the minimized equation is

$$f(A, B, C, D) = ABC\overline{D} + \overline{B}\,\overline{C} + \overline{B}\,\overline{D} + \overline{A}D + \overline{A}\,\overline{B}.$$

MINIMISATION OF BOOLEAN FUNCTIONS NOT IN MINTERMS/MAXTERMS

One way to minimize such functions is to convert them into standard forms *i.e.*,SOP or POS, then make the *k*-map and obtain the minimized function.

Another way is to directly prepare the *k*-map using the following algorithm :

1. Enter ones for minterms and zeros for maxterms.

2. Enter a pair of ones/zeros for each of the terms with one variable less than the total number of variables.

3. Enter a four adjacent ones/zeros for terms with two variables less than the total number of variables.

4. Repeat for other terms in similar way.

After preparing the *k*-map, the minimization procedure is same.

Example 21. *Minimize the four variable logic function*

$$f(A, B, C, D) = AB\overline{C}D + \overline{A}BCD + \overline{A}\,\overline{B}\,\overline{C} + \overline{A}\,\overline{B}\,\overline{D} + A\overline{C} + A\overline{B}C + \overline{B}.$$

Sol. The *k*-map is obtained by following way

(*a*) Enter 1 in the cell with A = 1, B = 1, C = 0, D = 1 corresponding to the minterm $AB\overline{C}D$.

(*b*) Enter 1 in the cell with A = 0, B = 1, C = 1, D = 1 corresponding to the minterm $\overline{A}BCD$.

(*c*) Enter 1's in the two cells with A = 0, B = 0, C = 0 corresponding to the term $\overline{A}\,\overline{B}\,\overline{C}$.

(*d*) Enter 1's in the two cells with A = 0, B = 0, D = 0 corresponding to the term $\overline{A}\,\overline{B}\,\overline{D}$.

(*e*) Enter 1's in the two cells with A = 1, B = 0, C = 1 corresponding to the term $A\overline{B}C$.

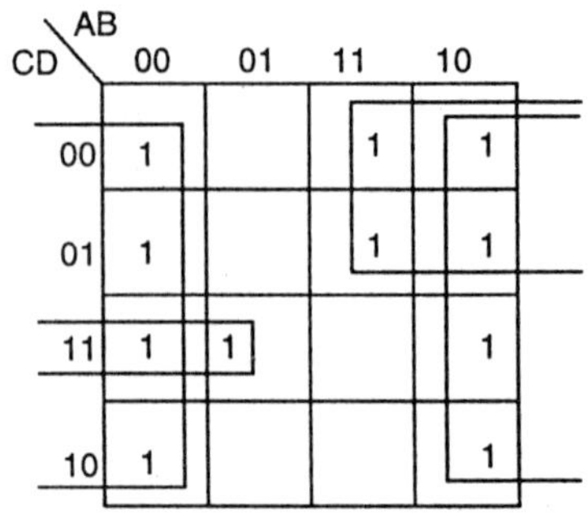

Fig. 20

(*f*) Enter 1's in the four cells with A = 1, C = 0 corresponding to the term $A\overline{C}$.

(*g*) Enter 1's in the eight cells with B = 0 corresponding to the term $\overline{B}$.

The minimized expression is $\overline{B} + A\overline{C} + \overline{A}CD$.

Example 22. *Given the Boolean expression* $f = ABC + B\overline{C}D + \overline{A}BC$.

(*i*) *Make a truth table* (*ii*) *Simplify using k-map*

(*iii*) *Make the switching circuit of the expression.*

Sol. (*i*) The truth table of the Boolean expression is shown in Fig. 21.

A	B	C	D	$ABC + B\overline{C}D + \overline{A}CD$
0	0	0	0	0
0	0	0	1	0
0	0	1	0	0
0	0	1	1	0
0	1	0	0	0
0	1	0	1	1
0	1	1	0	1
0	1	1	1	1
1	0	0	0	0
1	0	0	1	0
1	0	1	0	0
1	0	1	1	0
1	1	0	0	0
1	1	0	1	1
1	1	1	0	1
1	1	1	1	1

Fig. 21

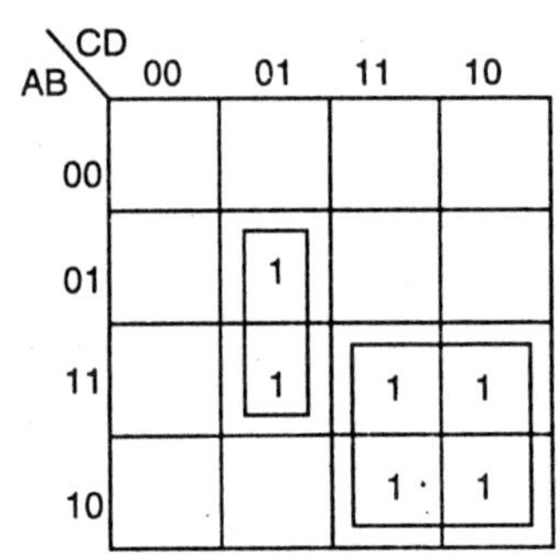

Fig. 22

(*ii*) Draw the 4-variable *k*-map and insert 1's in the corresponding cells as shown in Fig. 22. Now, group the 1's. The required simplified Boolean expression is as follows :

$$f = AC + BC\overline{D}.$$

(*iii*) The switching circuit of the Boolean expression is shown in Fig. 23.

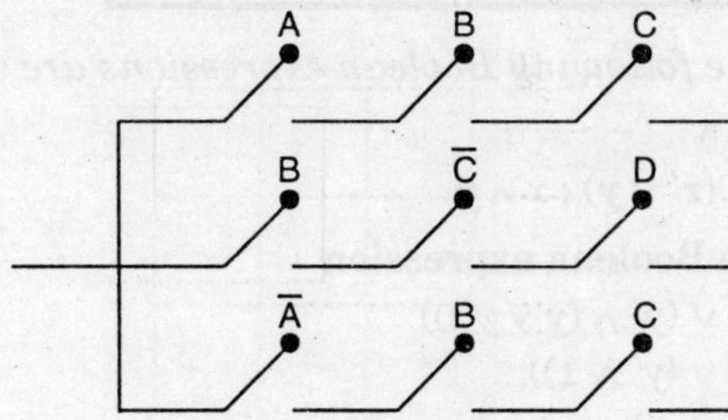

Fig. 23

EXERCISES

1. Using karnaugh maps minimize the following Boolean expressions :

   (*i*) $xyz + xyz' + x'yz' + x'y'z'$

   (*ii*) $yz' + xz'u + xyzu + x'y'z + x'y'u + x'u' + xy'z'u' + xyzu'$

2. Construct a minimal circuit for the following Boolean expression :

   $yz'(x + u') + yz'(x + u') + x'yz.$

3. For the logic circuit whose diagram is shown in figure. Write the Boolean expression.

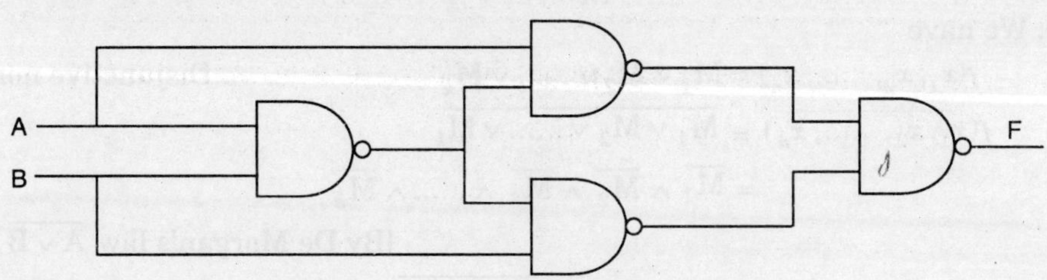

4. Represent the following expressions as switching circuits

   (*i*) $(xy')x$ (*ii*) $(x_1 + x_2)\, x_3 + x_4$

   (*iii*) $((a + b) . (a' + b') . (a' + b'))$.

5. Draw the logic diagram to represent the Boolean expressions

   (*i*) $(xy)' + xyz$ (*ii*) $(a_1 + a_2)\, a_3 + a_4$

   (*iii*) $(a' + b'c)'$ (*iv*) $((a + b')c)'$.

6. Use Nand gates to construct circuits with following outputs.

   (*i*) xy (*ii*) $x + y$

7. Draw Karnaugh map and simplify the following Boolean expressions :

   (*i*) $F(x, y, z) = \Sigma(0, 1, 3, 4, 5)$

   (*ii*) $F(A, B, C, D) = \Sigma(0, 2, 6, 8, 10, 12, 14, 15)$.

8. Show that $A \oplus B = ((A . B)')' . ((A' . B))'$ and hence design a logic circuit of XOR gate using NAND only.

9. Find the logic circuit corresponding to the Boolean expression

$$a'b'c + a'bc + ab'$$

SOLVED PROBLEMS

Problem 1. *Show that the following Boolean expressions are equivalent :*

(*i*) $x \wedge (y \vee (y' \wedge (y \vee y')))$; x

(*ii*) $(z' \vee x) \wedge ((x \wedge y) \vee z) \wedge (z' \vee y)$; $x \wedge y$.

Sol. (*i*) Take L.H.S. of the Boolean expression

$$= x \wedge (y \vee (y' \wedge (y \vee y')))$$
$$= x \wedge (y \vee (y' \wedge 1)) \qquad \{\because \; y \vee y' = 1\}$$
$$= x \wedge (y \vee y') \qquad \{\because \; y' \wedge 1 = y'\}$$
$$= x \wedge 1 \qquad \{\because \; y \vee y' = 1\}$$
$$= x \qquad \{\because \; x \wedge 1 = x\}$$

Hence, the Boolean expressions are equivalent.

(*ii*) Take the L.H.S. of the Boolean expression

$$= (z' \vee x) \wedge ((x \wedge y) \vee z) \wedge (z' \vee y)$$
$$= (z' \vee (x \wedge y)) \wedge ((x \wedge y) \vee z) \qquad \{\because \; x \wedge x = x, z \wedge z' = 0\}$$
$$= (x \wedge y) \vee (z' \wedge z) = (x \wedge y) \vee 0 \qquad \{\because \; z' \wedge z = 0\}$$
$$= x \wedge y.$$

Hence, the Boolean expressions are equivalent.

Problem 2. *Show that if* $M_1 \vee M_2 \vee \ldots\ldots \vee M_k$ *is the disjunctive normal form of* $f(x_1, x_2, \ldots\ldots, x_n)$*, then* $\overline{M_1} \wedge \overline{M_2} \wedge \ldots\ldots \wedge \overline{M_k}$ *is the conjunctive normal form of* $f\overline{(x_1, x_2, \ldots\ldots, x_n)}$.

Sol. We have

$$f(x_1, x_2, \ldots\ldots, x_n) = M_1 \vee M_2 \vee \ldots\ldots \vee M_k \qquad \text{...Disjunctive normal form}$$
$$f\overline{(x_1, x_2, \ldots\ldots, x_n)} = \overline{M_1 \vee M_2 \vee \ldots\ldots \vee M_k}$$
$$= \overline{M_1} \wedge \overline{M_2} \wedge \overline{M_3} \wedge \ldots\ldots \wedge \overline{M_k}$$

{By De Morgan's law $\overline{A \vee B} = \overline{A} \wedge \overline{B}$}

which is the conjunctive normal form of $f\overline{(x_1, x_2, \ldots\ldots, x_n)}$

Problem 3. *Show that the following switching circuits are equivalent (Fig. 24).*

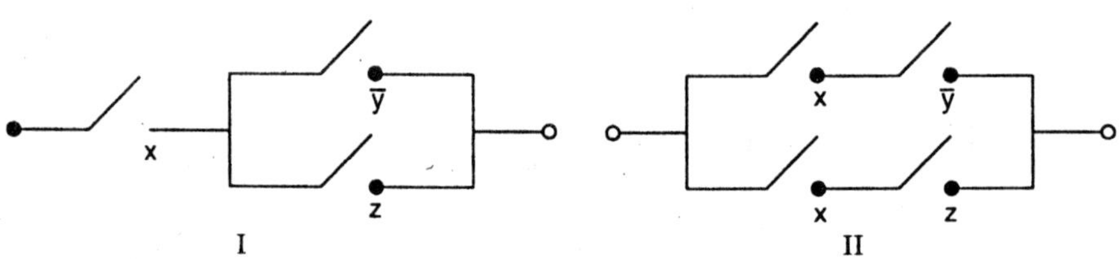

Fig. 24

Sol. Two switching circuits are equivalent if the Boolean expressions that represent them are equal.

The Boolean expression of both the switching circuits are as follows :

$$x \wedge (\bar{y} \vee z) \qquad \text{...(i)}$$

and

$$(x \wedge \bar{y}) \vee (x \wedge z) \qquad \text{...(ii)}$$

Now take the Boolean expression (*ii*), we have

$$= (x \wedge \bar{y}) \vee (x \wedge z) = x \wedge (\bar{y} \vee z) \quad \text{\{Distributive law\}}$$

which is equal to the Boolean expression (*i*). Hence both the switching circuits are equivalent.

Problem 4. *In any Boolean algebra B prove that*

$$a' + b = 1 \iff a + b = b \quad \forall \quad a, b \in B.$$

Sol. Let us assume that $a + b = b$

then $a' + (a + b) = (a' + a) + b$ {By associative law}

$= (a + a') + b$ {By commutative law}

$= 1 + b$ $\{\because \; a + a' = 1\}$

$= 1$ $\{\because \; 1 + b = 1\}$

$\therefore \quad a' + b = 1$

Conversely, let us assume that $a' + b = 1$

then $a + b = (a + b) \,.\, 1$

$= (a + b) \,.\, (a' + b)$ $\{\because \; a' + b = 1\}$

$= (b + a) \,.\, (b + a')$ {By commutative law}

$= b + aa'$

$= b + 0 \;\; = b$

Hence $a' + b = 1 \iff a + b = b.$

Problem 5. *Show that for any Boolean algebra B,*

$$(a + b) \,.\, (b + c) \,.\, (c + a) = a \,.\, b + b \,.\, c + c \,.\, a.$$

Sol. Take the L.H.S., $(a + b) \,.\, (b + c) \,.\, (c + a)$

$= ((a + b)\,(a + c))\,(b + c)$

$= (a + b \,.\, c)\,(b + c)$

$= (a + b \,.\, c)\, b + (a + b \,.\, c)\, c$

$= b(a + b \,.\, c) + c\,(a + b \,.\, c)$

$= b \,.\, a + b \,.\, b \,.\, c + c \,.\, a + c \,.\, b \,.\, c$

$= a \,.\, b + (bb + bb')\, c + ca + ccb + cc'b$ $\{\because \; bb' = 0, cc' = 0\}$

$= a\, b + b\,(b + b')c + ca + c(c + c')\, b$ $\{\because \; b + b' = c + c' = 1\}$

$= ab + bc + ca + cb$

$= ab + bc + ca =$ R.H.S.

Problem 6. *For a Boolean algebra B, show that*

$$a + (a'c + b) = (a + a'c) + b \quad \forall\, a, b, c \in B.$$

Sol. Take L.H.S., then we have

$a + (a'c + b) = (a + a'c) + b$ {By associativity}

$= (a + a')\,(a + c) + b$

$= 1\,(a + c) + b$

$= a + c + b$

$= a + b + c$ ...(*i*)

Similarly, By taking R.H.S. we can show that

$(a + a'c) + b = a + b + c$ ...(*ii*)

From (*i*) and (*ii*)

$$a + (a'c + b) = (a + a'c) + b.$$

Problem 7. *Show that for a Boolean algebra B,*

$$a + b + c = b(a + c) \quad \forall\, a, b, c \in B.$$

Sol. Let us take the L.H.S., then we have

$$\begin{aligned} a + b \,.\, c &= (a + b)\,(a + c) \\ &= (b + a)\,(a + c) && \{\text{By law of commutativity}\} \\ &= (b + a\,)\,.\,1\,.\,(a + c) \\ &= (b + a)\,(b + b')\,(a + c) \\ &= (b + ab')\,(a + c) \\ &= (b + 0)\,(a + c) && \{ab' = 0, \text{ when } a \le b\} \\ &= b(a + c) = \text{R.H.S.} \end{aligned}$$

Problem 8. *Show whether the following circuits are equivalent or not (Fig. 25).*

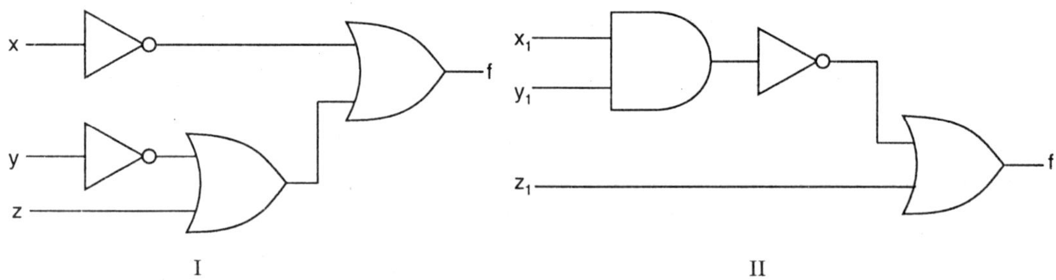

Fig. 25

Sol. Two combinatorial circuits, both having inputs $x_1, x_2, \ldots\ldots, x_n$ and a single output are equivalent if, whenever the circuits receive the same inputs, they produce the same outputs.

The truth table of both the circuits are as shown below :

x	y	z	f
0	0	0	1
0	0	1	1
0	1	0	0
0	1	1	1
1	0	0	1
1	0	1	1
1	1	0	1
1	1	1	1

x_1	y_1	z_1	f
0	0	0	1
0	0	1	1
0	1	0	1
0	1	1	1
1	0	0	1
1	0	1	1
1	1	0	0
1	1	1	1

Since the two combinatorial circuits have not the same logic tables, hence they are not equivalent.

Problem 9. *Write the Boolean expression that represents the following combinatorial circuit, write the logic table and also write the output of each gate symbolically on the figure.*

(i) (ii) (iii)

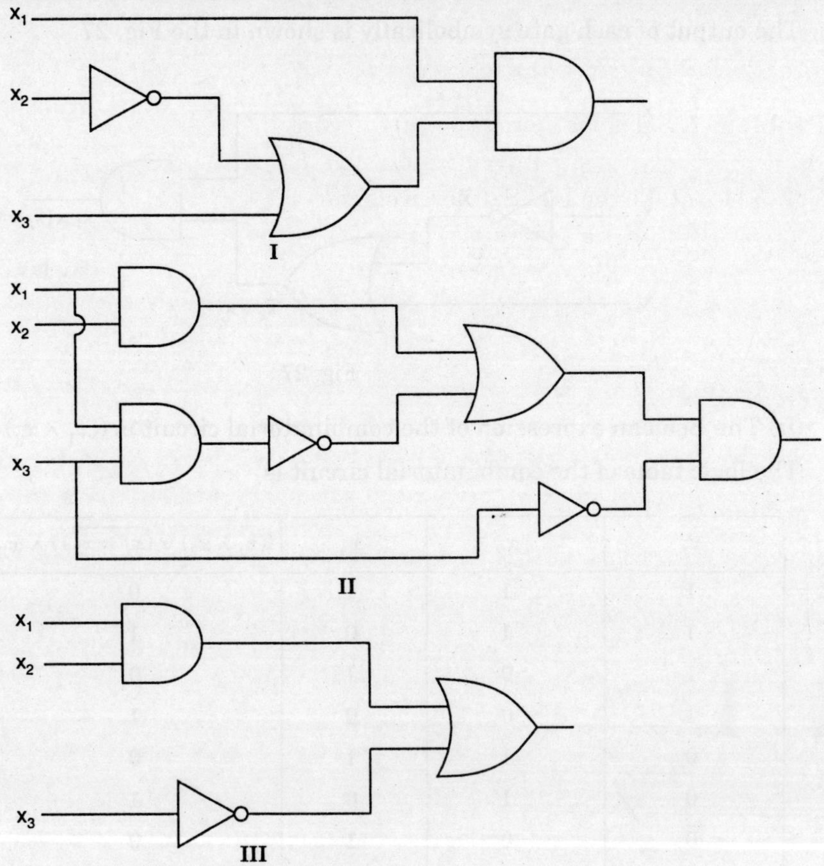

Fig. 26

Sol. (*i*) The Boolean expression of the combinatorial circuit is

$$x_1 \wedge (\bar{x}_2 \vee x_3).$$

The logic table of the combinatorial circuit is

x_1	x_2	x_3	$x_1 \wedge (\bar{x}_2 \vee x_3)$
1	1	1	1
1	1	0	0
1	0	1	1
1	0	0	1
0	1	1	0
0	1	0	0
0	0	1	0
0	0	0	0

The output of each gate symbolically is shown in the Fig. 27.

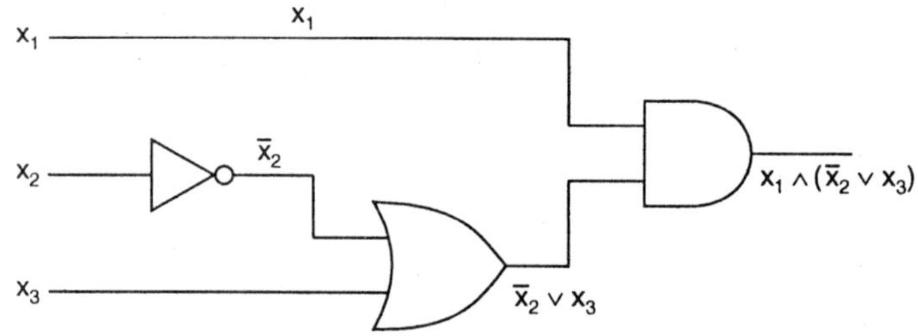

Fig. 27

(*ii*) The Boolean expression of the combinatorial circuit is $((x_1 \wedge x_2) \vee (\overline{x_1 \wedge x_3})) \wedge \bar{x}_3$.

The logic table of the combinatorial circuit is

x_1	x_2	x_3	$((x_1 \wedge x_2) \vee (\overline{x_1 \wedge x_3})) \wedge \bar{x}_3$
1	1	1	0
1	1	0	1
1	0	1	0
1	0	0	1
0	1	1	0
0	1	0	1
0	0	1	0
0	0	0	1

The output of each gate symbolically is shown in Fig. 28.

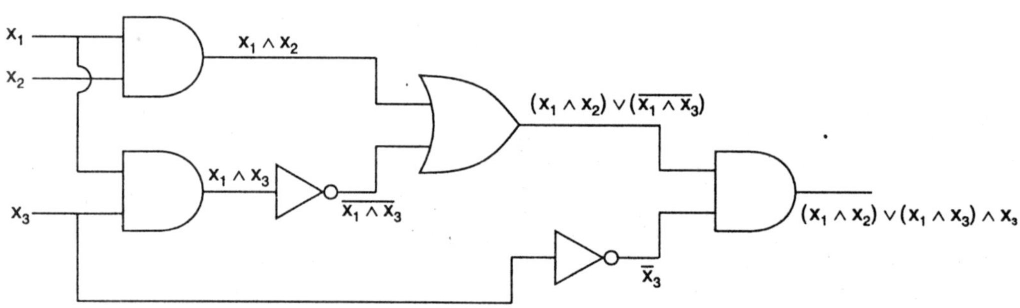

Fig. 28

(*iii*) The Boolean expression of the combinatorial circuit is

$$(x_1 \wedge x_2) \vee \bar{x}_3.$$

The logic table of the combinatorial circuit is

x_1	x_2	x_3	$(x_1 \wedge x_2) \vee \bar{x}_3$
1	1	1	1
1	1	0	1
1	0	1	0
1	0	0	1
0	1	1	0
0	1	0	1
0	0	1	0
0	0	0	1

The output of each gate is shown symbolically in Fig. 29.

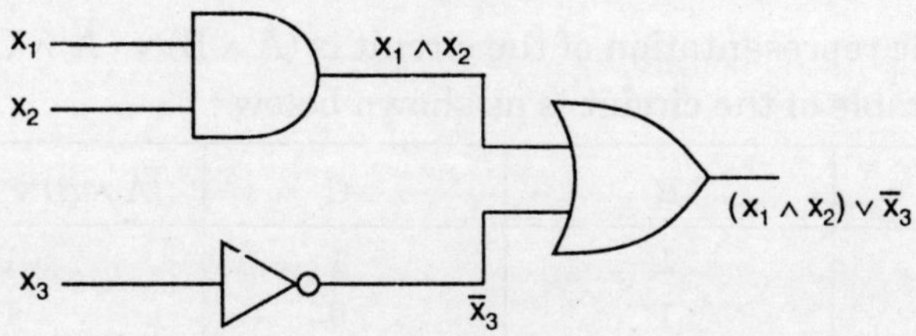

Fig. 29

Problem 10. *Represent the following circuits symbolically and also give its switching table.*

(*i*)

(*ii*)

(*iii*)

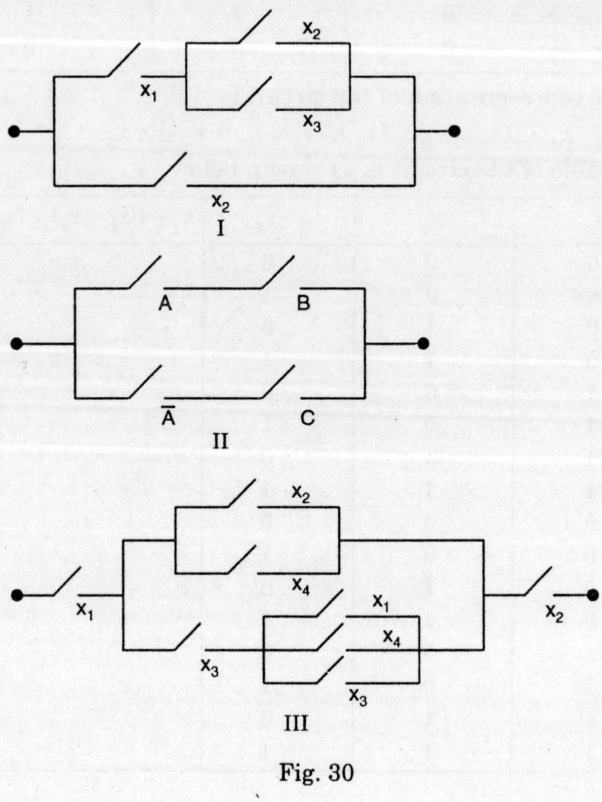

Fig. 30

Sol. (*i*) The symbolic representation of the circuit is $(x_1 \wedge (x_2 \vee x_3)) \vee \bar{x}_2$.

The switching table of the circuit is as shown below :

x_1	x_2	x_3	$(x_1 \wedge (x_2 \vee x_3)) \vee \bar{x}_2$
1	1	1	1
1	1	0	1
1	0	1	1
1	0	0	1
0	1	1	0
0	1	0	0
0	0	1	1
0	0	0	1

(*ii*) The symbolic representation of the circuit is $(A \wedge B) \vee (\bar{A} \wedge C)$.

The switching table of the circuit is as shown below :

A	B	C	$(A \wedge B) \vee (\bar{A} \wedge C)$
1	1	1	1
1	1	0	1
1	0	1	0
1	0	0	0
0	1	1	1
0	1	0	0
0	0	1	1
0	0	0	0

(*iii*) The symbolic representation of the circuit is

$$x_1 \wedge ((x_2 \vee x_4) \vee (x_3 \wedge (x_1 \vee x_4 \vee x_3))) \wedge x_2.$$

The switching table of the circuit is as shown below :

x_1	x_2	x_3	x_4	$x_1 \wedge ((x_2 \vee x_4) \vee (x_3 \wedge (x_1 \vee x_4 \vee x_3))) \wedge x_2$
0	0	0	0	0
0	0	0	1	0
0	0	1	0	0
0	0	1	1	0
0	1	0	0	0
0	1	0	1	0
0	1	1	0	0
0	1	1	1	0
1	0	0	0	0
1	0	0	1	0
1	0	1	0	0
1	0	1	1	0
1	1	0	0	1
1	1	0	1	1
1	1	1	0	1
1	1	1	1	1

Problem 11. *Find the combinatorial circuit corresponding to each of the following Boolean expressions :*

(*i*) $x_1 \wedge (x_2 \vee x_3)$ (*ii*) $(x_1 \wedge \bar{x}_2) \vee (x_1 \wedge \bar{x}_3)$

(*iii*) $\overline{(A \wedge B)}$ (*iv*) $(x_1 \wedge ((x_2 \wedge \bar{x}_3) \vee (\bar{x}_2 \wedge x_3))) \vee (\bar{x}_1 \wedge x_2 \wedge x_3)$.

Sol. (*i*) The combinatorial circuit of the Boolean expression is shown in Fig. 31.

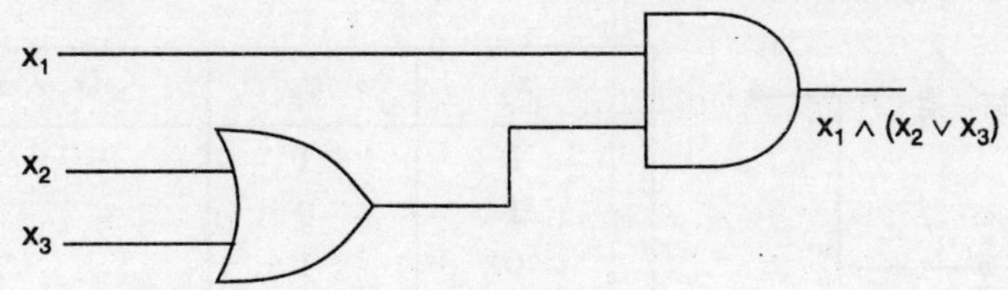

Fig. 31

(*ii*) The combinatorial circuit of the Boolean expression is shown in Fig. 32.

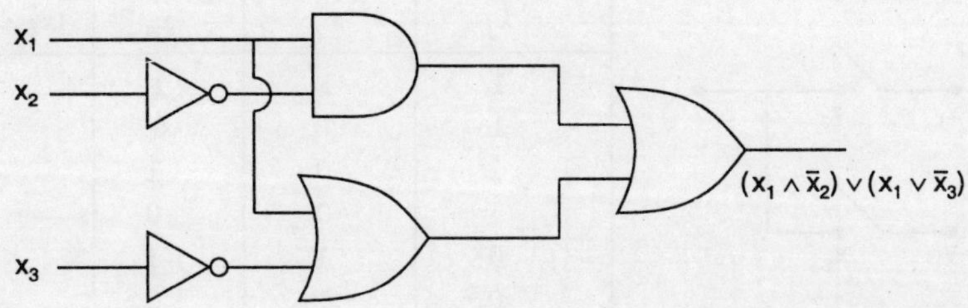

Fig. 32

(*iii*) The combinatorial circuit of the Boolean expression is shown in Fig. 33.

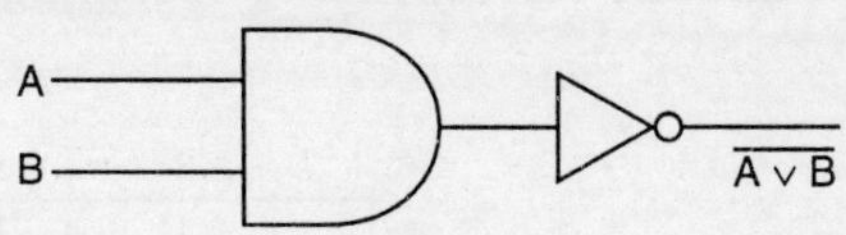

Fig. 33

(*iv*) The combinatorial circuit of the Boolean expression is shown in Fig. 34.

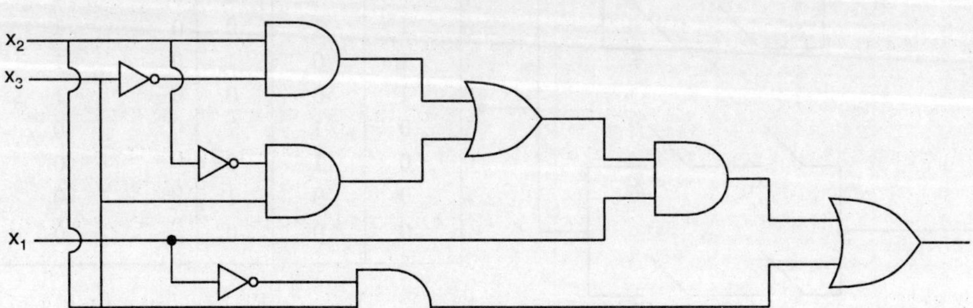

Fig. 34

Problem 12. *Represent the following expressions as switching circuits and also write the switching tables :*

(*i*) $(x_1 \wedge \bar{x}_2) \vee x_1$ (*ii*) $(x_1 \wedge x_3) \vee (\bar{x}_1 \wedge x_2)$

(*iii*) $(x_1 \wedge (x_2 \vee (x_1 \wedge \bar{x}_2))) \vee ((x_1 \wedge \bar{x}_2) \vee (x_1 \wedge \bar{x}_3))$.

Sol. (*i*) The switching circuit and the corresponding switching table of the Boolean expression is shown in Fig. 35.

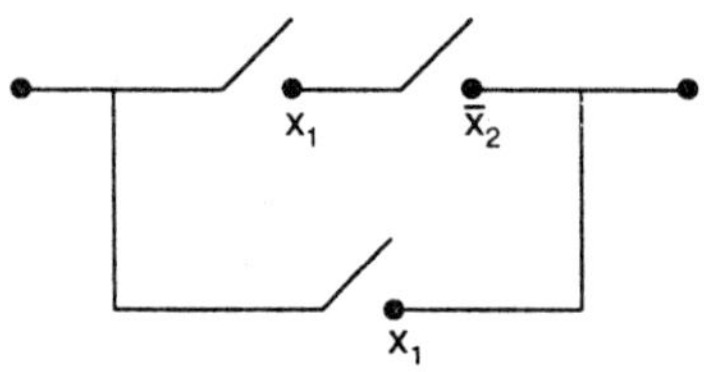

Fig. 35

x_1	x_2	$(x_1 \wedge \bar{x}_2) \vee x_1$
1	1	1
1	0	1
0	1	0
0	0	0

(*ii*) The switching circuit and the corresponding switching table of the Boolean expression is shown in Fig. 36.

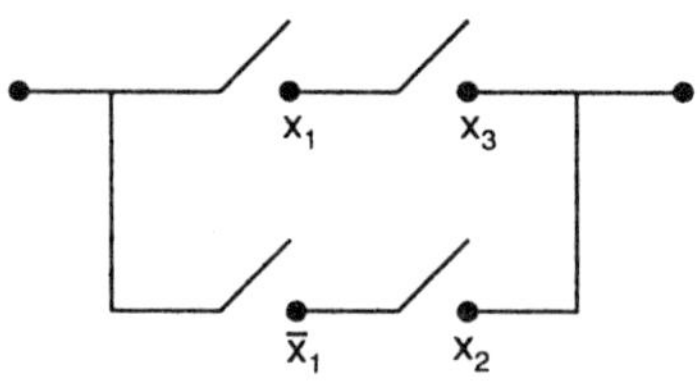

Fig. 36

x_1	x_2	x_3	$(x_1 \wedge x_3) \vee (\bar{x}_1 \wedge x_2)$
1	1	1	1
1	1	0	0
1	0	1	1
1	0	0	0
0	1	1	1
0	1	0	1
0	0	1	0
0	0	0	0

(*iii*) The switching circuit and the corresponding switching table of the Boolean expression is shown in Fig. 37.

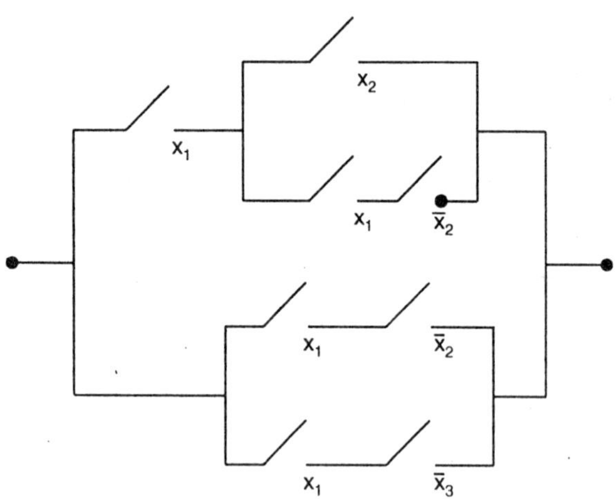

Fig. 37

x_1	x_2	x_3	$(x_1 \wedge (x_2 \vee (x_1 \wedge \bar{x}_2)))$ $\vee ((x_1 \wedge \bar{x}_2) \vee (x_1 \wedge \bar{x}_3))$
1	1	1	1
1	1	0	1
1	0	1	1
1	0	0	1
0	1	1	0
0	1	0	0
0	0	1	0
0	0	0	0

Problem 13. *Find the disjunctive normal form and also the corresponding combinatorial circuit of the following Boolean functions :*

(*i*)

A	*B*	*C*	*f(A, B, C)*
0	0	0	1
0	0	1	1
0	1	0	1
0	1	1	0
1	0	0	0
1	0	1	1
1	1	0	1
1	1	1	1

(*ii*)

A	*B*	*C*	*f(A, B, C)*
0	0	0	0
0	0	1	1
0	1	0	0
0	1	1	1
1	0	0	1
1	0	1	0
1	1	0	1
1	1	1	0

Sol. (*i*) The disjunctive normal form of the Boolean function is

$$f = \overline{A}\,\overline{B}\,\overline{C} + \overline{A}\,\overline{B}C + \overline{A}\,B\,\overline{C} + A\,\overline{B}\,C + A\,B\,\overline{C} + A\,B\,C$$

and the corresponding combinatorial circuit is shown in Fig. 38.

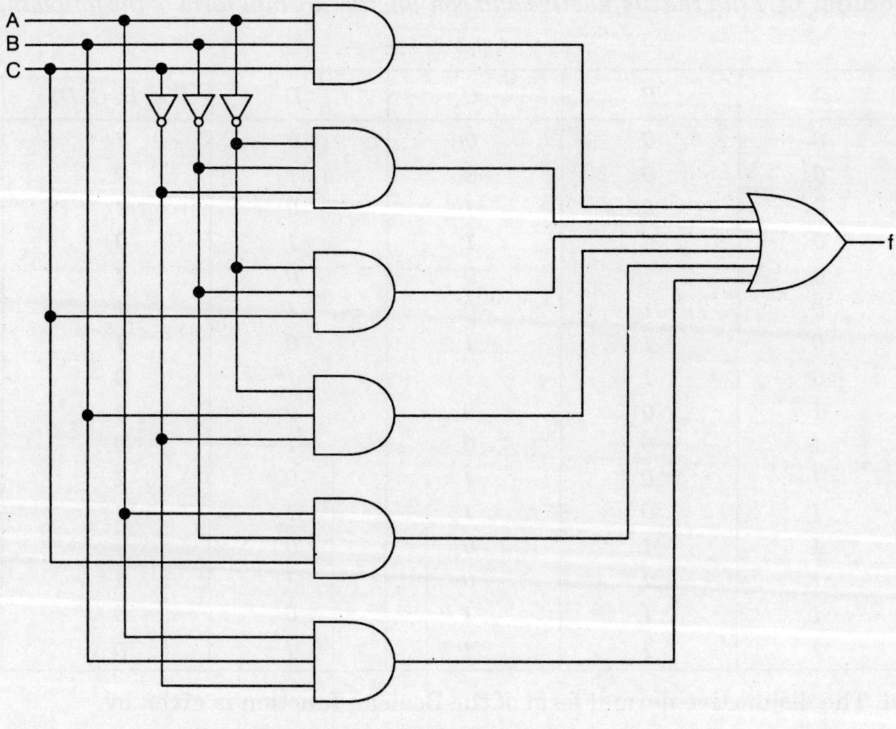

Fig. 38

(*ii*) The disjunctive normal form of the Boolean function is

$$f = \overline{A}\,\overline{B}\,\overline{C} + \overline{A}\,B\,C + A\overline{B}\,\overline{C} + A\,B\,\overline{C}$$

and the corresponding combinatorial circuit is shown in Fig. 39.

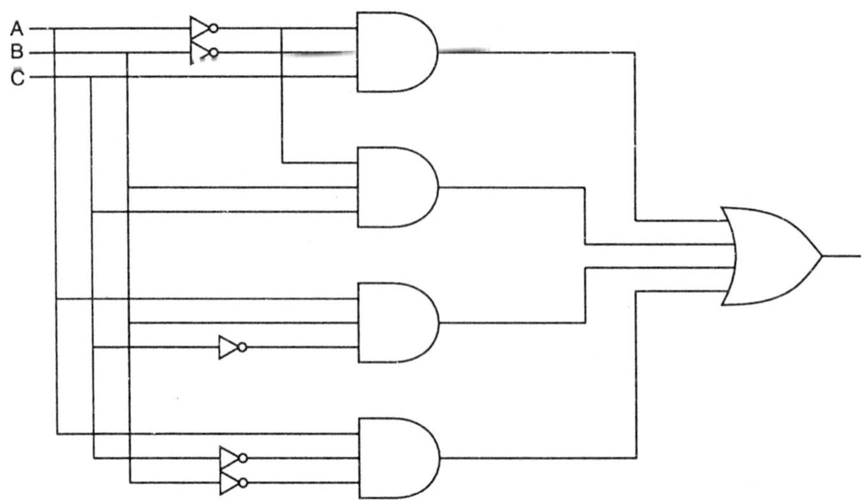

Fig. 39

Problem 14. *Find the disjunctive and conjunctive normal form of the following Boolean function :*

A	*B*	*C*	*D*	*F(A, B, C, D)*
0	*0*	*0*	*0*	*1*
0	*0*	*0*	*1*	*0*
0	*0*	*1*	*0*	*1*
0	*0*	*1*	*1*	*0*
0	*1*	*0*	*0*	*1*
0	*1*	*0*	*1*	*1*
0	*1*	*1*	*0*	*1*
0	*1*	*1*	*1*	*0*
1	*0*	*0*	*0*	*1*
1	*0*	*0*	*1*	*0*
1	*0*	*1*	*0*	*1*
1	*0*	*1*	*1*	*1*
1	*1*	*0*	*0*	*1*
1	*1*	*0*	*1*	*1*
1	*1*	*1*	*0*	*0*
1	*1*	*1*	*1*	*0*

Sol. The disjunctive normal form of the Boolean function is given by

$$F = \overline{A}\,\overline{B}\,\overline{C}\,\overline{D} + \overline{A}\,\overline{B}\,C\overline{D} + \overline{A}\,B\,\overline{C}\,\overline{D} + \overline{A}\,B\,\overline{C}\,D + \overline{A}\,B\,C\,\overline{D} + A\,\overline{B}\,\overline{C}\,\overline{D} + A\,\overline{B}\,C\,\overline{D}$$
$$+ A\,\overline{B}\,C\,D + A\,B\,\overline{C}\,\overline{D} + A\,B\,\overline{C}\,\overline{D} + A\,B\,\overline{C}\,D.$$

The conjunctive normal form of the Boolean function is given by

$$F = (A + B + C + \overline{D}) * (A + B + \overline{C} + \overline{D}) * (A + \overline{B} + \overline{C} + \overline{D}) * (\overline{A} + B + C + \overline{D}) *$$
$$(\overline{A} + \overline{B} + \overline{C} + D) * (\overline{A} + \overline{B} + \overline{C} + \overline{D}).$$

Problem 15. *Design a circuit that accepts a 3-bit number and gives an output 0 if input represents even decimal number and gives an output 1 if input represents an odd decimal number.*

Sol. The block diagram of the circuit and the corresponding logic table is shown in Fig. 40.

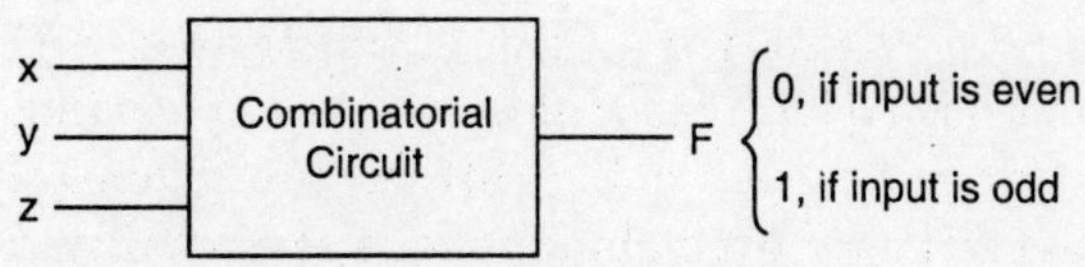

Fig. 40

Decimal Number	x	y	z	F
0	0	0	0	0
1	0	0	1	1
2	0	1	0	0
3	0	1	1	1
4	1	0	0	0
5	1	0	1	1
6	1	1	0	0
7	1	1	1	1

From the logic table we obtained the following Boolean expression :

$$F = \bar{x}\,\bar{y}z + \bar{x}\,yz + x\,\bar{y}\,z + xyz$$

The logic circuit corresponding to the above Boolean expression is shown in Fig. 41.

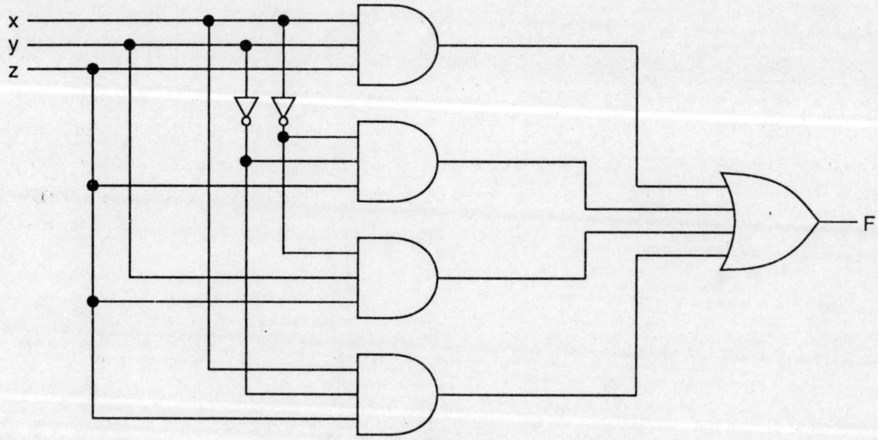

Fig. 41

SECTION IV

Chapter 10

Graphs

BASIC TERMINOLOGY

The graphs consist of points or nodes called vertices which are connected to each other by way of lines called edges. These lines may be directed or undirected.

UNDIRECTED GRAPHS

An undirected graph G consists of a set of vertices, V and a set of edges E. The edge set contains the unordered pair of vertices. If $(u, v) \in E$ then we say u and v are connected by an edge where u and v are vertices in the set V.

Example 1. *Let $V = \{1, 2, 3, 4\}$ and $E = \{(1, 2), (1, 4), (3, 4), (2, 3)\}$. Draw the graph.*

Sol. The graph can be drawn in several ways.

Two of which are as follows (Fig. 1 and Fig. 2).

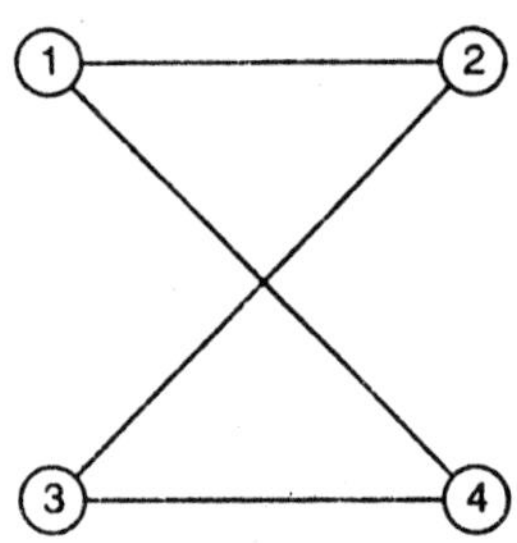

Fig. 1

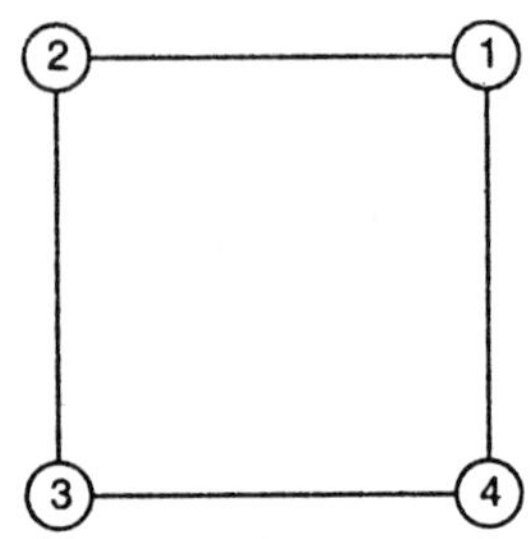

Fig. 2

Example 2. *Consider the graph shown in Fig. 3. Determine the edge set and the vertex set of the graph.*

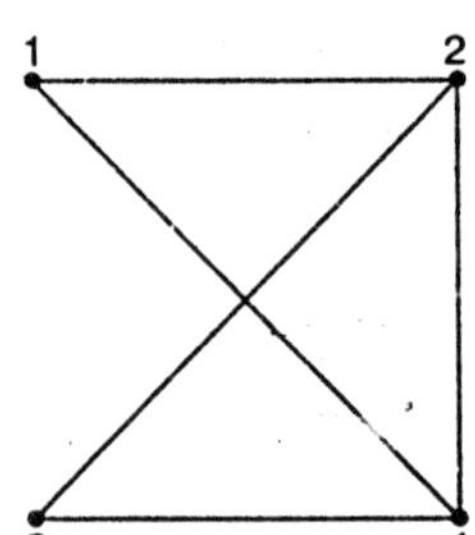

Fig. 3

Sol. The edge set is

$$E = \{(1, 2), (1, 4), (2, 3), (2, 4), (3, 4)\}$$

The vertex set is

$$V = \{1, 2, 3, 4\}.$$

Degree of a Vertex. The degree of a vertex is the number of edges incident on a vertex v. The self-loop is counted twice. The degree of vertex is denoted by $d(v)$.

Example 3. *Consider the graph G shown in Fig. 4. Determine the degree of each vertex of graph.*

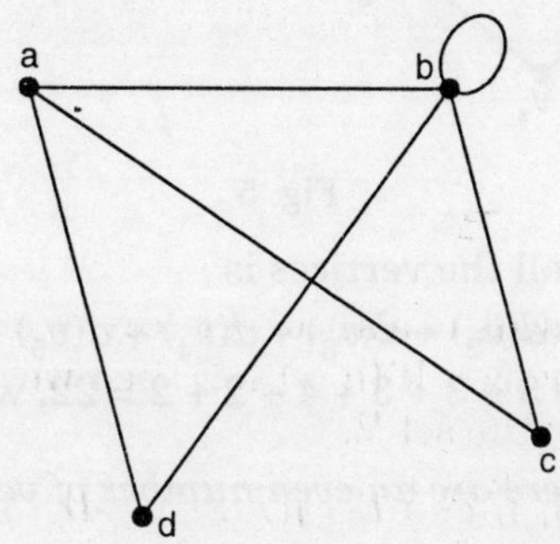

Fig. 4

Sol. The degree of each vertex is as follows :

$$d(a) = 3 \; ; d(b) = 5 \; ; d(c) = 2 \; ; d(d) = 2.$$

Theorem I. *Show that the sum of degree of all the vertices in a graph G, is even.*

Proof. Each edge contribute two degrees in a graph. Also, each edge contributes one degree to each of the vertices on which it is incident.

Hence, if there are N edges in G, then we have

$$2N = d(v_1) + d(v_2) + \ldots\ldots + d(v_N)$$

Thus, 2N is always even.

Theorem II. *Prove that in any graph, there are an even number of vertices of odd degree.*

Proof. Consider a graph having vertices of degree even and odd. Now, make two groups of vertices. One with even degree of vertices $v_1, v_2, \ldots, v_k$ and other with odd degree of vertices $u_1, u_2, \ldots, u_n$.

Suppose,

$$V = d(v_1) + d(v_2) + \ldots + d(v_k)$$

$$U = d(u_1) + d(u_2) + \ldots + d(u_n).$$

Now, we know that sum of degree of all the vertices is even (Theorem I). So, V + U is even.

Since, V is the sum of k even numbers. Hence, it is even. But U is the sum of n odd numbers. So, to be U an even number, n must be even. Hence proved.

Example 4. *Verify that the sum of the degree of all the vertices is even for the graph shown in Fig. 5.*

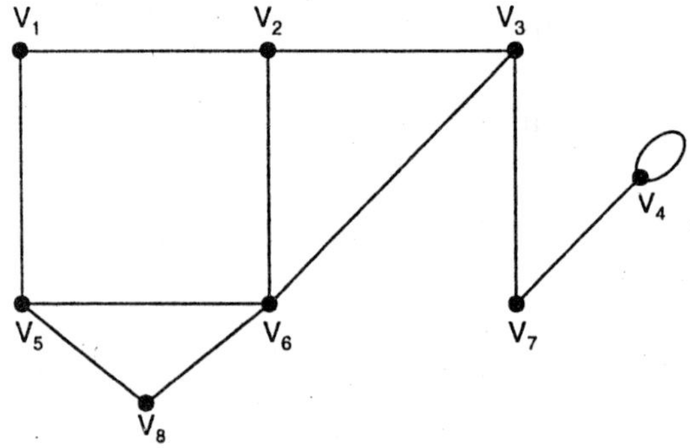

Fig. 5

Sol. The sum of degree of all the vertices is

$$= d(v_1) + d(v_2) + d(v_3) + d(v_4) + d(v_5) + d(v_6) + d(v_7) + d(v_8)$$
$$= 2 + 3 + 3 + 3 + 3 + 4 + 2 + 2 = 22, \text{ which is even.}$$

Example 5. *Verify that there are an even number of vertices of odd degree in the graph shown in Fig. 6.*

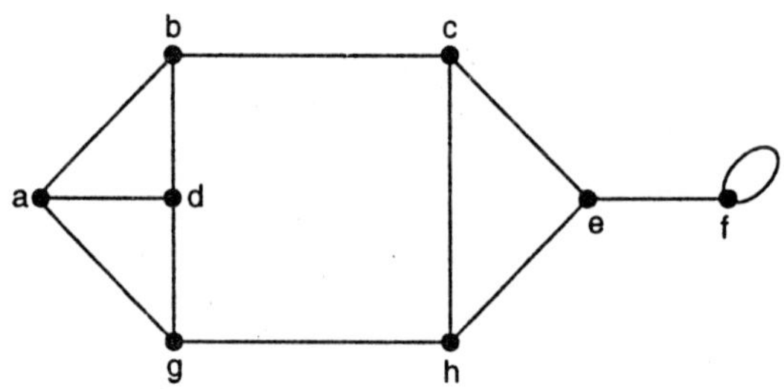

Fig. 6

Sol. The number of vertices of degree odd are 8 and each have degree three in the above graph. Hence, we have even number of vertices of odd degree.

Path. A path of length n is a sequence of $n + 1$ vertices of a graph in which each pair of vertices is an edge of the graph.

1. **A Simple Path.** The path is called simple one if no edge is repeated in the path *i.e.,* all the vertices are distinct except that first vertex equal to last vertex.

2. **An Elementary Path.** The path is called elementary one if no vertex is repeated in the path *i.e.,* all the vertices are distinct.

3. **Circuit or Closed Path.** The circuit or closed path is a path which starts and ends at the same vertex *i.e.,* $v_o = v_n$.

4. **Simple Circuit Path.** The simple circuit is a simple path which is a circuit.

Theorem III. *Suppose a graph G contains two distinct paths from a vertex u to a vertex v. Show that G has a cycle.*

Proof. Consider two distinct paths from u to v be $P_1 = (e_1, e_2, e_3, \ldots\ldots, e_n)$ and $P_2 = (e_1', e_2', e_3', \ldots\ldots, e_n')$.

Now delete from the paths P_1 and P_2 all the initial edges which are identical *i.e.*, of we have $e_1 = e_1', e_2 = e_2', e_3 = e_3', ..., e_k = e_k'$ but $e_{k+1} \neq e'_{k+1}$. We will delete all the first k edges of both the paths P_1 and P_2.

Now, after deleting the k edges both the paths start from the same vertex, (let u_1) and end at v.

Now, to construct a cycle, start from vertex u, and follow the left over path of P_1 until we first meet any vertex of the left over path of P_2.

If this vertex is u_2, then the remaining cycle is computed by following the left over path of P_2 which starts from u_2 and ends at v.

Example 6. *Consider the graph shown in Fig. 7. Give an example of the following :*

(*i*) *A simple path from V_1 to V_6.*

(*ii*) *An elementary path from V_1 to V_6.*

(*iii*) *A simple path which is not elementary from V_1 to V_6.*

(*iv*) *A path which is not simple and starting from V_2.*

(*v*) *A simple circuit starting from V_1.*

(*vi*) *A circuit which is not simple and starting from V_2.*

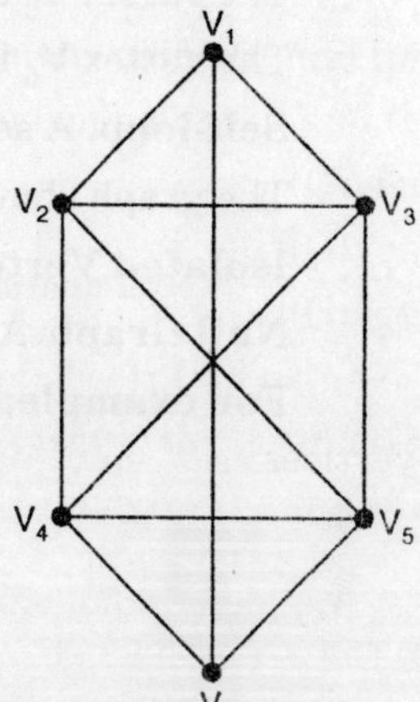

Fig. 7

Sol. (*i*) A simple path from V_1 to V_6 is

$$V_1, V_2, V_3, V_4, V_5, V_6.$$

(*ii*) An elementary path from V_1 to V_6 is

$$V_1, V_2, V_3, V_5, V_4, V_6.$$

(*iii*) A simple path which is not elementary from V_1 to V_6 is

$$V_1, V_2, V_3, V_5, V_2, V_4, V_6.$$

(*iv*) A path which is not simple and starting from V_2 is

$$V_2, V_3, V_4, V_5, V_3, V_4, V_6.$$

(*v*) A simple circuit starting from V_1 is

$$V_1, V_2, V_4, V_6, V_5, V_3, V_1.$$

(*vi*) A circuit which is not simple and starting from V_2 is

$$V_2, V_3, V_1, V_2, V_5, V_4, V_2.$$

Pendent Vertex. A vertex with degree one is called a pendent vertex.

Pendent Edge. The only edge which is incident with a pendent vertex is called the pendent edge.

Odd Vertex. A vertex having degree odd is called an odd vertex.

Even Vertex. A vertex having degree even is called an even vertex.

Incident Edge. An edge is called incident with the vertices is connects.

Adjacent Vertices. Two vertices are called adjacent if they are connected by an edge. If there is an edge (u, v), then we can say vertex u is adjacent to vertex v and vertex v is adjacent to vertex u.

Example 7. *Consider the graph as shown in Fig. 8.*

Determine the following :

(*i*) *Pendent vertices*

(*ii*) *Pendent edges*

(*iii*) *Odd vertices*

(*iv*) *Even vertices*

(*v*) *Incident edges*

(*vi*) *Adjacent vertices.*

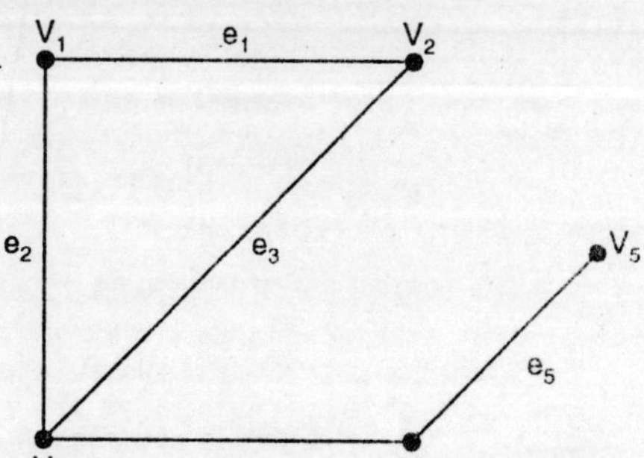

Fig. 8

Sol. (*i*) The vertex V_5 is the pendent vertex.

(*ii*) The edge (V_4, V_5) or e_5 is the pendent edge.

(*iii*) The vertices V_3 and V_5 are odd vertices.

(*iv*) The vertices V_1, V_2 and V_4 are even vertices.

(*v*) The edge e_1 is incident on V_1 and V_2. The edge e_2 is incident on V_1 and V_3.
The edge e_3 is incident on V_2 and V_3. The edge e_4 is incident on V_3 and V_4.
The edge e_5 is incident on V_4 and V_5.

(*vi*) The vertex V_1 is adjacent to V_2 and V_3. The vertex V_2 is adjacent to V_1 and V_3.
The vertex V_3 is adjacent to V_1 and V_4. The vertex V_4 is adjacent to V_3 and V_5.
The vertex V_5 is adjacent to V_4.

Self-loop. A self-loop is an edge *e* if it has a same end point.

The graph shown in Fig. 4 contains the self-loop at vertex *b i.e., e* = (*b, b*).

Isolated Vertex. An isolated vertex is a vertex having no edges incident to it.

Null Graph. A null graph is defined as a graph which consists only the isolated vertices.

For example: The graph shown in Fig. 9 is a null graph and the vertices are isolated vertices.

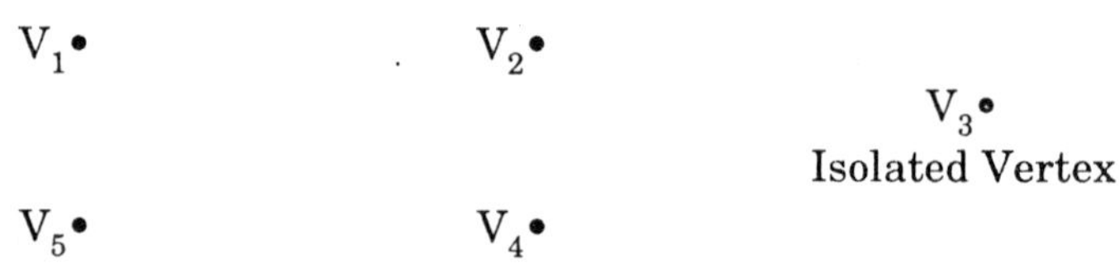

Fig. 9. Null Graph.

Complementary Graph. The complement of a graph G is defined to be a graph which has the same number of vertices as in graph G and has two vertices connected if and only if they are not connected in graph.

Example 8. *Consider the graph G shown in Fig. 10. Find the complement of this graph.*

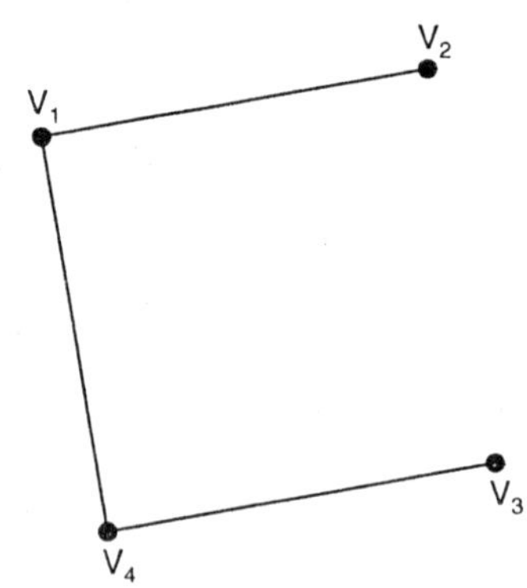

Fig. 10

Sol. The complement of above graph is shown in Fig. 11.

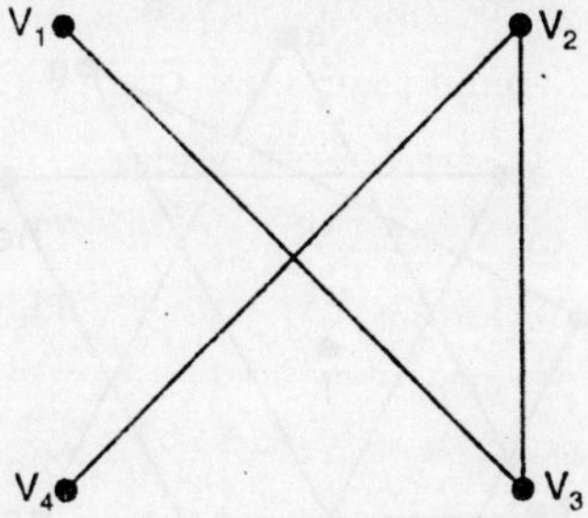

Fig. 11. Complement of Graph G

Undirected Complete Graph. An undirected complete graph G = (V, E) of n vertices is a graph in which each vertex is connected to every other vertex *i.e.,* and edge exists between every pair of distinct vertices. It is denoted by K_n. A complete graph with n vertices will have $n(n-1)/2$ edges.

Example 9. *Draw undirected complete graphs K_4 and K_6.*

Sol. The undirected complete graph of K_4 is shown in Fig. 12 and that of K_6 is shown in Fig. 13.

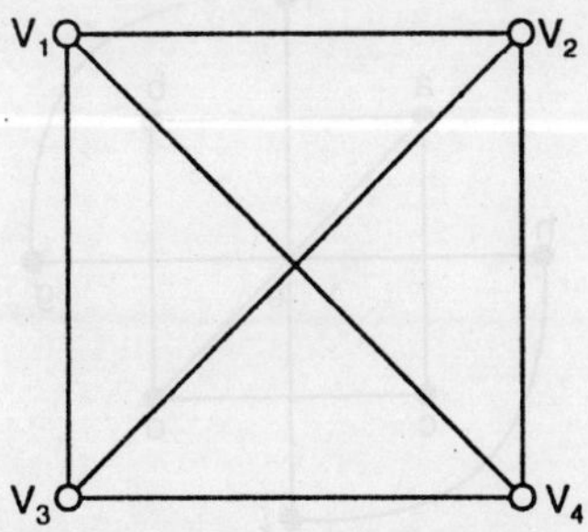

Fig. 12. K_4

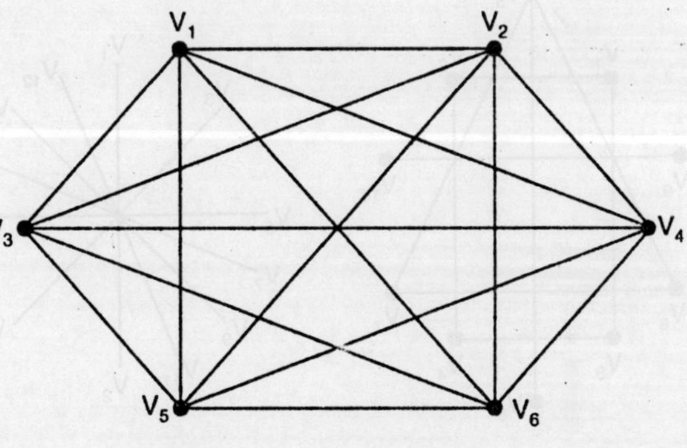

Fig. 13. K_6

Connected Graph. A graph is called connected if there is a path from any vertex u to v or vice-versa.

Disconnected Graph. A graph is called disconnected if there is no path between any two of its vertices.

Connected Component. A subgraph of graph G is called the connected component of G, if it is not contained in any bigger subgraph of G, which is connected. It is defined by listing its vertices.

Example 10. *Consider the graph shown in Fig. 14. Determine its connected components.*

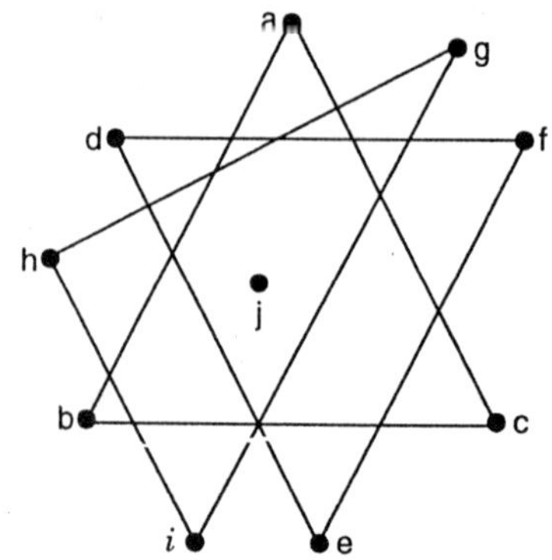

Fig. 14

Sol. The connected components of this graph is {*a, b, c*}, {*d, e, f*}, {*g, h, i*} and {*j*}.

Example 11. *Consider the graphs shown in Figs. 15, 16 and 17. Determine whether the graphs are (a) Connected graphs or (b) Disconnected graphs.*

Also write their connected components.

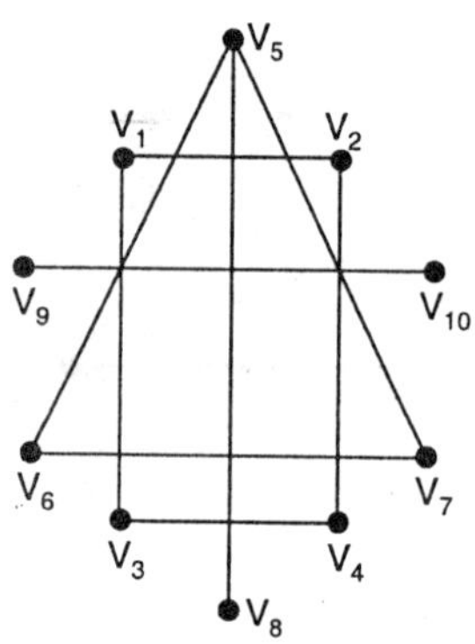

Fig. 15

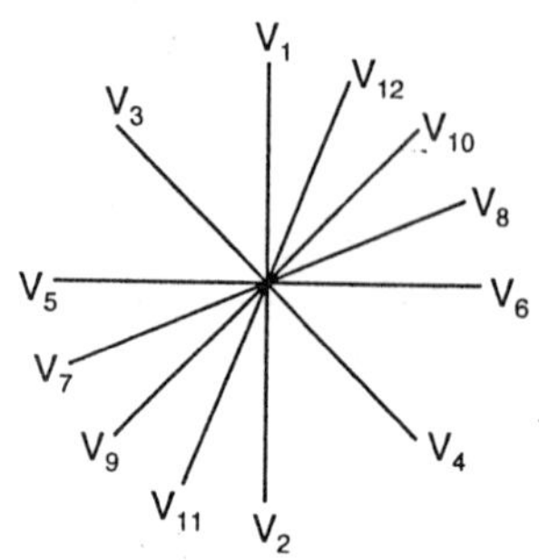

Fig. 16

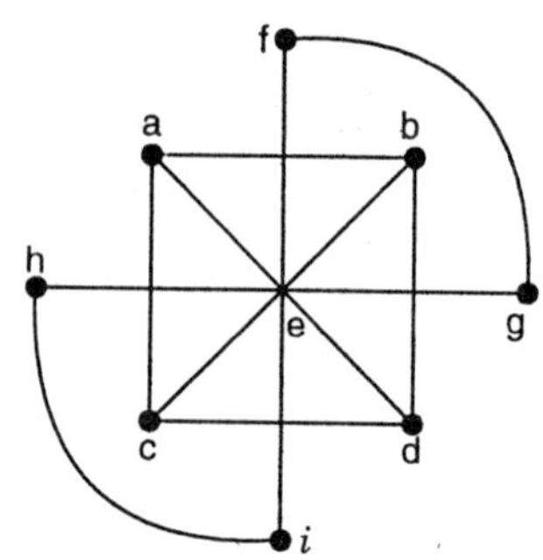

Fig. 17

Sol. (*i*) The graph shown in Fig. 15 is a disconnected graph and its connected components are

$$\{V_1, V_2, V_3, V_4\}, \{V_5, V_6, V_7, V_8\} \text{ and } \{V_9, V_{10}\}.$$

(*ii*) The graph shown in Fig. 16 is a disconnected graph and its connected components are

$$\{V_1, V_2\}, \{V_3, V_4\}, \{V_5, V_6\}, \{V_7, V_8\}, \{V_9, V_{10}\} \text{ and } \{V_{11}, V_{12}\}.$$

(*iii*) The graph shown in Fig. 17 is a connected graph.

Subgraph. A subgraph of a graph G = (V, E) is a graph G′ = (V′, E′) in which V′ ⊆ V and E′ ⊆ E and each edge of G′ has the same end vertices in G′ as in graph G.

Note. A single vertex is a subgraph.

Example 12. *Consider the graph G shown in Fig. 18. Show the different subgraphs of this graph.*

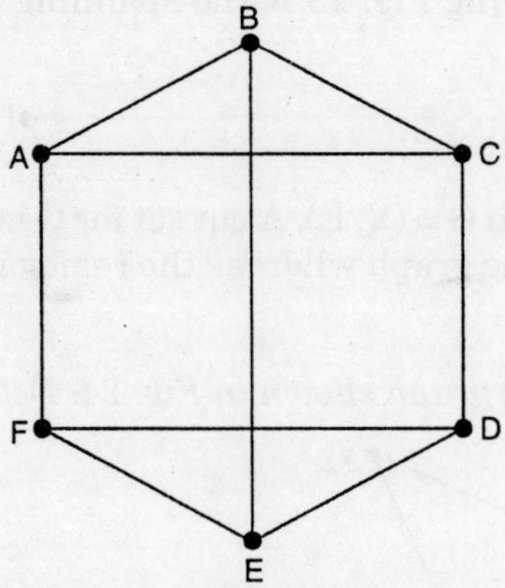

Fig. 18

Sol. The following are all subgraphs of the above graph (shown in Figs. 19, 20, 21, 22).

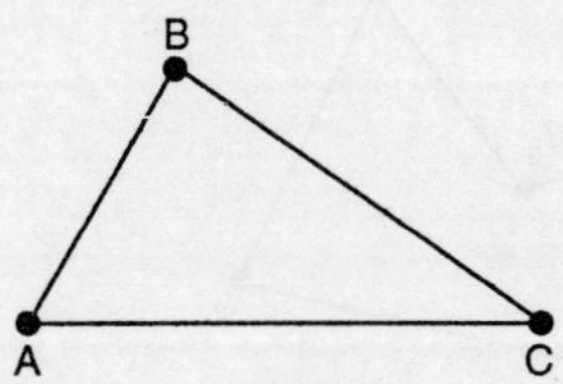

Fig. 19

Fig. 20

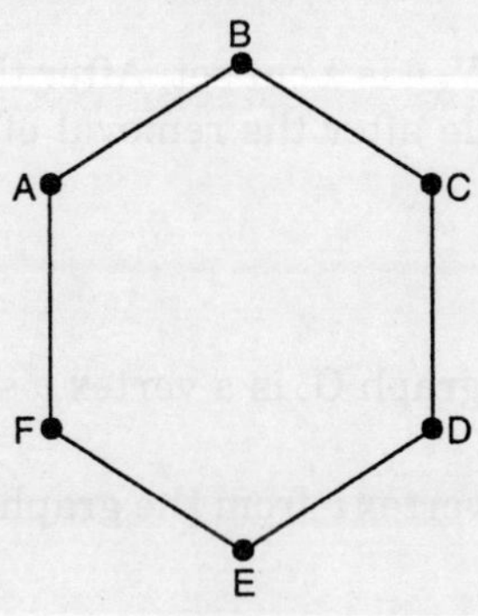

Fig. 21

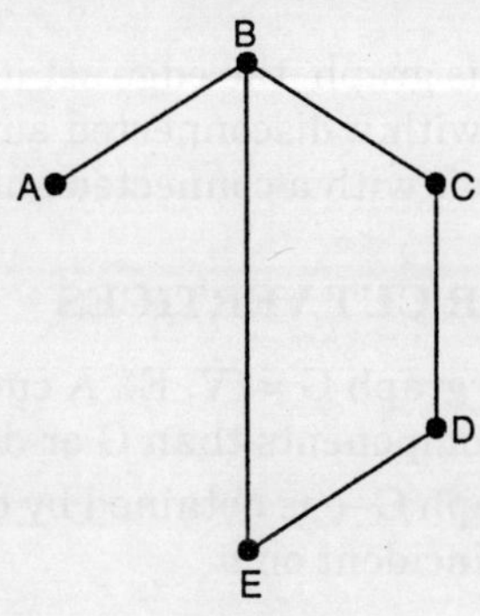

Fig. 22

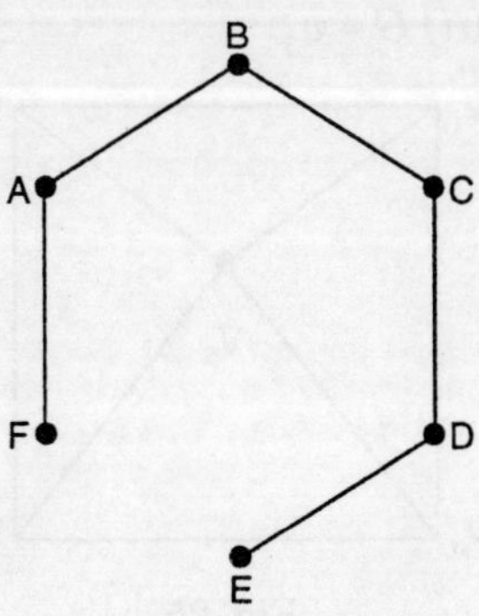

Fig. 23. Spanning Subgraph

Spanning Subgraph. A graph G_1 is called a spanning subgraph of G if G_1 contains all the vertices of G.

For example : The following Fig. 23 is the spanning subgraph of the graph shown in Fig. 18.

CUT SET

Consider a connected graph G = (V, E). A cut set for G is a smallest set of edges such that removal of the set, disconnects the graph whereas the removal of any proper subset of this set, left a connected subgraph.

Example 13. *Consider the graph shown in Fig. 24. Determine the cut set for this graph.*

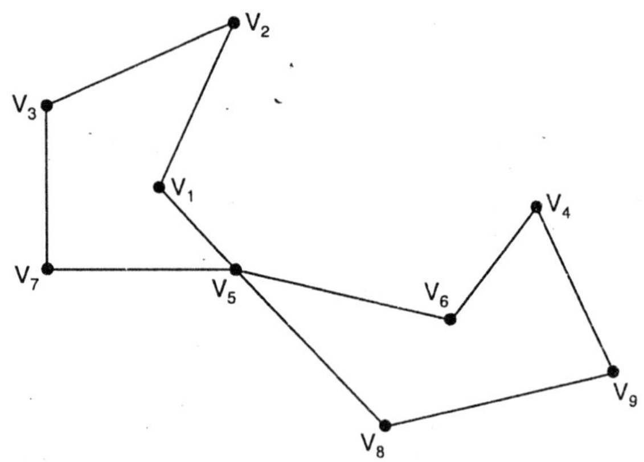

Fig. 24

Sol. For this graph, the edge set $\{(V_1, V_5), (V_7, V_5)\}$ is a cut set. After the removal of this set, we have left with a disconnected subgraph. While after the removal of any of its proper subset, we have left with a connected subgraph.

CUT POINTS OR CUT VERTICES

Consider a graph G = (V, E). A cut point for a graph G, is a vertex v such that $G-v$ has more connected components than G or disconnected.

The subgraph $G-v$ is obtained by deleting the vertex v from the graph G and also deleting all the edges incident on v.

Example 14. *Consider the graph shown in Fig. 25. Determine the subgraphs*

(*i*) $G - v_1$ (*ii*) $G - v_3$ (*iii*) $G - v_5$.

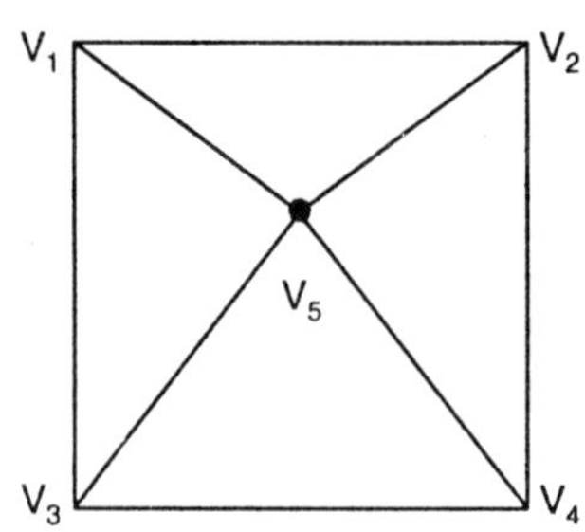

Fig. 25

Sol. (*i*) The subgraph G – v_1 is as shown in Fig. 26.

(*ii*) The subgraph G–v_3 is as shown in Fig. 27.

(*iii*) The subgraph G–v_5 is as shown in Fig. 28.

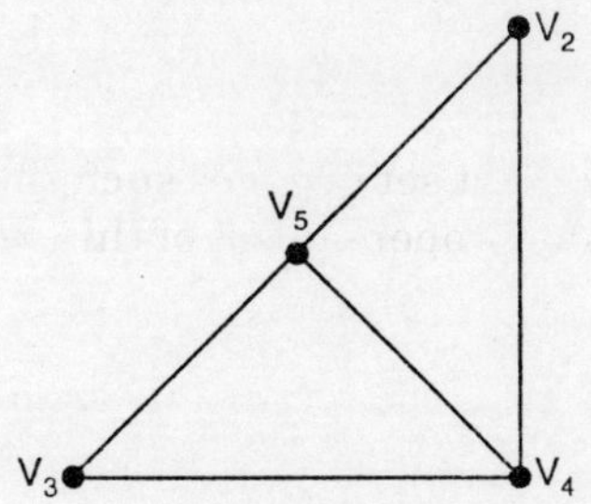

Fig. 26

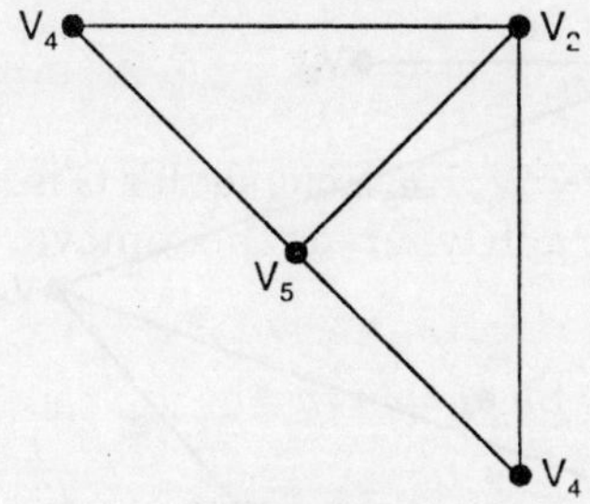

Fig. 27

Fig. 28

Example 15. *Consider the graph G shown in Fig. 29. Determine all the cut points of G.*

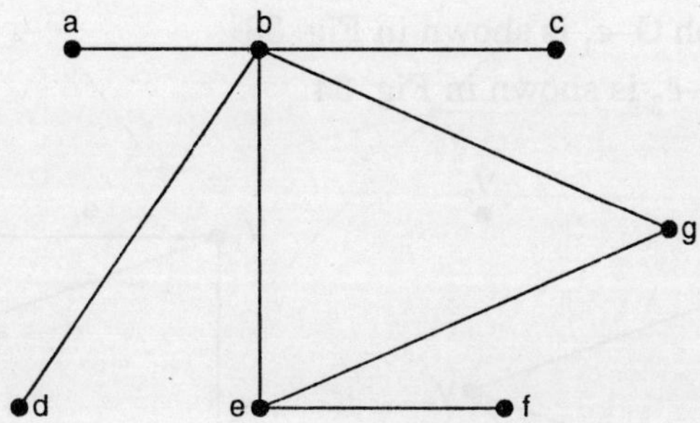

Fig. 29

Sol. (*a*) The vertex *b* is cut point for G. Since, G–*b* has more than one connected components as shown in Fig. 30.

(*b*) The vertex *e* is also a cut point for G. Since G–*e* has more than one connected components as shown in Fig. 31.

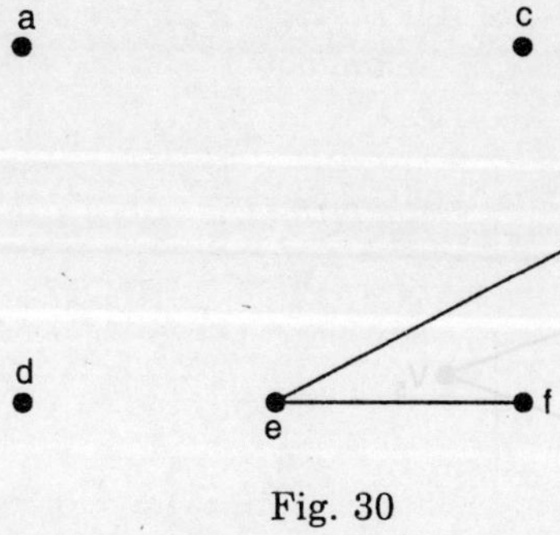

Fig. 30

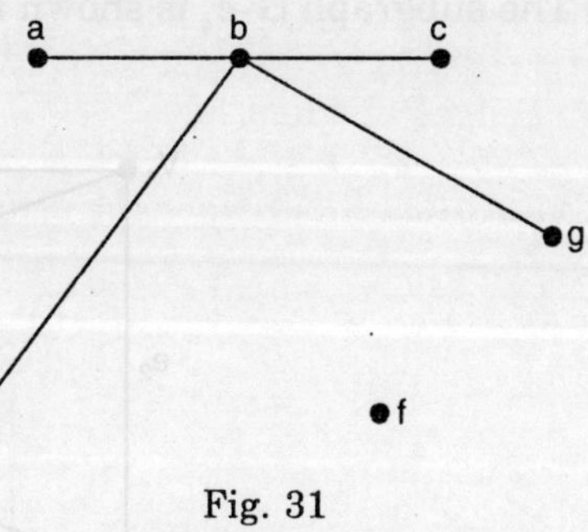

Fig. 31

BRIDGE (CUT EDGES)

Consider a graph G = (V, E). A bridge for a graph G, is an edge *e* such that G–*e* has more connected components than G or disconnected.

Example 16. *Consider the graph shown in Fig. 32. Determine the subgraphs*

(*i*) $G - e_1$ (*ii*) $G - e_3$ (*iii*) $G - e_4$.

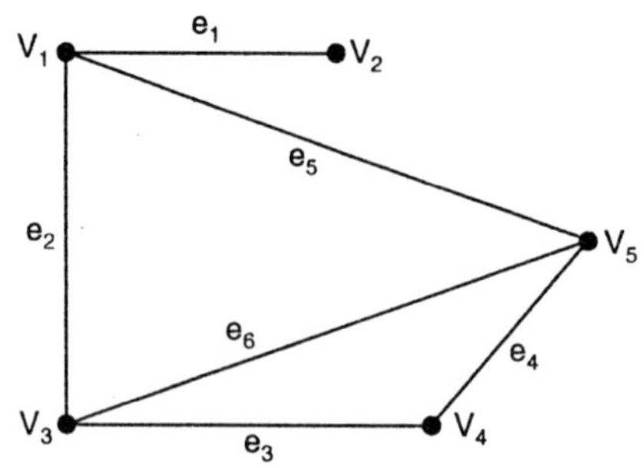

Fig. 32

Sol. (*i*) The subgraph $G–e_1$ is shown in Fig. 33.

(*ii*) The Subgraph $G–e_3$ is shown in Fig. 34.

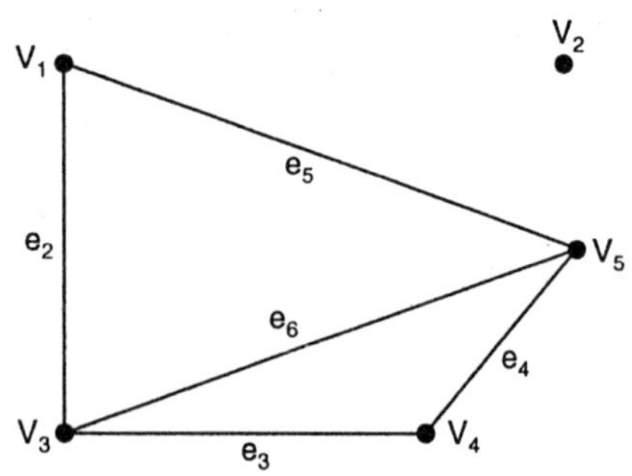

Fig. 33

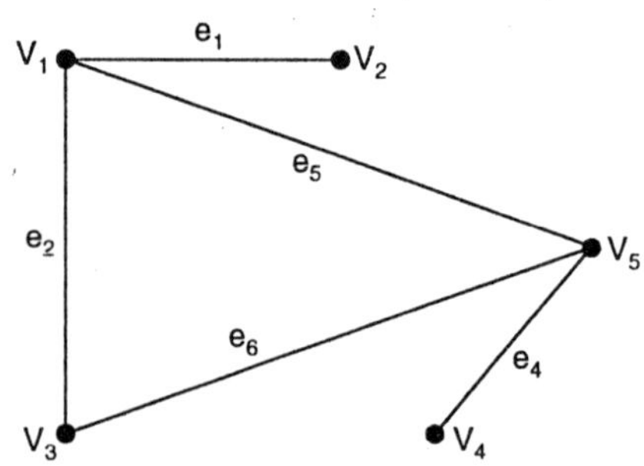

Fig. 34

(*iii*) The subgraph $G–e_4$ is shown in Fig. 35.

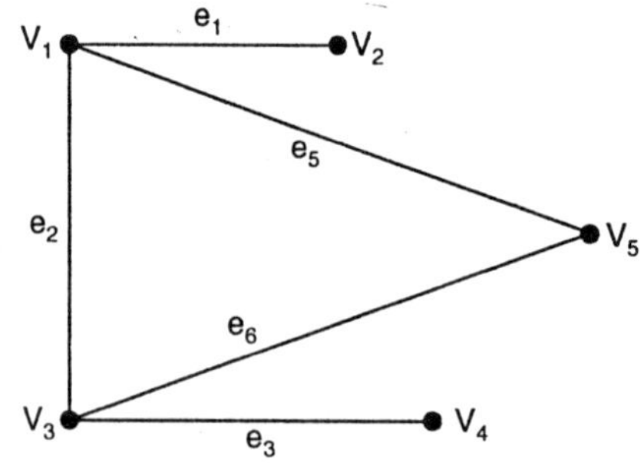

Fig. 35

Example 17. *Consider the graph G shown in Fig. 36. Determine all the bridges of G.*

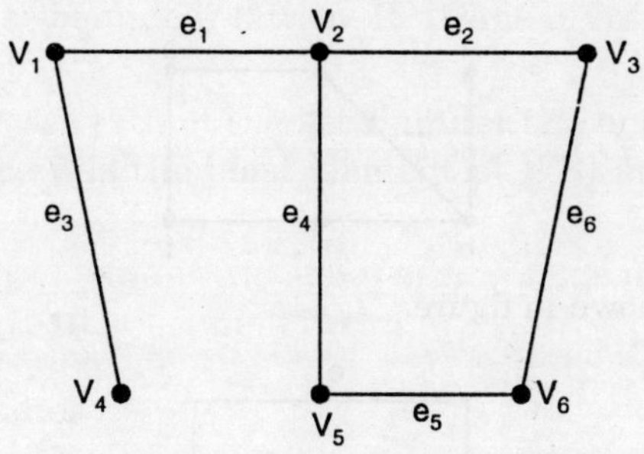

Fig. 36

Sol. (*a*) The edge e_1 is a bridge for G. Since G–e_1 has more than one connected components as shown in Fig. 37.

(*b*) The edge e_3 is a bridge for G. Since G–e_3 has more than one connected components as shown in Fig. 38.

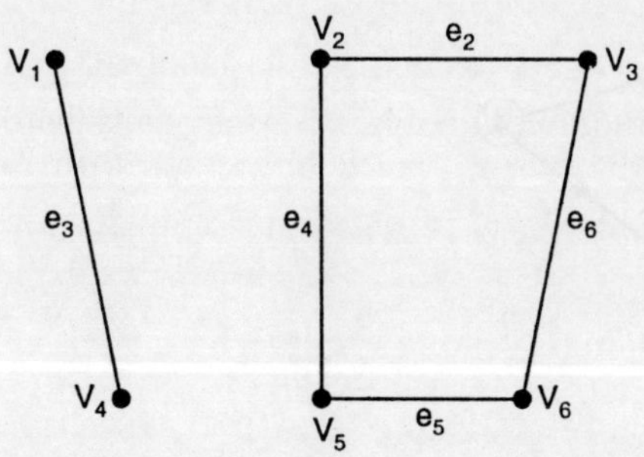

Fig. 37

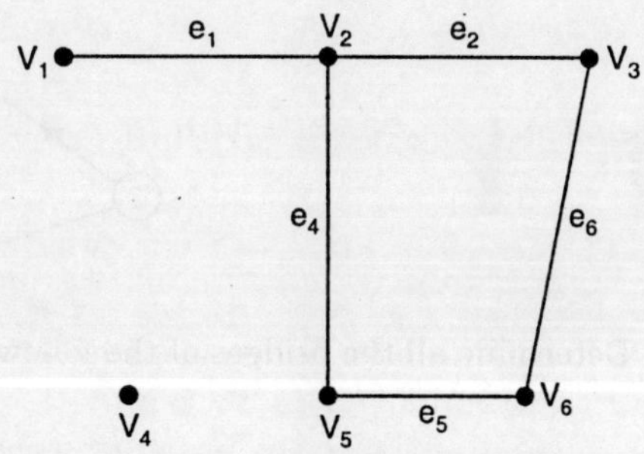

Fig. 38

EXERCISES

1. Let G be the connected graph with three or more vertices. Show that G has at least two vertices which are not cut vertices.
2. Draw a graph having the given properties or explain why no such graph exists.
   (*i*) Graph with five vertices of degree 3, 3, 3, 3, 2.
   (*ii*) Simple graph with four vertices of degree 1, 1, 3, 3.
   (*iii*) Graph with five vertices of degree 0, 1, 2, 2, 3.
3. Draw the complementary graphs of the following graphs.

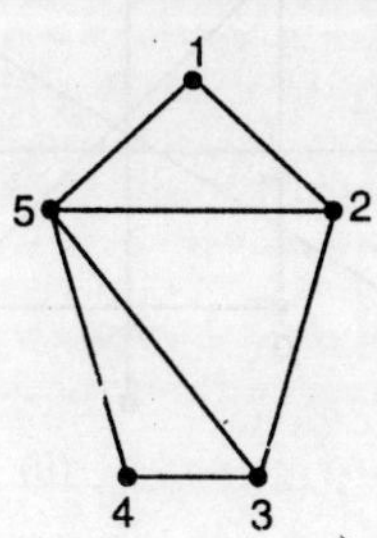

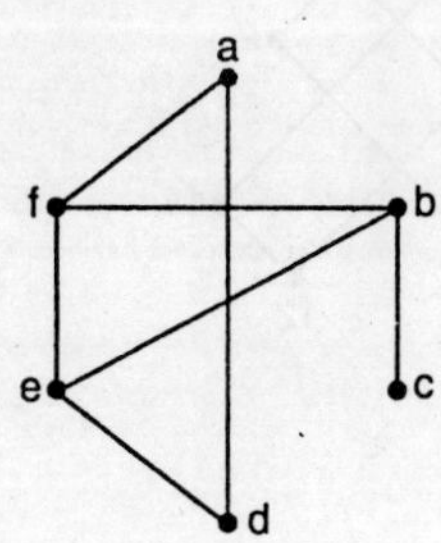

4. If the simple graph G has V vertices and e edges, how many edges does G have ?
5. Determine all the cut vertices of the following graph.

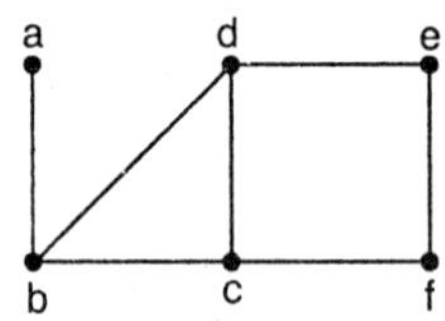

6. Consider the graph as shown in figure.

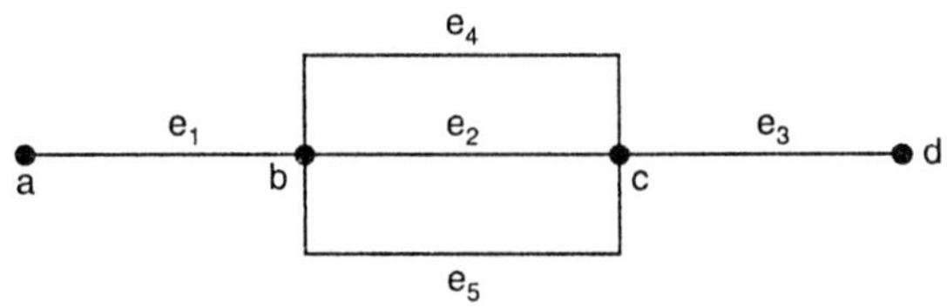

(*i*) How many paths are there from a to d ?

(*ii*) How many simple paths are there from a to d ?

7. Determine whether the following graph is strongly connected ?

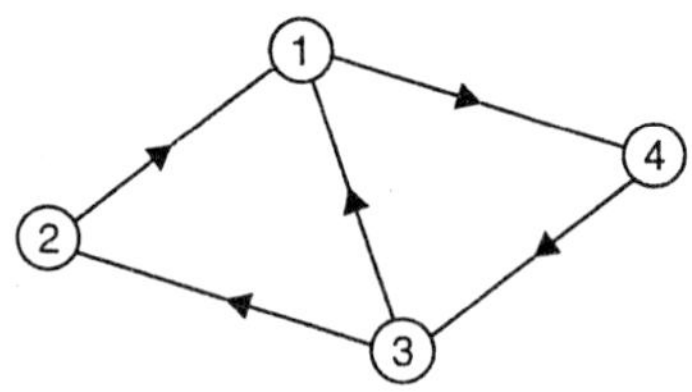

8. Determine all the bridges of the following graph.

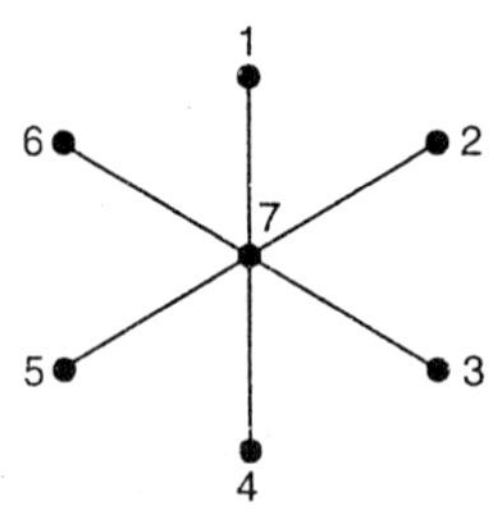

9. Determine the number of connected components of each of the following graphs.

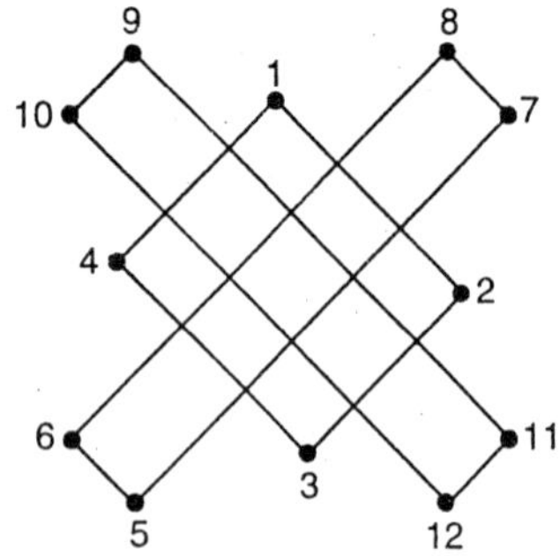

(*i*)

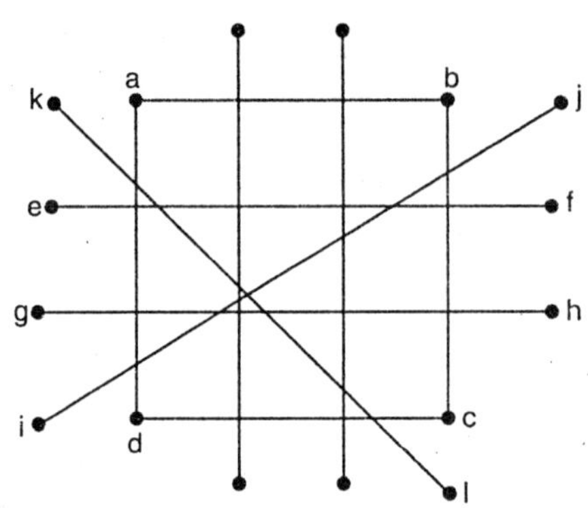

(*ii*)

10. Consider the graph as shown in figure. Determine which of the following sequences are paths, simple paths, cycle and simple cycle.

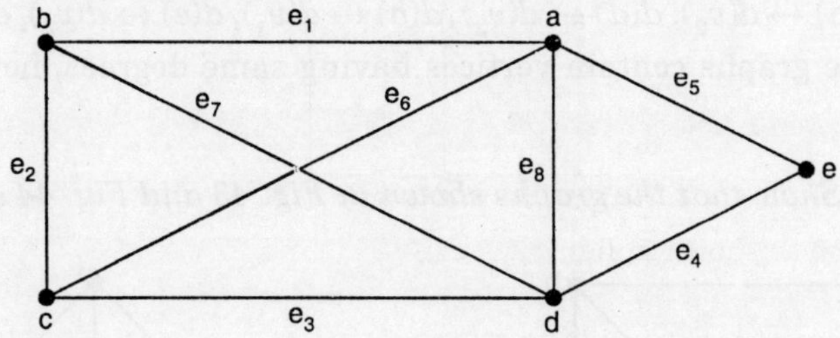

(i) $a\ e_8\ d\ e_3\ c\ e_7\ a\ e_8\ d$ (ii) $b\ e_2\ c\ e_3\ d\ e_4\ e\ e_5\ a\ e_1\ b$

ISOMORPHIC GRAPHS

Two graphs G_1 and G_2 are called isomorphic graphs if there is a one-to-one correspondence between their vertices and between their edges *i.e.,* the graphs have identical representation except that the vertices may have different labels.

For example : Consider the following graphs shown in Fig. 39 and Fig. 40. They are isomorphic graphs.

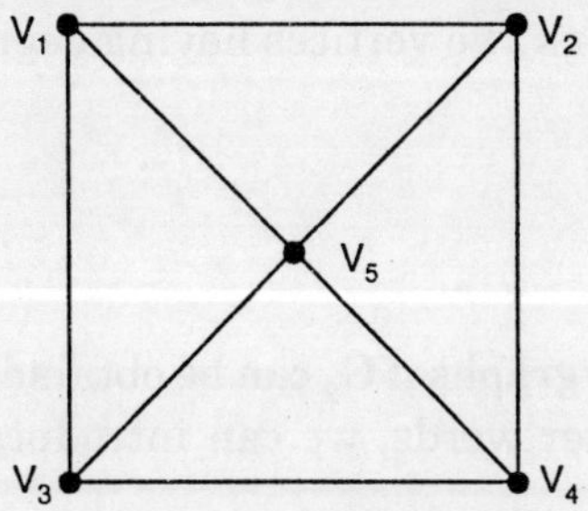

Fig. 39

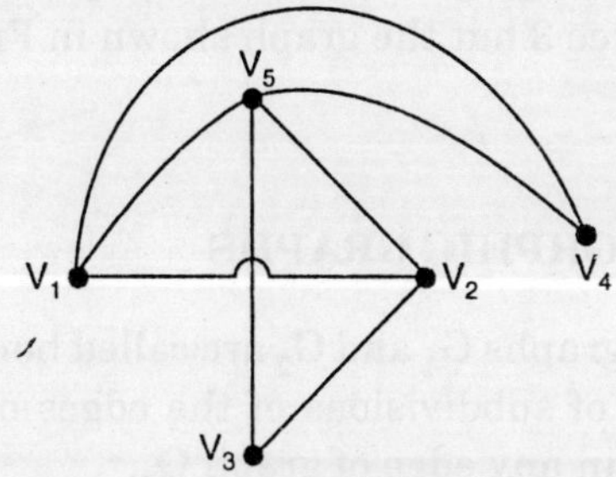

Fig. 40

Example 18. *Show that the graphs shown in Fig 41 and Fig. 42 are isomorphic.*

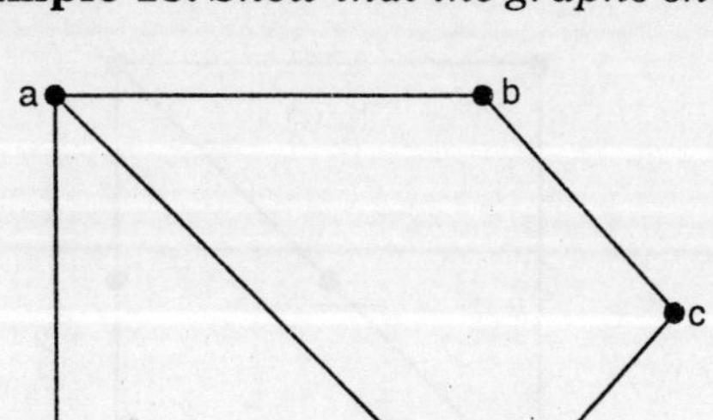

Fig. 41

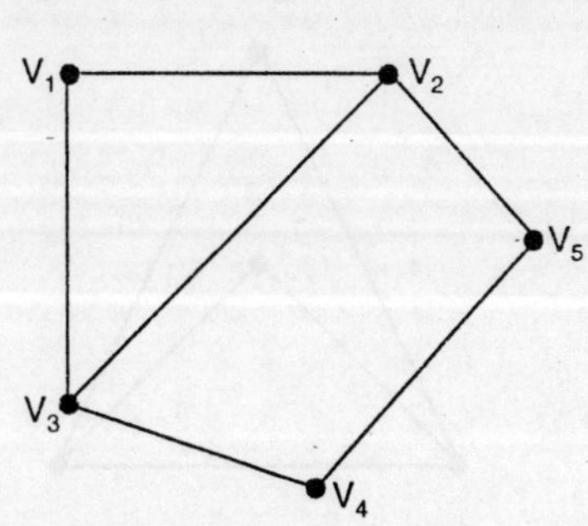

Fig. 42

Sol. Compare the degrees of vertices of two graphs and find the vertices from both the graphs having same degrees and make the pairs of the vertices in decreasing order of degree.

If both the graphs contain vertices having same degree, then they are isomorphic otherwise not. The pairs of vertices in decreasing order of degree are as follows :

$$d(a) \leftrightarrow d(v_2), d(d) \leftrightarrow d(v_3), d(b) \leftrightarrow d(v_1), d(e) \leftrightarrow d(v_4), d(c) \leftrightarrow d(v_5).$$

Since, both the graphs contain vertices having same degrees, hence they are isomorphic.

Example 19. *Show that the graphs shown in Fig. 43 and Fig. 44 are not isomorphic.*

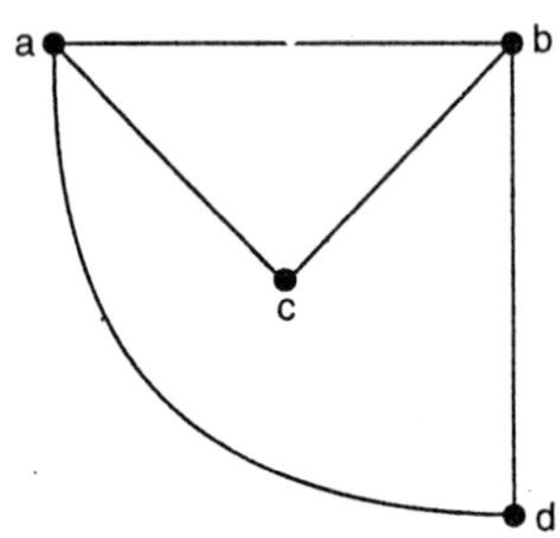

Fig. 43

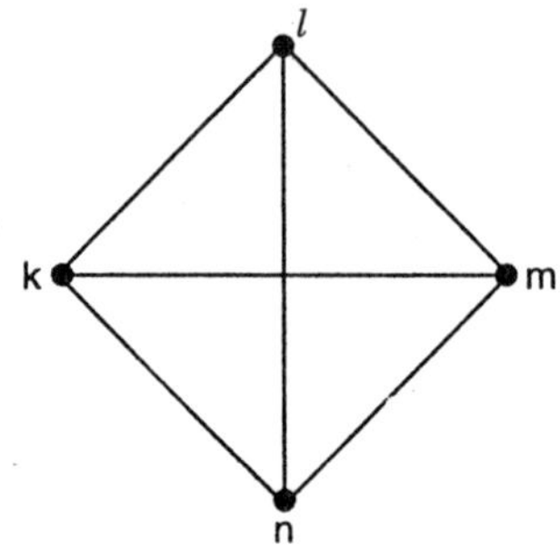

Fig. 44

Sol. The graphs are not isomorphic because the vertices of the graph shown in Fig. 44 is having degree 3 but the graph shown in Fig. 43 contains two vertices having degree less than three.

HOMEOMORPHIC GRAPHS

Two graphs G_1 and G_2 are called homeomorphic graphs if G_2 can be obtained from G_1 by a sequence of subdivisions of the edges of G_1. In other words, we can introduce vertices of degree two in any edge of graph G_1.

For example : Consider the graph shown in Fig. 45 and Fig. 46. They are homeomorphic graphs.

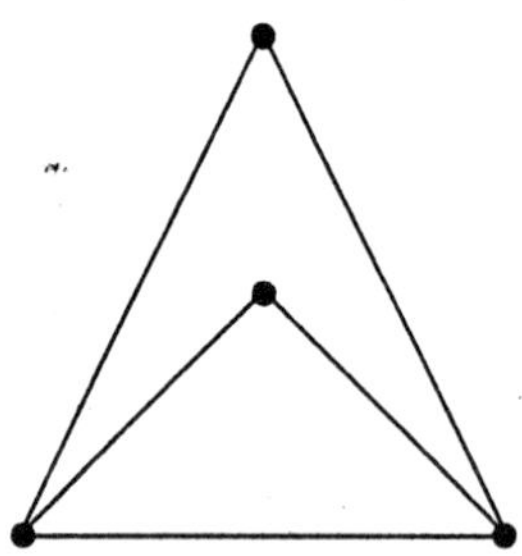

Fig. 45. G_1

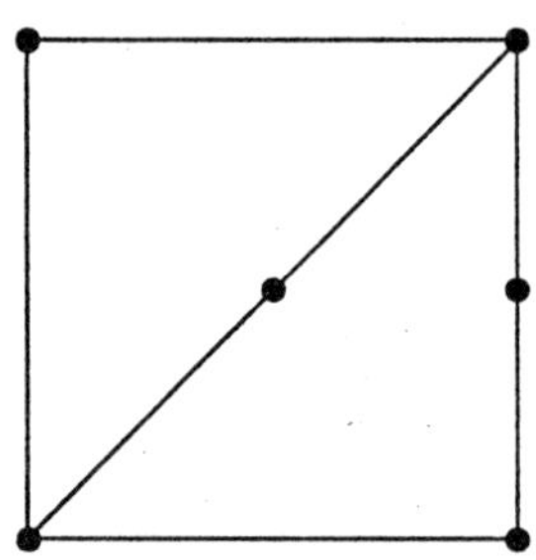

Fig. 46. G_2

Example 20. *Show that the graphs shown in Figs. 47 and 48 are homeomorphic.*

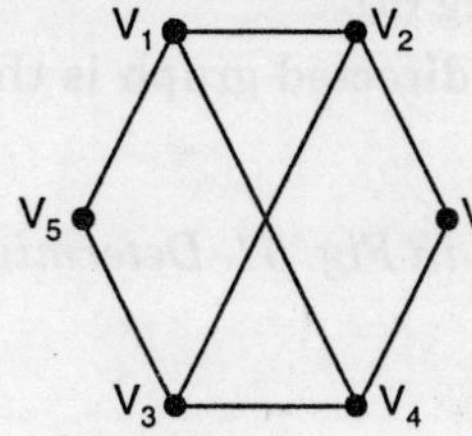

Fig. 47. G_1

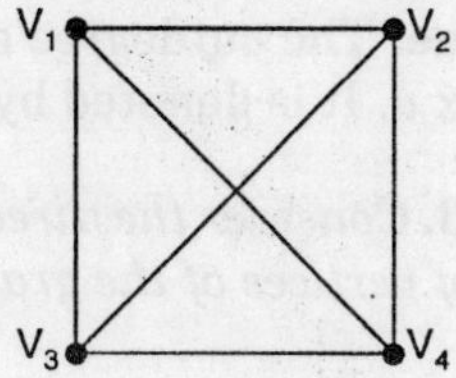

Fig. 48. G_2

Sol. The two graphs are homeomorphic because G_1 can be obtained from G_2 by introducing vertices of degree 2 on edges (V_1, V_3) and (V_2, V_4).

DIRECTED GRAPHS

A directed graphs or digraph G is defined as an ordered pair (V, E), where V is the set of points called vertices and E is the set of edges. Each edge in the graph G is assigned a direction and is identified with an ordered pair (u, v), where u is the initial vertex and v is the end vertex.

Example 21. *Consider the directed graph G = (V, E) as shown in Fig. 49. Determine the vertex set and edge set of graph G.*

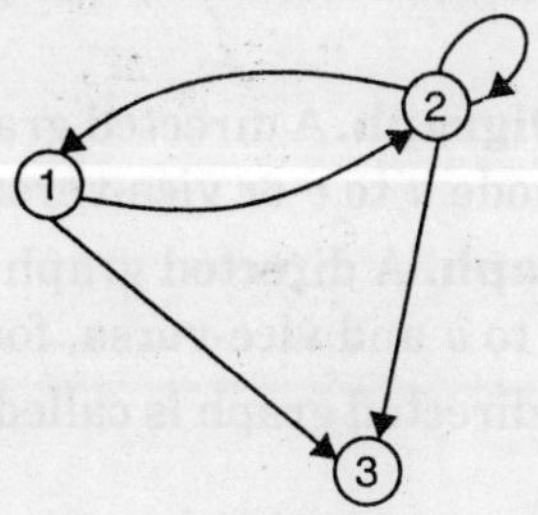

Fig. 49

Sol. The vertex and edge set of graph G = (V, E) is as follows

$$G = \{\{1, 2, 3\}, \{(1, 2), (2, 1), (2, 2), (2, 3), (1, 3)\}\}.$$

Example 22. *Let G = {{a, b, c, d}, {(a, b), (b, c), (c, c), (d, d), (d, a)}}. Draw the graph G.*

Sol. The graph of G = (V, E) is shown in Fig. 50.

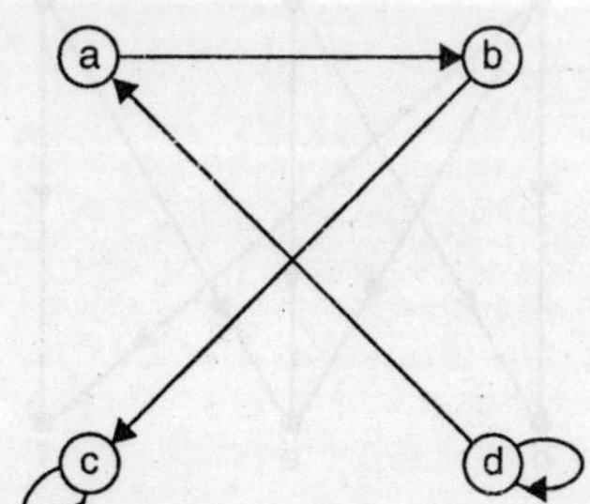

Fig. 50

1. **Indegree.** The indegree of a vertex v in a directed graph is the number of edges ending or terminating at vertex v. It is denoted by indeg (v).

2. **Outdegree.** The outdegree of a vertex v in a directed graph is the number of edges beginning at vertex v. It is denoted by outdeg (v).

Example 23. *Consider the directed graph shown in Fig. 51. Determine the indegree and outdegree of each of vertices of the graph.*

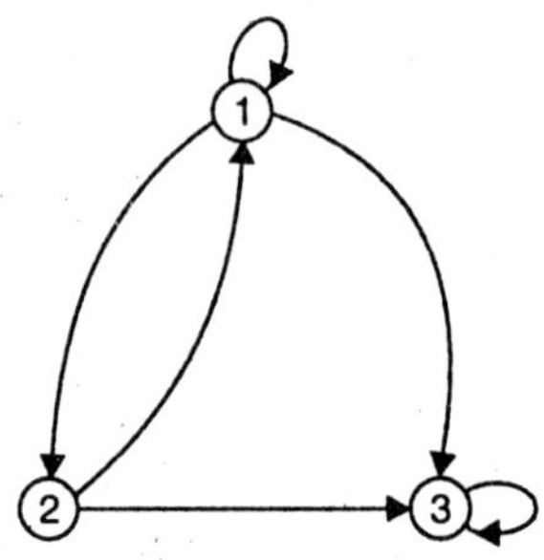

Fig. 51

Sol. The indegree of digraph is indeg (1) = 2, indeg (2) = 1, indeg (3) = 3

The outdegree of digraph is outdeg (1) = 3, outdeg (2) = 2, outdeg (3) = 1.

Weakly Connected or Connected Digraph. A directed graph is called weakly connected if its undirected graph is connected *i.e.,* the graph obtained after neglecting the direction.

Unilaterally Connected Digraph. A directed graph is called unilaterally connected if there is a directed path from any node u to v or vice-versa, for any pair of nodes of the graph.

Strongly Connected Digraph. A directed graph is called strongly connected if there is a directed path from any node u to v and vice-versa, for any pair of nodes of the graph.

Disconnected Digraph. A directed graph is called disconnected if its undirected graph is disconnected.

Example 24. *Consider the graphs shown in Figs. 52, 53 and 54 which of the graphs are*

(*i*) *Unilaterally connected digraph* (*ii*) *Weakly connected digraph*

(*iii*) *Strongly connected digraph*

(*iv*) *Disconnected digraph (also find its connected components).*

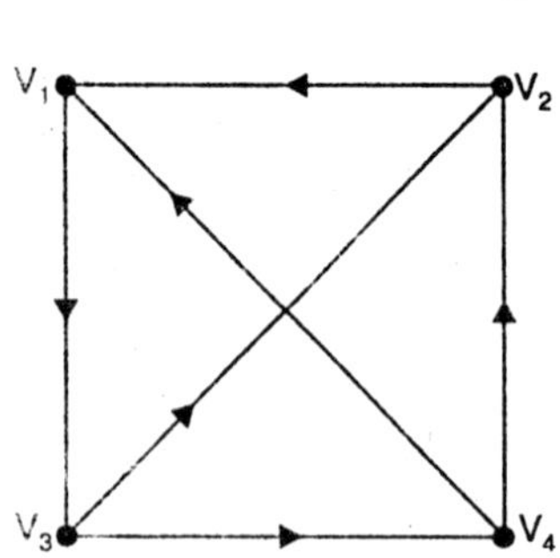

Fig. 52

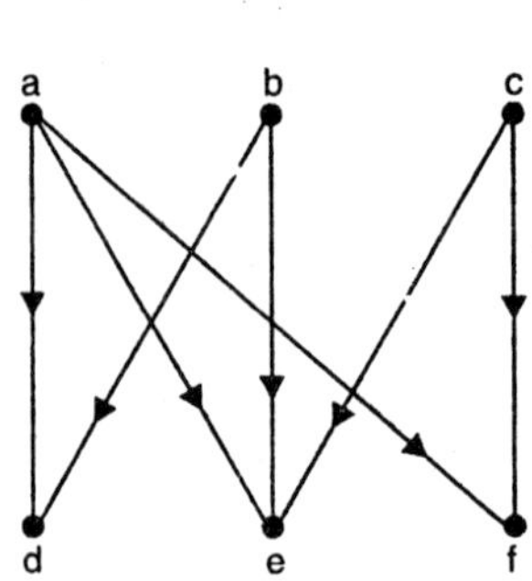

Fig. 53

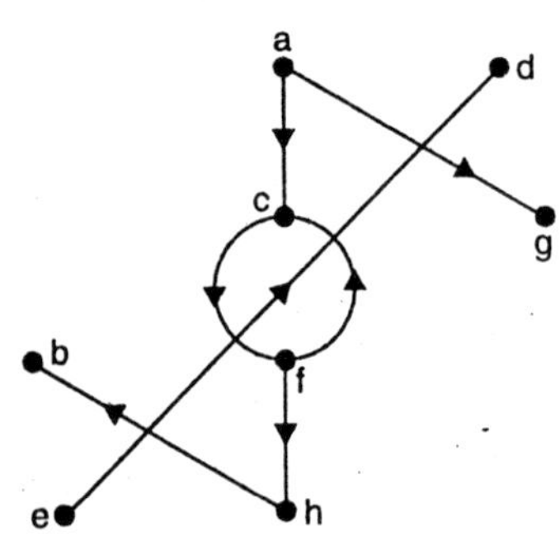

Fig. 54

Sol. The graph shown in Fig. 52 is strongly connected because there is a path from every vertex u to v and also there is a path from v to u. It is also weakly and unilaterally connected because a strongly connected digraph is both weakly and unilaterally connected.

The graph shown in Fig. 53 is weakly connected but not unilaterally connected because there is no directed path from vertex a to b or a to c etc. but its undirected graph is connected.

The graph shown in Fig. 54 is a disconnected graph. The components of this graph are $\{g, a, c, f, h, b\}$ and $\{e, d\}$.

Directed Complete Graph. A directed complete graph G = (V, E) on n vertices is a graph in which each vertex is connected to every other vertex by an arrow. It is denoted by K_n.

Example 25. *Draw directed complete graphs K_3 and K_5.*

Sol. Place the number of vertices at appropriate place and then draw an arrow from each vertex to every other vertex as shown in Figs. 55 and 56.

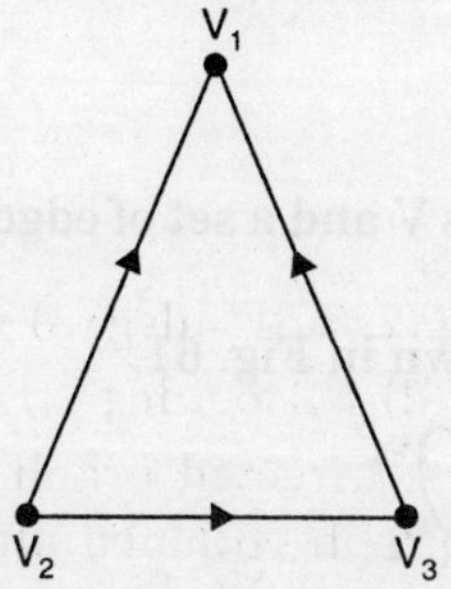

Fig. 55. K_3

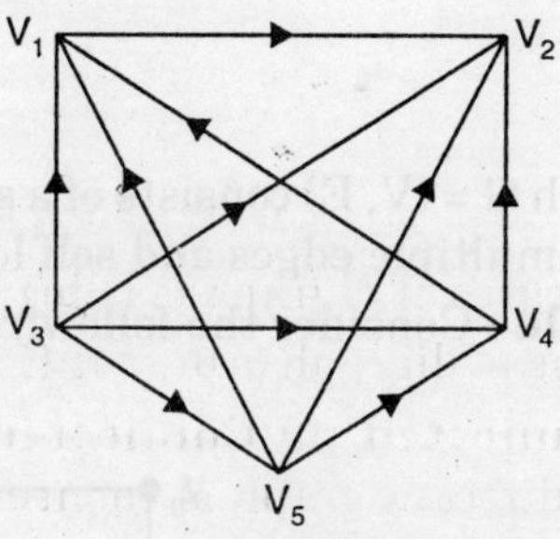

Fig. 56. K_5

LABELED GRAPHS

A graph G = (V, E) is called a labeled graph if its edges are labeled with some name or data. So, we can write these labels in place of an ordered pair in the edge set.

For example : The graphs shown in Figs. 57 and 58 are labeled graphs.

G = {{1, 2, 3, 4, 5}, {e_1, e_2, e_3, e_4, e_5}}.

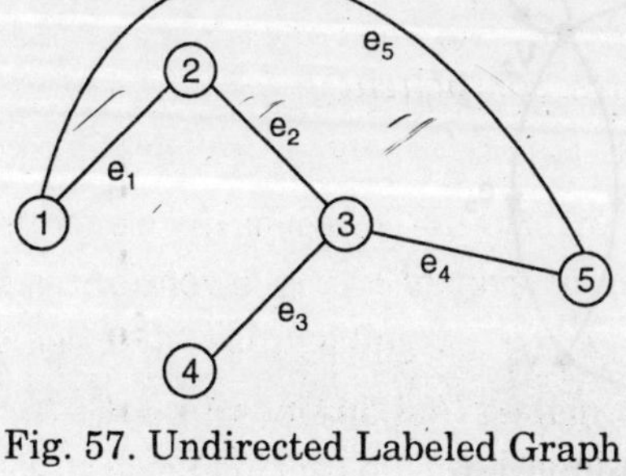

Fig. 57. Undirected Labeled Graph

G = {{a, b, c, d}, {e_1, e_2, e_3, e_4}}.

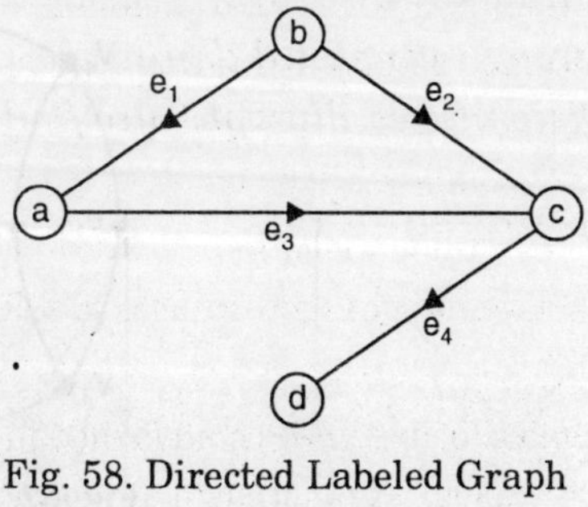

Fig. 58. Directed Labeled Graph

WEIGHTED GRAPHS

A graph G = (V, E) is called a weighted graph if each edge of graph G is assigned a positive number w called the weight of the edge e.

For example : The graph shown in Figs. 59 and 60 is a weighted graph.

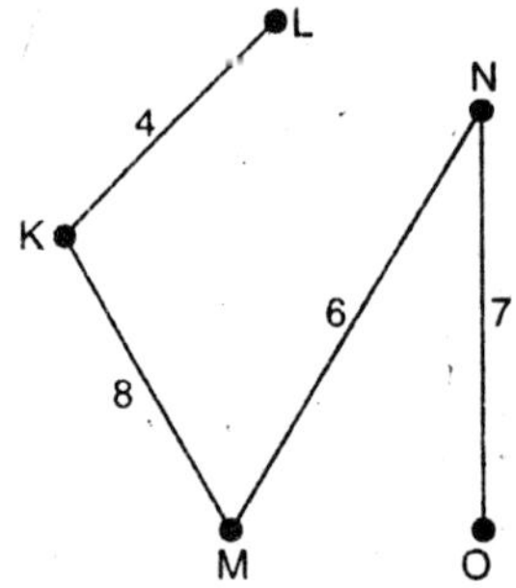

Fig. 59

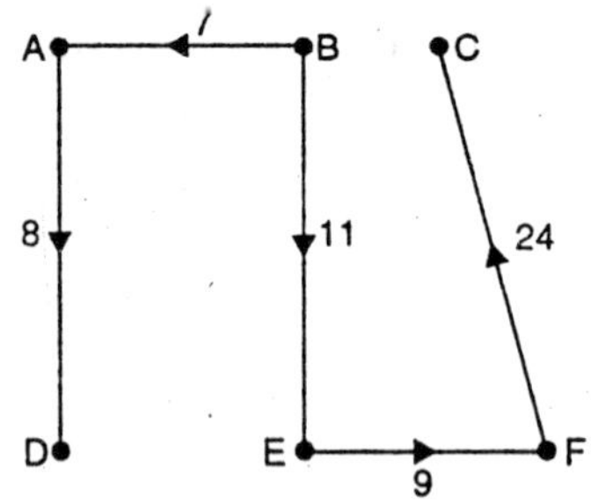

Fig. 60

Multiple Edges. Two edges e_1 and e_1' which are distinct are said to be multiple edges if they connect the end points *i.e.*, if $e_1 = (u, v)$ and $e_1' = (u, v)$ then e_1 and e_1' are multiple edges.

MULTIGRAPH

A multigraph G = (V, E) consists of a set of vertices V and a set of edges E such that edge set E may contain multiple edges and self loops.

For example : Consider the following graph shown in Fig. 61.

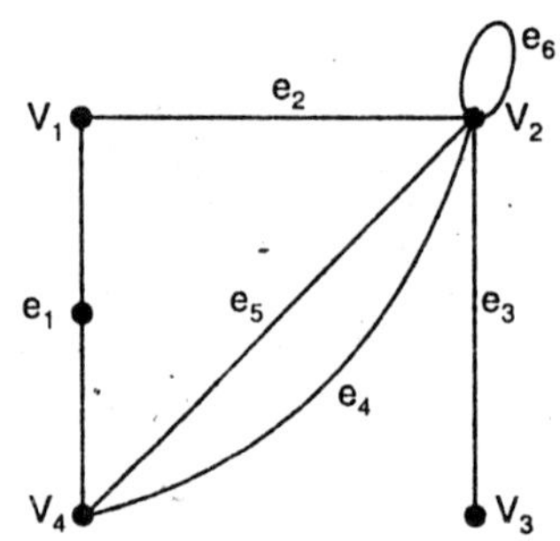

Fig. 61. Undirected Multigraph

In the above Fig. 61, e_4 and e_5 are multiple edges, e_6 is a self-loop.

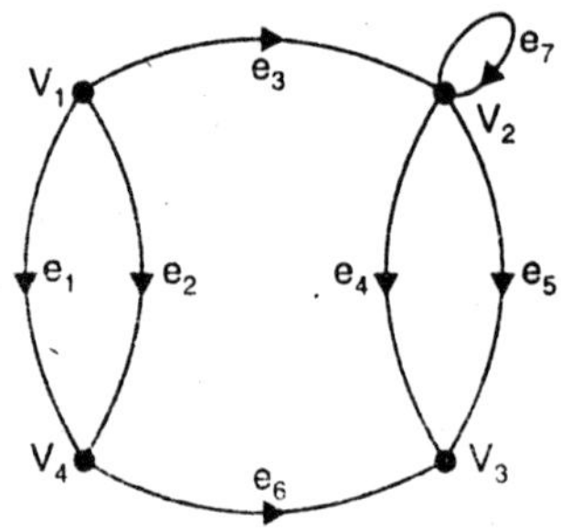

Fig. 62. Directed Multigraph

In the graph shown in Fig. 62, the edges e_1, e_2 and e_4, e_5 are multiple edges. e_7 is a loop.

Traversable Multigraphs. Consider a multigraph G = (V, E). If the multigraph G consists of a path which includes all vertices and whose edge list contains each edge of graph exactly once. Then, the multigraph G is called a traversable multigraph.

The sufficient and necessary condition for a multigraph to be traversable is that it should be connected and have either zero or two vertices of odd degree.

For example : Consider the multigraph shown in Fig. 63.

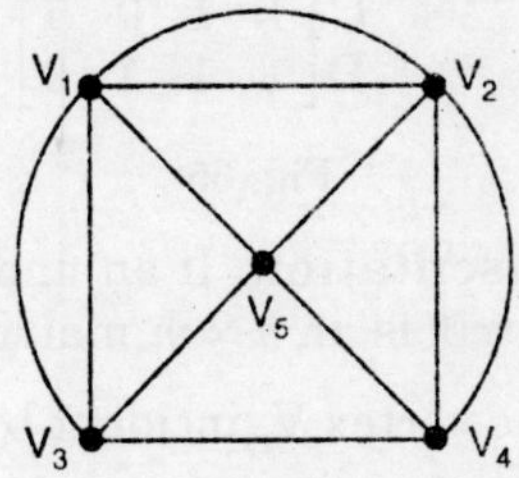

Fig. 63

The multigraph has three even degree vertices *i.e.*, V_3, V_4 and V_5 and two odd degree vertices *i.e.*, V_1 and V_2. Hence it is a traversable multigraph.

REPRESENTATION OF GRAPHS

There are two important ways to represent a graph G with the matrices *i.e.*, adjacency matrix and incidence matrix representation.

(*a*) Representation of Undirected Graph

(*i*) **Adjacency matrix representation.** If an undirected graph G consists of n vertices then the adjacency matrix of graph is an $n \times n$ matrix $A = [a_{ij}]$ and defined by

$$a_{ij} = \begin{cases} 1, \text{ if } \{V_i, V_j\} \text{ is an edge } i.e., v_i \text{ is adjacent to } v_j \\ 0, \text{ if there is no edge between } v_i \text{ and } v_j \end{cases}$$

If there exists an edge between vertex v_i and v_j, where i is a row and j is a column then value of $a_{ij} = 1$.

If there is no edge between vertex v_i and v_j, then value of $a_{ij} = 0$.

Example 26. *Find the adjacency matrix M_A of graph G shown in Fig. 64.*

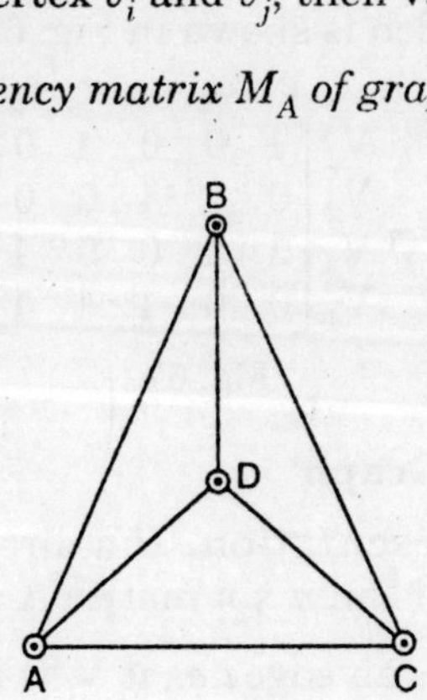

Fig. 64

Sol. Since the graph G consists of four vertices. Therefore, the adjacency matrix will be a 4 × 4 matrix. The adjacency matrix is as follows in Fig. 65.

$$M_A = \begin{array}{c} \\ A \\ B \\ C \\ D \end{array}\begin{array}{c} \begin{array}{cccc} A & B & C & D \end{array} \\ \begin{bmatrix} 0 & 1 & 1 & 1 \\ 1 & 0 & 1 & 1 \\ 1 & 1 & 0 & 1 \\ 1 & 1 & 1 & 0 \end{bmatrix} \end{array}$$

Fig. 65

(*ii*) **Incidence matrix representation.** If an undirected graph consists of n vertices and m edges, then the incidence matrix is an $n \times m$ matrix $C = [c_{ij}]$ defined by

$$c_{ij} = \begin{cases} 1, \text{ if the vertex } V_i \text{ incident by edge } e_j \\ 0, \text{ otherwise} \end{cases}$$

There is a row for every vertex and a column for every edge in the incidence matrix.

The number of one's in an incidence matrix of undirected graph (without loops) is equal to the sum of degrees of all the vertices of the graph.

Example 27. *Consider the undirected graph G as shown in Fig. 66. Find its incidence matrix M_I.*

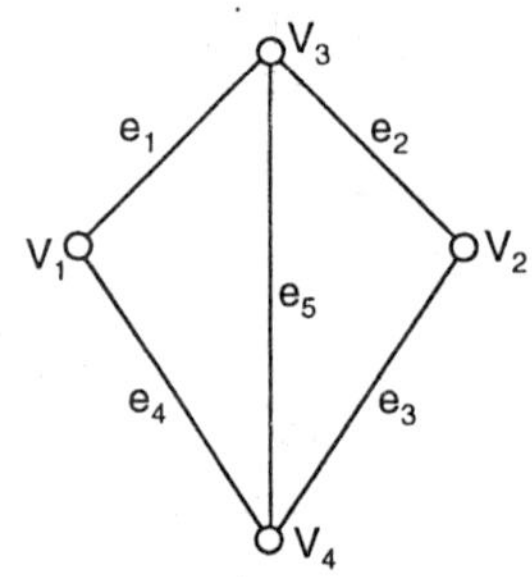

Fig. 66

Sol. The undirected graph consists of four vertices and five edges. Therefore, the incidence matrix is an 4×5 matrix, which is shown in Fig. 67.

$$M_I = \begin{array}{c} \\ V_1 \\ V_2 \\ V_3 \\ V_4 \end{array}\begin{array}{c} \begin{array}{ccccc} e_1 & e_2 & e_3 & e_4 & e_5 \end{array} \\ \begin{bmatrix} 1 & 0 & 0 & 1 & 0 \\ 0 & 1 & 1 & 0 & 0 \\ 1 & 1 & 0 & 0 & 1 \\ 0 & 0 & 1 & 1 & 1 \end{bmatrix} \end{array}$$

Fig. 67

(*b*) Representation of Directed Graph

(*i*) **Adjacency matrix representation.** If a directed graph G consists of n vertices, then the adjacency matrix of graph is an $n \times n$ matrix $A = [a_{ij}]$ defined by

$$a_{ij} = \begin{cases} 1, \text{ if } V_i, V_j \text{ is an edge } i.e., \text{ if } V_i \text{ is initial vertex and } V_j \text{ is final vertex} \\ 0, \text{ if there is no edge between } V_i \text{ and } V_j \end{cases}$$

If there exists an edge between vertex V_i and V_j with V_i as initial vertex and V_j as final vertex, then value of $a_{ij} = 1$.

If there is no edge between vertex V_i and V_j then value of $a_{ij} = 0$.

The number of one's in the adjacency matrix of a directed graph is equal to the number of edges.

Example 28. *Consider the directed graph shown in Fig. 68. Determine its adjacency matrix* M_A.

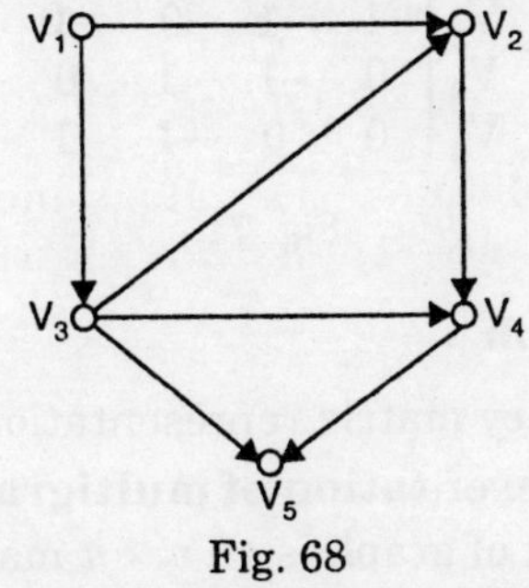

Fig. 68

Sol. Since the directed graph G consists of five vertices. Therefore, the adjacency matrix will be a 5 × 5 matrix. The adjacency matrix of the directed graph is as follows in Fig. 69.

$$M_A = \begin{array}{c} \\ V_1 \\ V_2 \\ V_3 \\ V_4 \\ V_5 \end{array} \begin{array}{c} \begin{array}{ccccc} V_1 & V_2 & V_3 & V_4 & V_5 \end{array} \\ \begin{bmatrix} 0 & 1 & 1 & 0 & 0 \\ 0 & 0 & 0 & 1 & 0 \\ 0 & 1 & 0 & 1 & 1 \\ 0 & 0 & 0 & 0 & 1 \\ 0 & 0 & 0 & 0 & 0 \end{bmatrix} \end{array}$$

Fig. 69

(ii) **Incidence matrix representation.** If a directed graph consists of n vertices and m edges then the incidence matrix is an $n \times m$ matrix $C = [c_{ij}]$, defined by

$$c_{ij} = \begin{cases} 1, & \text{if } V_i \text{ is initial vertex of edge } e_j \\ -1, & \text{if } V_i \text{ is final vertex of edge } e_j \\ 0, & \text{if } V_i \text{ is not incident on edge } e_j \end{cases}$$

The number of one's in the incidence matrix is equal to the number of edges in the graph.

Example 29. *Consider the directed graph G shown in Fig. 70. Find its incidence matrix* M_I.

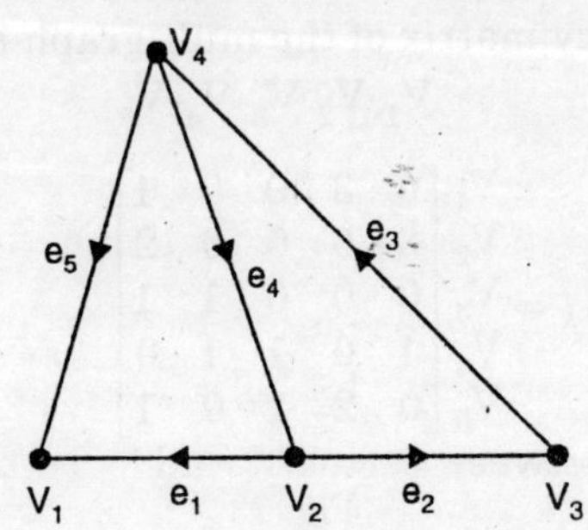

Fig. 70

Sol. The directed graph consists of four vertices and five edges. Therefore, the incidence matrix is a 4 × 5 matrix which is shown in Fig. 71.

$$M_I = \begin{array}{c} \\ V_1 \\ V_2 \\ V_3 \\ V_4 \end{array} \begin{array}{c} \begin{array}{ccccc} e_1 & e_2 & e_3 & e_4 & e_5 \end{array} \\ \begin{bmatrix} -1 & 0 & 0 & 0 & -1 \\ 1 & 1 & 0 & -1 & 0 \\ 0 & -1 & 1 & 0 & 0 \\ 0 & 0 & -1 & 1 & 1 \end{bmatrix} \end{array}$$

Fig. 71

(c) Representation of Multigraph

Represented only by adjacency matrix representation.

(*i*) **Adjacency matrix representation of multigraph.** If a multigraph G consists of n vertices, then the adjacency matrix of graph is an $n \times n$ matrix $A = [a_{ij}]$ and is defined by

$$a_{ij} = \begin{cases} N, \text{ If there are one or more than one edges between vertex } v_i \text{ and } v_j\text{, where} \\ \quad N \text{ is the number of edges.} \\ 0, \text{ otherwise.} \end{cases}$$

If there exists one or more than one edges between vertex v_i and v_j then $a_{ij} = N$, where N is the number of edges between v_i and v_j.

If there is no edge between vertex v_i and v_j then value of $a_{ij} = 0$.

Example 30. *Consider the multigraph shown in Fig. 72. Determine its adjacency matrix.*

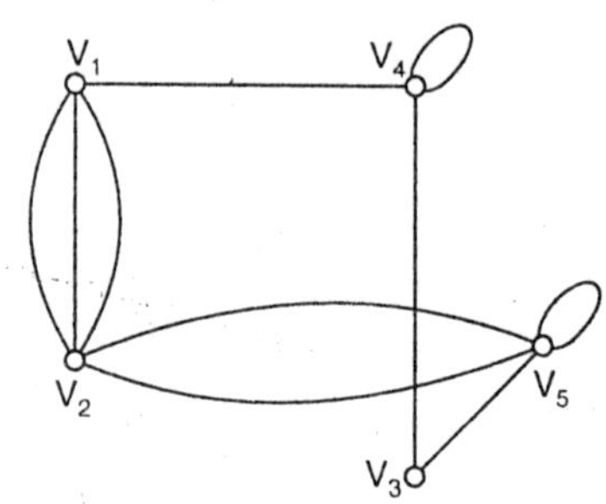

Fig. 72

Sol. Since the multigraph consists of five vertices. Therefore, the adjacency matrix will be an 5 × 5 matrix. The adjacency matrix of the multigraph is as follows in Fig. 73.

$$M_A = \begin{array}{c} \\ V_1 \\ V_2 \\ V_3 \\ V_4 \\ V_5 \end{array} \begin{array}{c} \begin{array}{ccccc} V_1 & V_2 & V_3 & V_4 & V_5 \end{array} \\ \begin{bmatrix} 0 & 3 & 0 & 0 & 1 \\ 3 & 0 & 0 & 0 & 2 \\ 0 & 0 & 0 & 1 & 1 \\ 1 & 0 & 1 & 1 & 0 \\ 0 & 2 & 1 & 0 & 1 \end{bmatrix} \end{array}$$

Fig. 73

EXERCISES

1. Determine whether the following pairs of graphs are isomorphic.

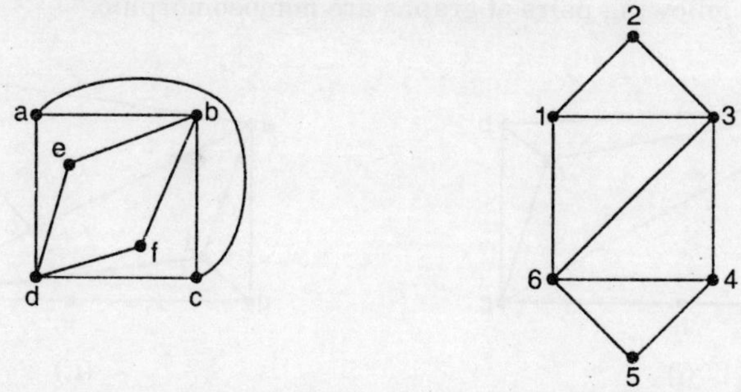

2. Determine whether the graphs shown in figure is a simple graph, a multigraph.

(*i*) (*ii*)

3. Draw the graph represented by the given adjacency matrix.

$$\begin{bmatrix} 0 & 1 & 0 & 1 & 0 \\ 1 & 0 & 1 & 0 & 1 \\ 0 & 1 & 0 & 1 & 1 \\ 1 & 0 & 1 & 0 & 0 \\ 0 & 1 & 1 & 0 & 0 \end{bmatrix}$$

4. Determine the adjacency matrix of each of the following graphs.

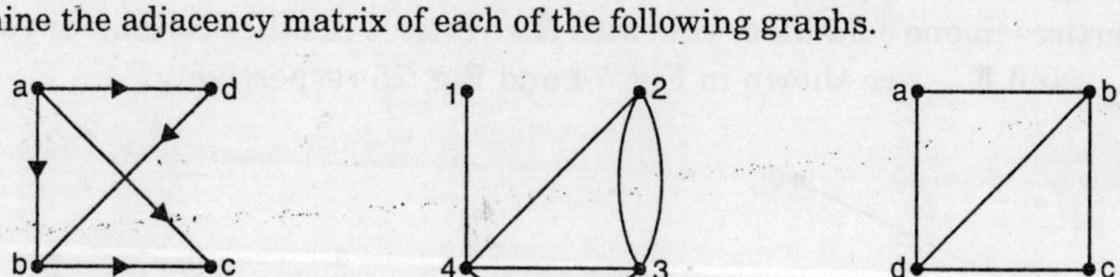

(*i*) (*ii*) (*iii*)

5. Draw the graph represented by the incidence matrix.

$$\begin{matrix} 1 \\ 2 \\ 3 \\ 4 \\ 5 \end{matrix}\begin{bmatrix} 1 & 0 & 0 & 0 & 0 \\ 0 & 1 & 1 & 0 & 1 \\ 1 & 0 & 0 & 1 & 0 \\ 0 & 1 & 0 & 1 & 0 \\ 0 & 0 & 1 & 0 & 1 \end{bmatrix}$$

6. If the simple graph G has V vertices and e edges, how many edges does $\overline{G}$ have ?

7. Draw an undirected graph represented by the given adjacency matrix.

$$\begin{bmatrix} 1 & 4 & 2 \\ 4 & 0 & 3 \\ 2 & 3 & 0 \end{bmatrix}$$

8. Suppose that G and H are isomorphic simple graphs. Show that their complementary graph $\overline{G}$ and $\overline{H}$ are also isomorphic.

9. Show that a simple graph G with n vertices is connected if it has more than $(n - 1)(n - 2)/2$ edges.

10. Show whether the following pairs of graphs are homoeomorphic.

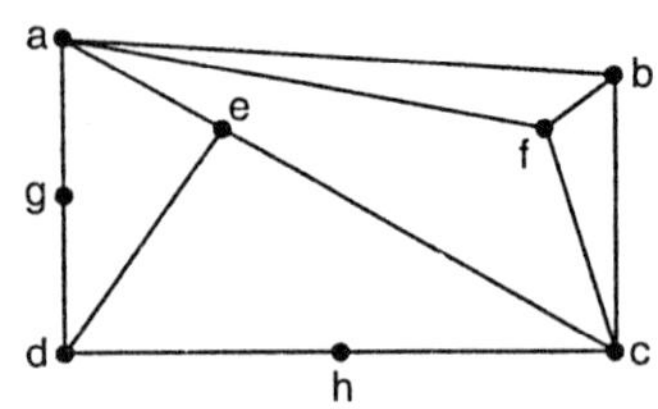

(*i*)

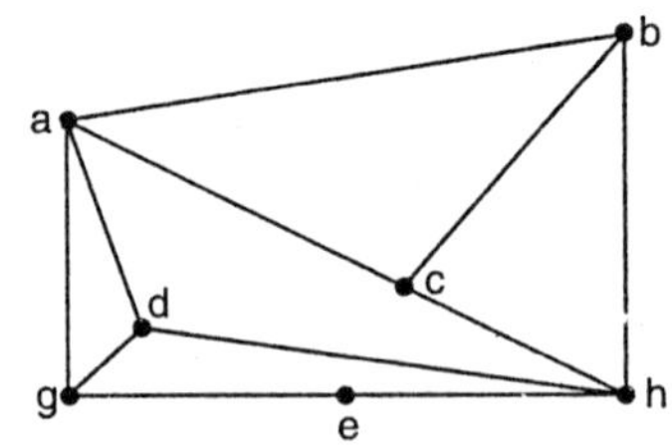

(*ii*)

OTHER IMPORTANT GRAPHS

(*a*) **Bipartite Graph.** A graph G = (V, E) is called a bipartite graph if its vertices V can be partitioned into two subsets V_1 and V_2 such that each edge of G connects a vertex of V_1 to a vertex of V_2. In other words, no edge joining two vertices in V_1 or two vertices in V_2. It is denoted by $K_{m, n}$, where m and n are number of vertices in V_1 and V_2 respectively.

(*b*) **Complete Bipartite Graph.** A graph G = (V, E) is called a complete bipartite graph if its vertices V can be partitioned into two subsets V_1 and V_2, such that each vertex of V_1 is connected to each vertex of V_2. The number of edges in a complete bipartite graph is $m. n$ as each of the m vertices is connected to each of the n vertices. It is denoted by $K_{m, n}$ and $m \leq n$.

Example 31. *Draw the bipartite graphs $K_{2,4}$ and $K_{3, 4}$. Assuming any number of edges.*

Sol. First draw the appropriate number of vertices on two parallel columns or rows and connect the vertices in one column or row with the vertices in other column or row. The bipartite graphs $K_{2, 4}$ and $K_{3, 4}$ are shown in Fig. 74 and Fig. 75 respectively.

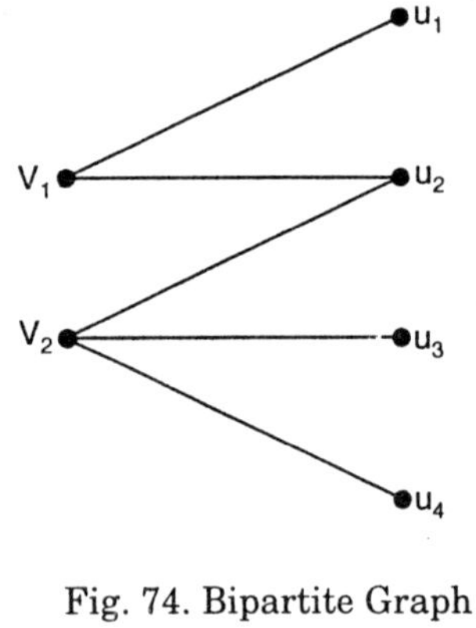

Fig. 74. Bipartite Graph $K_{2, 4}$

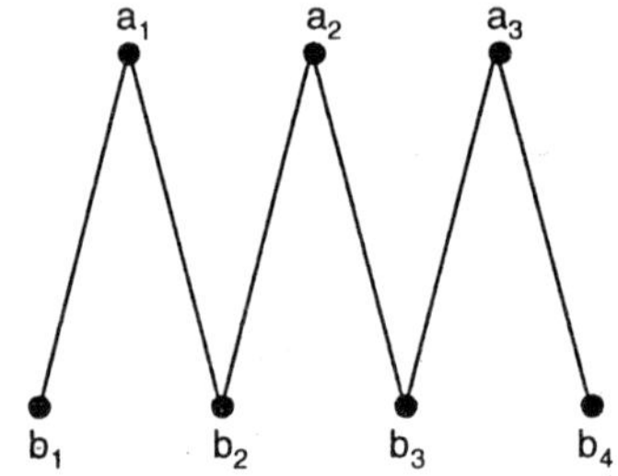

Fig. 75. Bipartite Graph $K_{3, 4}$

Example 32. *Draw the complete bipartite graphs $K_{3,4}$ and $K_{1,5}$.*

Sol. First draw the appropriate number of vertices in two parallel columns or rows and connect the vertices in first column or row with all the vertices in second column or row. The graphs $K_{3,4}$ and $K_{1,5}$ are shown in Fig. 76 and Fig. 77 respectively.

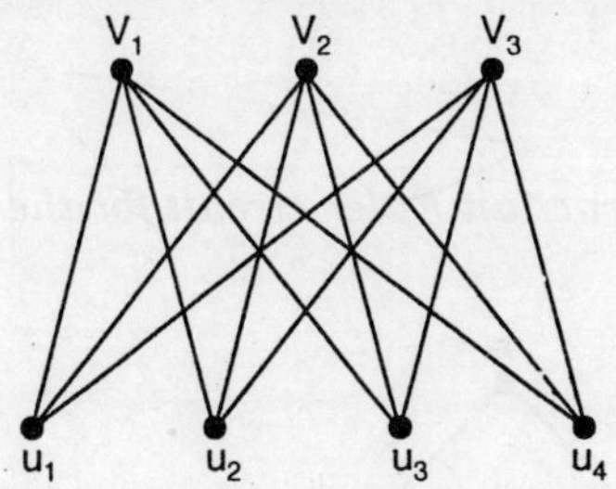

Fig. 76. $K_{3,4}$

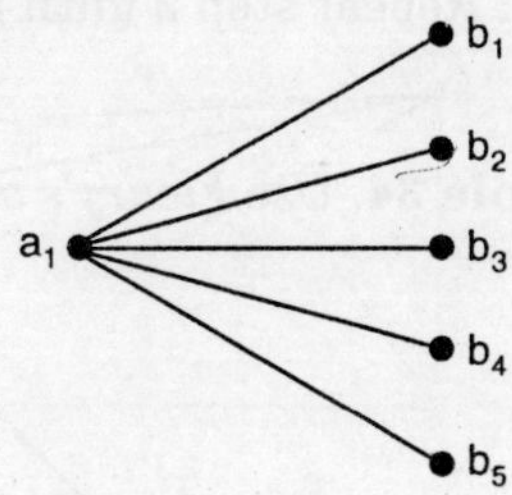

Fig. 77. $K_{1,5}$

Euler Path. An Euler path through a graph is a path whose edge list contains each edge of the graph exactly once.

Euler Circuit. An Euler circuit is a path through a graph, in which the initial vertex appears second time as the terminal vertex.

(*c*) **Euler Graph.** An Euler graph is a graph that possesses an Euler circuit. An Euler circuit uses every edge exactly once but vertices may be repeated.

Example 33. *The graph shown in Fig. 78 is an Euler graph. Determine Euler circuit for this graph.*

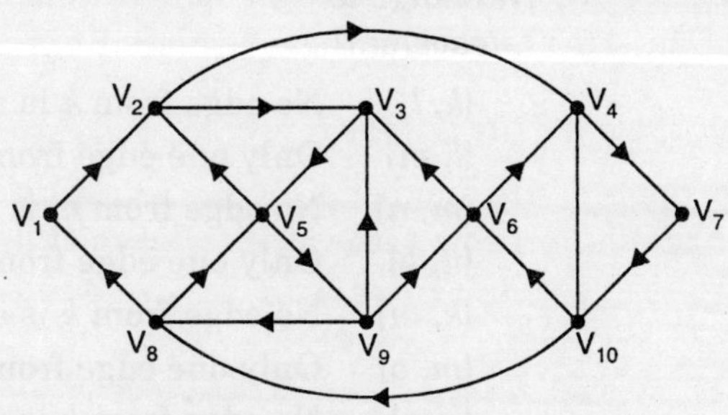

Fig. 78. Euler Graph

Sol. The Euler circuit for this graph is

$$V_1, V_2, V_3, V_5, V_2, V_4, V_7, V_{10}, V_6, V_3, V_9, V_6, V_4, V_{10}, V_8, V_5, V_9, V_8, V_1.$$

We can produce an Euler circuit for a connected graph with no vertices of odd degree. The following algorithm called Fleury's algorithm can do this—

FLEURY'S ALGORITHM

Input

Let G = (V, E) be a connected graph with every vertex of even degree.

Step 1. Select a vertex v_0 of V as the starting vertex to construct the circuit and designate P : v_0 as the starting of the path to be constructed.

Step 2. Consider that P : $v_0, v_1, v_2, ..., v_k$ as been constructed so far. If at v_k there is only one edge say $\{v_k, v_{k+1}\}$, then extend P to P : $v_0, v_1, v_2, ..., v_{k+1}$. Remove $\{v_k, v_{k+1}\}$ from edge set E and v_{k+1} from V. If at v_k there are several edges, choose one that is not a bridge to the remaining graph, say $\{v_k, v_{k+1}\}$. Extend P to P : $v_0, v_1, v_2, ..., v_k, v_{k+1}$ and remove $\{v_k, v_{k+1}\}$ from edge set E.

Step 3. Repeat step 2 until no edges left in E.

END.

Example 34. *Use Fleury's algorithm to construct an Euler circuit for the graph shown in Fig. 78(a).*

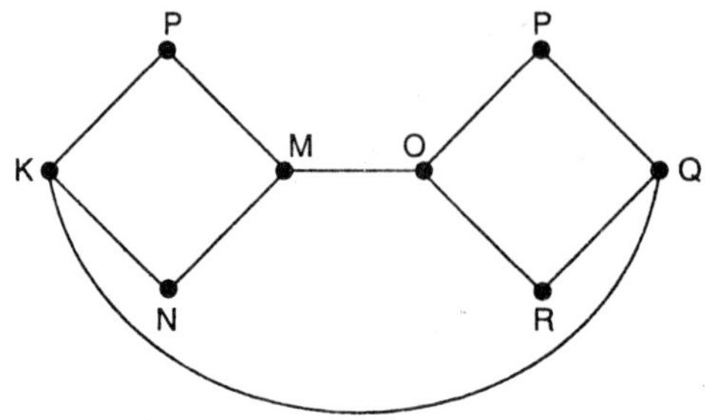

Fig. 78(*a*)

Sol. Start from any vertex, as per step 1. So, choose vertex *k* as the starting vertex. The summary of the results applying step 2 repeatedly is shown in the table—

Current path	*Next edge to traverse*	*Reason*
P : *k*	{*k*, *l*}	No edge from *k* is a bridge. Choose any one.
P : *k*, *l*	{*l*, *m*}	Only one edge from *l*.
P : *k*, *l*, *m*	{*m*, *n*}	No edge from *m* is a bridge. Choose any one.
P : *k*, *l*, *m*, *n*	{*n*, *k*}	Only one edge from *n*.
P : *k*, *l*, *m*, *n*, *k*	{*k*, *m*}	No edge from *k* is a bridge. Choose any one.
P : *k*, *l*, *m*, *n*, *k*, *m*	{*m*, *o*}	Only one edge from *m* remains.
P : *k*, *l*, *m*, *n*, *k*, *m*, *o*	{*o*, *p*}	No edge from *k* is a bridge. Choose any one.
P : *k*, *l*, *m*, *n*, *k*, *m*, *o*, *p*	{*p*, *q*}	Only one edge from *k* remains.
P : *k*, *l*, *m*, *n*, *k*, *m*, *o*, *p*, *q*	{*q*, *r*}	No edge from *k* is a bridge. Choose any one.
P : *k*, *l*, *m*, *n*, *k*, *m*, *o*, *p*, *q*, *r*	{*r*, *o*}	Only one edge from *r* remains.
P : *k*, *l*, *m*, *n*, *k*, *m*, *o*, *p*, *q*, *r*, *o*	{*o*, *q*}	Only one edge from *o* remains.
P : *k*, *l*, *m*, *n*, *k*, *m*, *o*, *p*, *q*, *r*, *o*, *q*	{*q*, *k*}	Only one edge from *q* remains.
P : *k*, *l*, *m*, *n*, *k*, *m*, *o*, *p*, *q*, *r*, *o*, *q*, *k*		

Hamiltonian Path. A Hamiltonian path through a graph is a path whose vertex list contain each vertex of the graph exactly once, except if path is a circuit.

Hamiltonian Circuit. A Hamiltonian circuit is a path in which the initial vertex appears a second time as the terminal vertex.

(*d*) **Hamiltonian Graph.** A Hamiltonian graph is a graph that possesses a Hamiltonian path. A Hamiltonian path uses each vertex exactly once but edges may not be included.

Example 35. *The graph shown in Fig. 79 is a Hamiltonian graph. Determine Hamiltonian circuit for this graph.*

Sol. The Hamiltonian circuit is shown in Fig. 80.

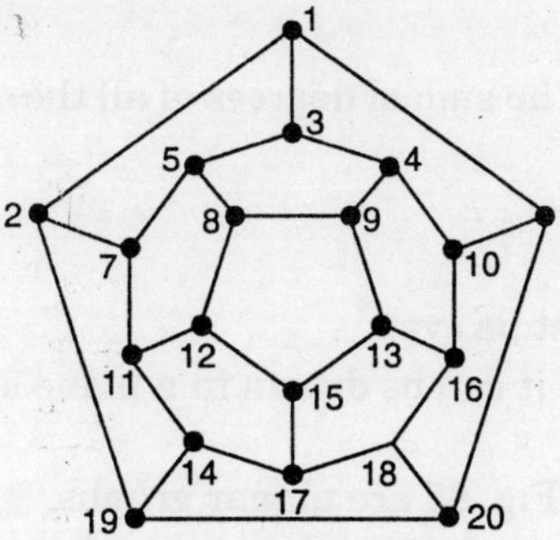

Fig. 79. Hamiltonian Graph

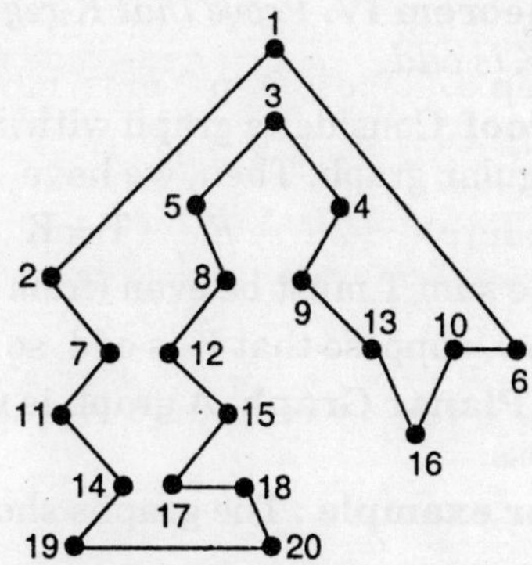

Fig. 80. Hamiltonian Circuit

(*e*) **Regular Graph.** A graph is said to be regular or K-regular if all its vertices have the same degree K. A graph whose all vertices have degree 2 is called 2-regular graph. A complete graph K_n is regular of degree $n - 1$.

Example 36. *Draw regular graphs of degree 2 and 3.*

Sol. The regular graphs of degree 2 and 3 are shown in Figs. 81 and 82.

Fig. 81. 2-regular Graph

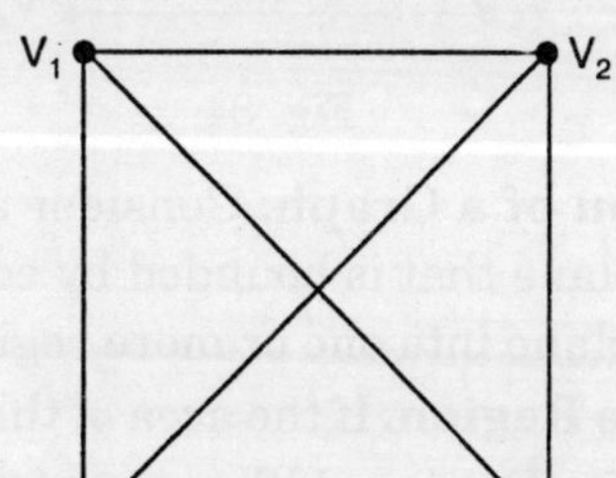

Fig. 82. Regular Graph

Example 37. *Draw a 2-regular graph of five vertices.*

Sol. The 2-regular graph of five vertices is shown in Fig. 83.

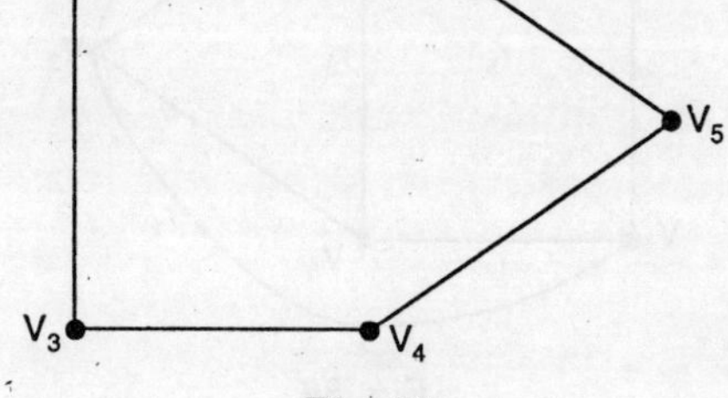

Fig. 83

Example 38. *Draw a 3-regular graph of five vertices.*

Sol. It is not possible to draw 3-regular graph of five vertices. The 3-regular graph must have an even number of vertices.

Theorem IV. *Prove that K-regular graph must have even number of vertices when the value of K is odd.*

Proof. Consider a graph with n vertices. Let T is the sum of degrees of all the n vertices of a K-regular graph. Then, we have

$$T = K \cdot n$$

The sum T must be even (from the theorem I).

Now, suppose that K is odd, so the value of n must be even.

(*f*) **Planar Graph.** A graph is said to be planar if it can be drawn in a plane so that no edges cross.

For example : The graphs shown in Fig. 84 and Fig. 85 are planar graphs.

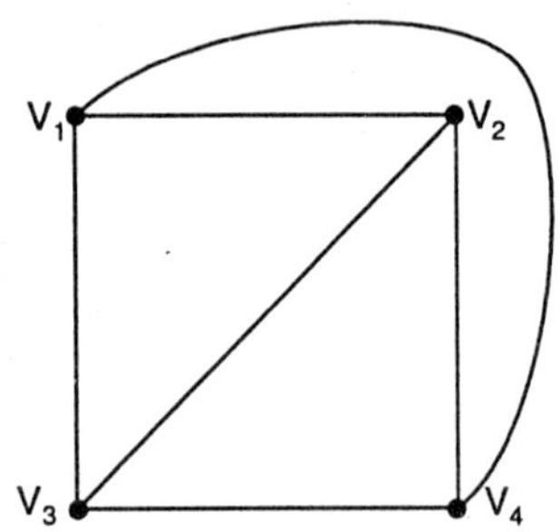

Fig. 84

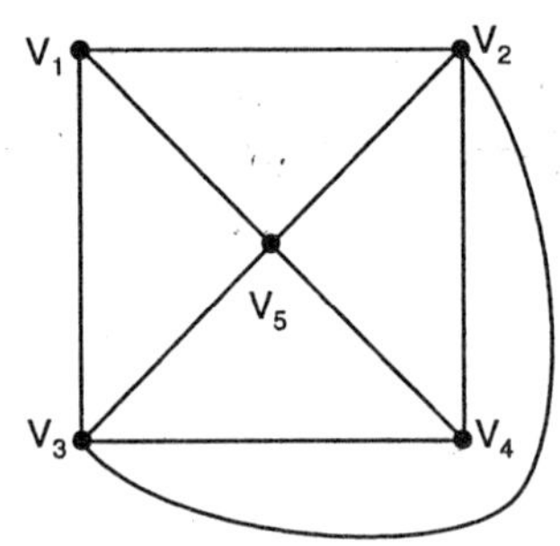

Fig. 85

Region of a Graph. Consider a planar graph G = (V, E). A region is defined to be an area of the plane that is bounded by edges and cannot be further subdivided. A planar graph divides the plane into one or more regions. One of these regions will be infinite.

Finite Region. If the area of the region is finite, then that region is called finite region.

Infinite Region. If the area of the region is infinite, that region is called infinite region. A planar graph has only one infinite region.

Example 39. *Consider the graph shown in Fig. 86. Determine the number of regions, finite regions and an infinite region.*

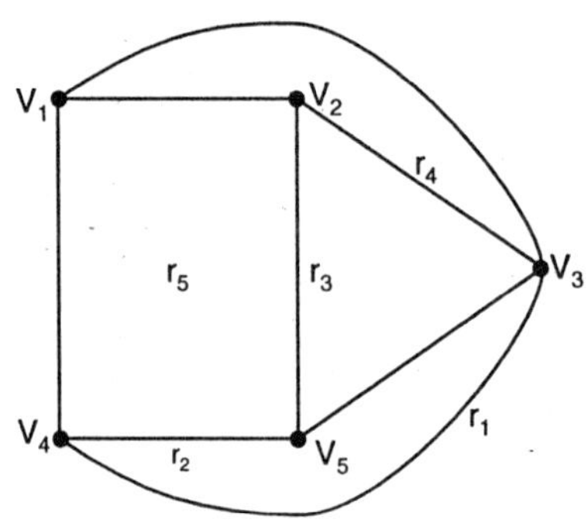

Fig. 86

Sol. There are five regions in the above graph *i.e.*, r_1, r_2, r_3, r_4 and r_5.

There are four finite regions in the graph *i.e.*, r_2, r_3, r_4 and r_5.

There is only one infinite region *i.e.*, r_1.

PROPERTIES OF PLANAR GRAPHS

1. If a connected planar graph G has e edges and r regions, then $r \leq \frac{2}{3}e$.

2. If a connected planar graph G has e edges, v vertices and r regions, then $v - e + r = 2$. [Euler's rule]

3. If a connected planar graph G has e edges and v vertices, then $3v - e \geq 6$.

4. A complete graph K_n is planar if and only if $n < 5$.

5. A complete bipartite graph $K_{m, n}$ is planar if and only if $m < 3$ or $n > 3$.

Example 40. *Prove that complete graph* K_4 *is planar.*

Sol. The complete graph K_4 contains 4 vertices and 6 edges.

We know that for a connected planar graph $3v - e \geq 6$. Hence for K_4, we have

$$3 \times 4 - 6 = 6 \text{ which satisfies the property (3).}$$

Thus K_4 is a planar graph. Hence proved.

STATE AND PROVE EULER'S THEOREM

Statement. Consider any connected planar graph G = (V, E) having R regions, V vertices and E edges. Then

$$V + R - E = 2.$$

Proof. Use induction on the number of edges to prove this theorem.

Basis of Inductions. Assume that the edges $e = 1$. Then we have two cases, graphs of which are shown in Figs. 87 and 88.

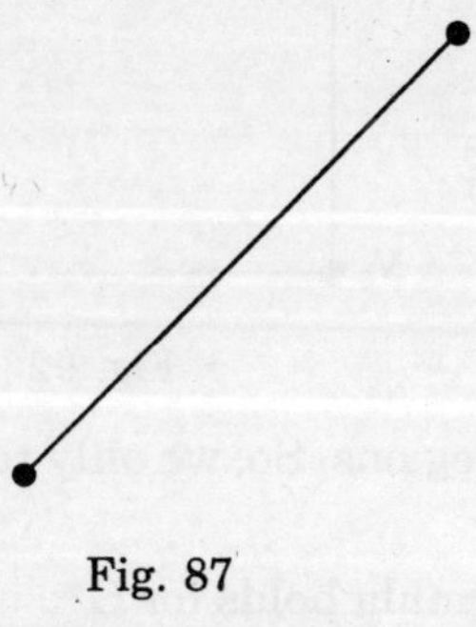

Fig. 87

Fig. 88

In Fig. 87 we have V = 2 and R = 1. Thus 2 + 1 – 1 = 2

In Fig. 88 we have V = 1 and R = 2. Thus 1 + 2 – 1 = 2.

Hence, the basis of induction is verified.

Induction Step. Let us assume that the formula holds for connected planar graphs with K edges.

Let G be a graph with K + 1 edges.

Firstly, we suppose that G contains no circuits. Now, take a vertex v and find a path starting at v. Since G is circuit free, whenever we find an edge, we have a new vertex. At last we will reach a vertex v with degree 1. So we cannot move further as shown in Fig. 89.

Now remove vertex v and the corresponding edge incident on v. So, we are left with a graph G* having K edges as shown in Fig. 90.

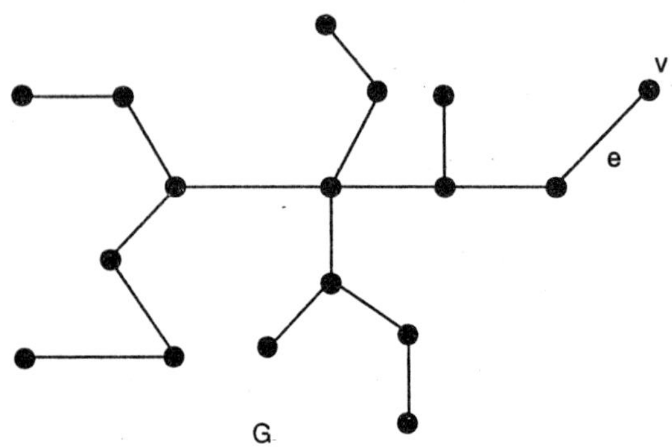

Fig. 89. G

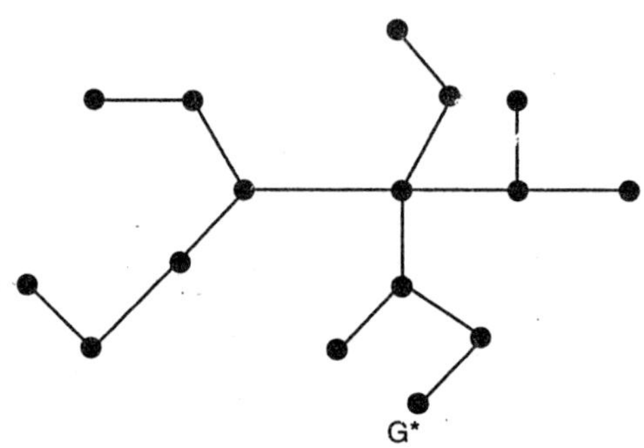

Fig. 90. G*

Hence, by inductive assumption, Euler's formula holds for G*.

Now, since G has one more edge than G*, one more vertex than G* with same number of regions as in G*. Hence, the formula also holds for G.

Secondly, we assume that G contains a circuit and *e* is an edge in the circuit shown in Fig. 91.

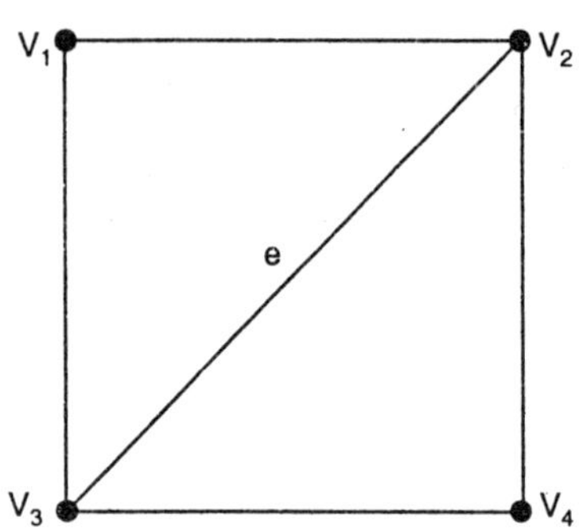

Fig. 91

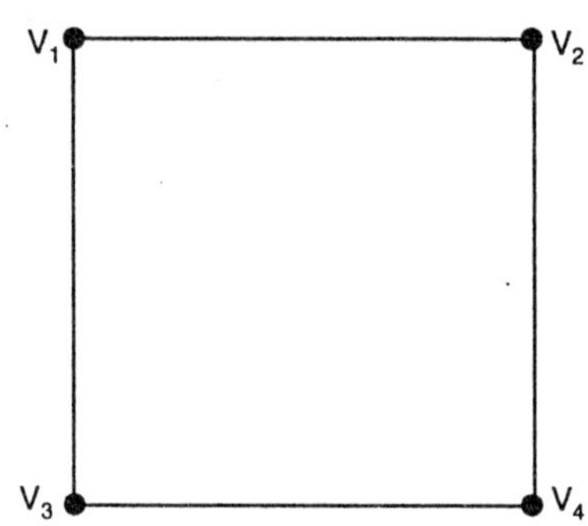

Fig. 92

Now, as *e* is the part of a boundary for two regions. So, we only remove the edge and we are left with graph G* having K edges (Fig. 92).

Hence, by inductive assumption, Euler's formula holds for G*.

Now, since G has one more edge than G*, one more region than G* with same number of vertices as G*. Hence the formula also holds for G which, verifies the inductive step and hence proves the theorem.

(*g*) **Non Planar Graphs.** A graph is said to be non planar if it cannot be drawn in a plane so that no edges cross.

For example : The graphs shown in Figs. 93 and 94 are non planar graphs.

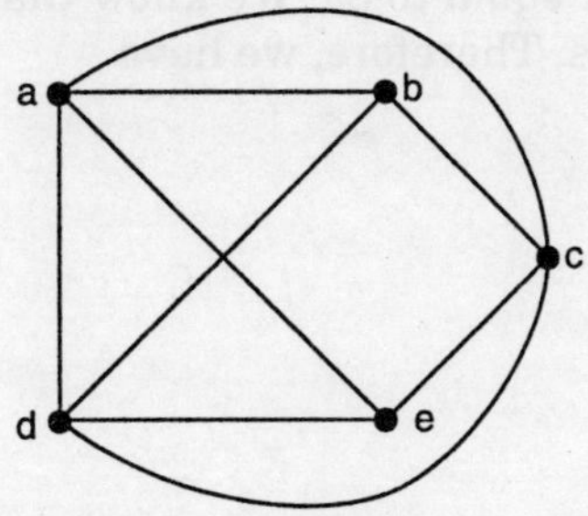

Fig. 93

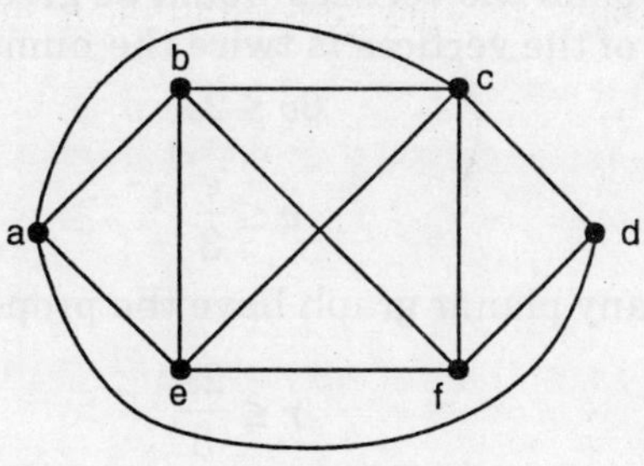

Fig. 94

These graphs cannot be drawn in a plane so that no edges cross hence they are non planar graphs.

Properties of Non planar Graphs. A graph is non-planar if and only if it contains a subgraph homeomorphic to K_5 or $K_{3,3}$ [KURATOWSKI'S THEOREM].

Example 41. *Show that K_5 is non-planar.*

Sol. The complete graph K_5 contains 5 vertices and 10 edges.

Now, for a connected planar graph $3v - e \geq 6$.

Hence, for K_5 we have $3 \times 5 - 10 = 5$ (which does not satisfy property 3 because it must be greater than or equal to 6).

Thus, K_5 is a non planar graph.

Example 42. *Show that the graphs shown in Figs. 95 and 96 are non-planar by finding a subgraph homeomorphic to K_5 or $K_{3,3}$.*

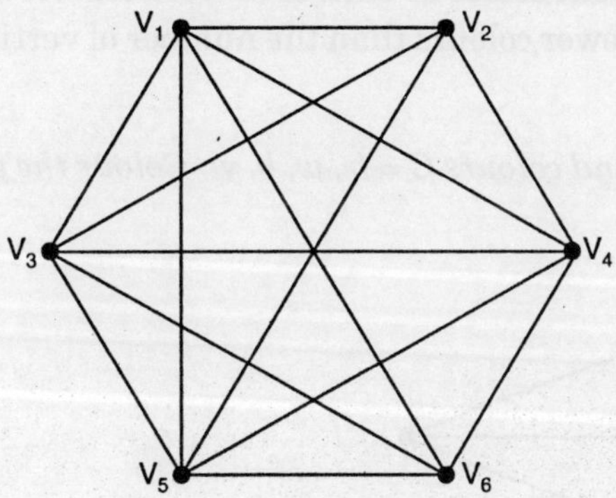

Fig. 95. G_1

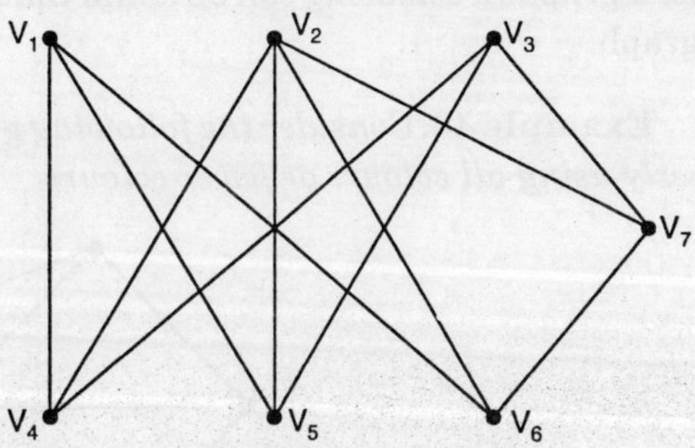

Fig. 96. G_2

Sol. If we remove the edges (V_1, V_4), (V_3, V_4) and (V_5, V_4) the graph G_1, becomes homeomorphic to K_5. Hence it is non-planar.

If we remove the edge (V_2, V_7) the graph G_2 becomes homeomorphic to $K_{3,3}$. Hence it is non-planar.

Theorem V. *Prove that every planar graph has at least one vertex of degree 5 or less than 5.*

Proof. Consider a graph G, whose all vertices are of degree 6 or more, then the sum of the degrees of all the vertices would be greater than or equal to $6v$. We know that the sum of the degrees of the vertices is twice the number of edges. Therefore, we have

$$6v \leq 2e$$

or $$v \leq \frac{e}{3} \quad \text{...(i)}$$

But, any planar graph have the property,

$$r \leq \frac{2e}{3} \quad \text{...(ii)}$$

Also, from Euler's formula, we have

$$2 = v - e + r \quad \text{...(iii)}$$

Now, putting the value of v and r from (*i*) and (*ii*) in (*iii*), we have

$$2 \leq \frac{e}{3} - e + \frac{2e}{3} = 0$$

Since, the statement $2 \leq 0$ is not true, hence we conclude that there must exist some vertex in G with degree 5 or less than 5.

GRAPH COLOURING

Suppose that G = (V, E) is a graph with no multiple edges. A vertex colouring of G is an assignment of colours to the vertices of G such that adjacent vertices have different colours. A graph G is M-colourable if there exists a colouring of G which uses M-colours.

Proper Colouring. A colouring is proper if any two adjacent vertices u and v have different colours otherwise it is called improper colouring.

A graph can be coloured by assigning a different colour to each of its vertices. However, for most graphs a colouring can be found that uses fewer colours than the number of vertices in the graph.

Example 43. *Consider the following graph and colours C = {r, w, b, y}. Colour the graph properly using all colours or fewer colours.*

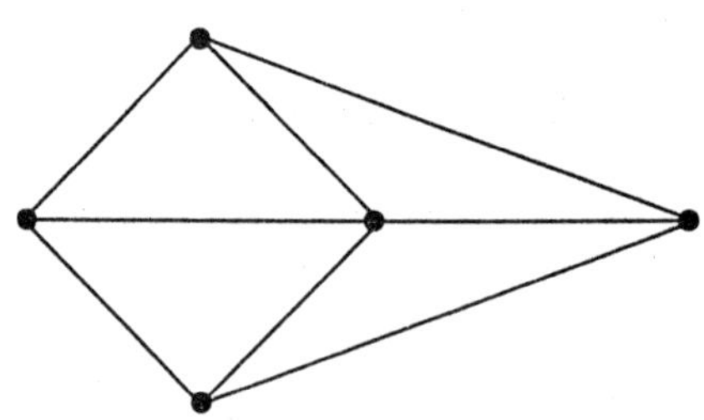

Fig. 97

Sol. Fig. 98 shows the graph properly coloured with all the four colours.

Fig. 99 shows the graph properly coloured with three colours.

It is not possible to colour this graph properly with two colours.

Chromatic number of G. The minimum number of colours needed to produce a proper colouring of a graph G is called the chromatic number of G and is denoted by $\chi(G)$.

The graph shown in Fig. 97 is minimum 3-colourable, hence $\chi(G) = 3$.

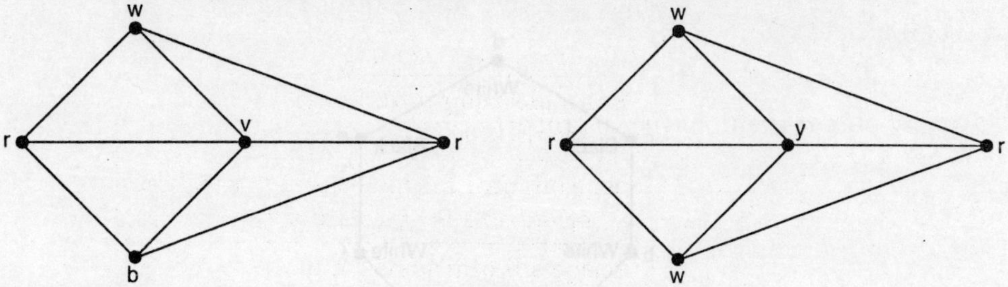

Fig. 98

Example 44. *The chromatic number of k_n is n.*

Sol. A colouring of k_n can be constructed using n colours by assigning a different colour to each vertex. No two vertices can be assigned the same colour, since every two vertices of this graph are adjacent. Hence the chromatic number of $k_n = n$.

Example 45. *The chromatic number of complete bipartite graph $k_{m,\,n}$, where m and n are positive integers is two.*

Sol. The number of colours needed does not depend upon m and n. However, only two colours are needed colour the set of m vertices with one colour and the set of n vertices with a second colour. Since, edges connect only a vertex from the set of m vertices and a vertex from the set of n vertices, no two adjacent vertices have the same colour.

Note 1. Every connected bipartite simple graph has a chromatic number of 2 or 1.

2. Conversely, every graph with a chromatic number of 2 is bipartite.

Example 46. *The chromatic number of graph c_n, where c_n is the cycle with n vertices is either 2 or 3.*

Sol. Two colours are needed to colour c_n, where c_n is even. To construct such a colouring, simply pick a vertex and colour it black. Then move around the graph in clockwise direction colouring the second vertex white, the third vertex black, and so on. The nth vertex can be coloured white since the two vertices adjacent to it, namely the $(n-1)$th and the first are both coloured black as shown in Fig. 99.

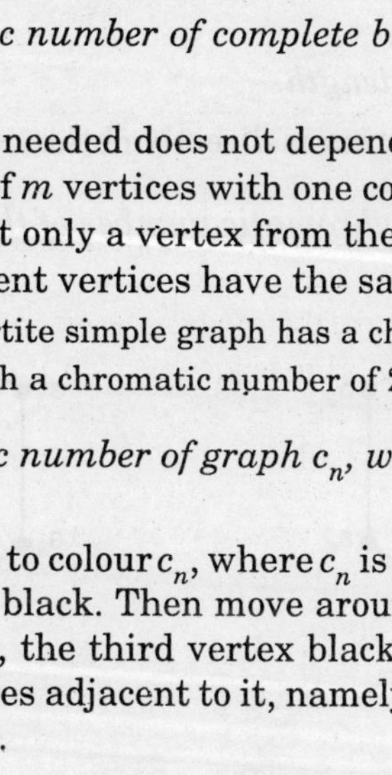

Fig. 99

When n is odd and $n > 1$, the chromatic number of c_n is 3. To construct such a colouring, pick an initial vertex. First use only two colours and alternate colours as the graph is traversed in a clockwise direction. However, the nth vertex reached is adjacent to two vertices of different colours, the first and $(n - 1)$th. Hence, a third colour is needed.

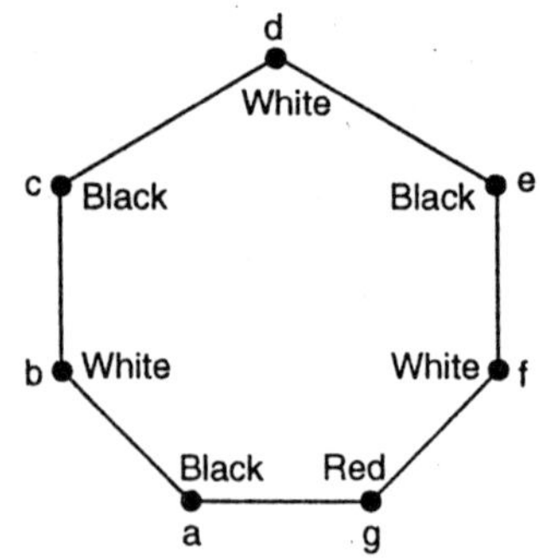

Fig. 100

Theorem VI. *The following are equivalent for a graph G :*

(*i*) *G is 2-colourable*

(*ii*) *G is bipartite*

(*iii*) *Every cycle of G has even length.*

Proof. The proof of this theorem is directly obtained from the example 44, 45 and 46.

Example 47. *Determine the chromatic number of the graphs shown in Fig. 101.*

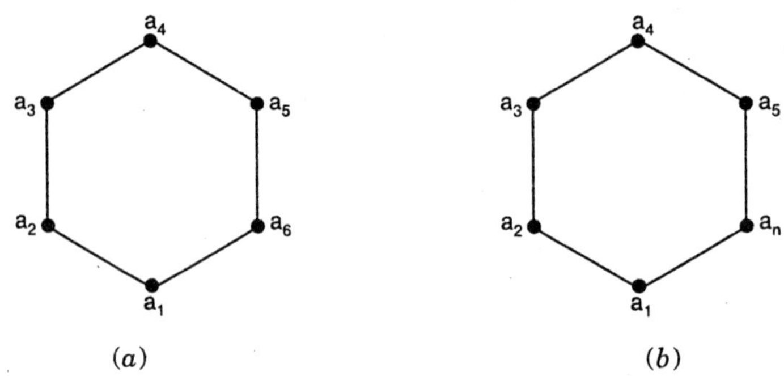

Fig. 101

Sol. The graphs shown in Fig. 101(*a*), has the chromatic number $\chi(G) = 2$.

The graph shown in Fig. 101(*b*) has the chromatic number

$\chi(G) = 2$, when n is an even number and

$\chi(G) = 3$, where n is odd.

Theorem VII. *The vertices of every planar graph can be properly coloured with five colours.*

Proof. We will prove this theorem by induction. All the graphs with 1, 2, 3, 4 or 5 vertices can be properly coloured with five colours. Now let us assume that every planar graph with $n - 1$ vertices can be properly coloured with five colours. Next, if we prove that any planar graph G with n vertices will require no more than five colours, we have done.

Consider the planar graph G with n vertices.

Since G is planar, it must have at least one vertex with degree five or less as shown in theorem V. Assume this vertex to be 'u'.

Let G_1 be a graph of $n - 1$ vertices obtained from G by deleting vertex 'u'. The G_1 graph requires no more than five colours (Induction hypothesis). Consider that the vertices in G_1 have been properly coloured and now add to it 'u' and all the edges incident on u. If the degree of u is 1, 2, 3, or 4, a proper colour to u can be easily assigned.

Now, we have one case left, in which the degree of u is 5, and all the 5 colours have been used in colouring the vertices adjacent to u, as shown in Fig. 102.

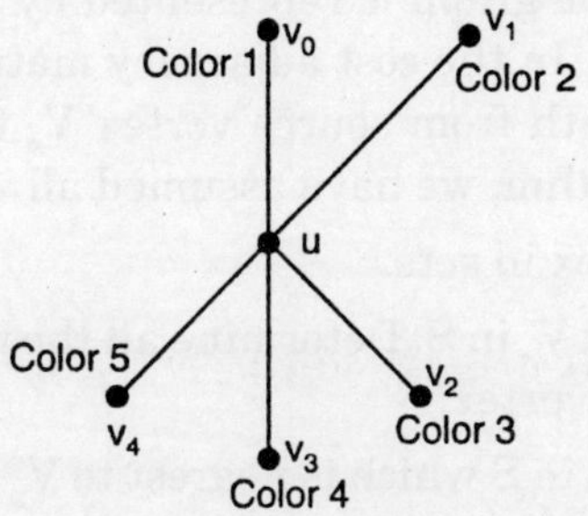

Fig. 102

Suppose that there is a path in G_1 between vertices v_0 and v_3 coloured alternately with colours 1 and 4 as shown in Fig. 103.

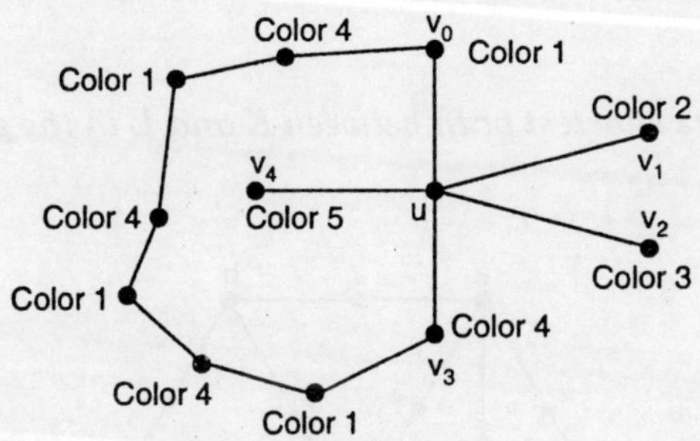

Fig. 103

Then a similar path between v_4 and v_2, coloured alternately with colours 5 and 3, cannot exist ; otherwise, these two paths will intersect and cause G to be non-planar.

Thus, if there is no path between v_4 and v_2 coloured alternately with colour 5 and 3 of all vertices connected to v_2 through vertices of alternating colours 5 and 3. This interchange will colour vertex v_2 with colour 5 and yet keep G, properly coloured. As vertex v_4 is still with colour 5, the colour 3 is left over with which to colour vertex u which proves the theorem.

SHORTEST PATH IN WEIGHTED GRAPHS

Weighted graphs can be used to represent highways connecting the different cities. The weighted edges represent the distance between different cities and the vertices represent the cities. A common problem with this type of graph is to find the shortest path from one city to another city. There are many ways to tackle this problem one of which is as follows :

Shortest Paths from Single Source. We will find shortest paths from a single vertex to all other vertices of the graph. The first algorithm was proposed by E. Dijkstra in 1959. Some common terms related with this algorithm are as follows :

Path Length. The length of a path is the sum of the weights of the edges on that path.

Source. The starting vertex of the graph from which we have to start to find the shortest path.

Destination. The terminal or last vertex upto which we have to find the path.

Dijkstra's Algorithm. This algorithm maintains a sets of vertices whose shortest path from source is already known. The graph is represented by its cost adjacency matrix, where cost being the weight of the edge. In the cost adjacency matrix of the graph, all the diagonal values are zero. If there is no path from source vertex V_s to any other vertex V_i then it is represented by $+ \infty$. In this algorithm, we have assumed all weights are positive.

1. Initially there is no vertex in sets.

2. Include the source vertex V_s in S. Determine all the paths from V_s to all other vertices without going through any other vertex.

3. Now, include that vertex in S which is nearest to V_s and find shortest paths to all the vertices through this vertex and update the values.

4. Repeat the step 3 until $n - 1$ vertices are not included in S if there are n vertices in the graph.

After completion of the process, we got the shortest paths to all the vertices from the source vertex.

Example 48. *Find the shortest path between K and L in the graph shown in Fig. 104 by using Dijkstra's Algorithm.*

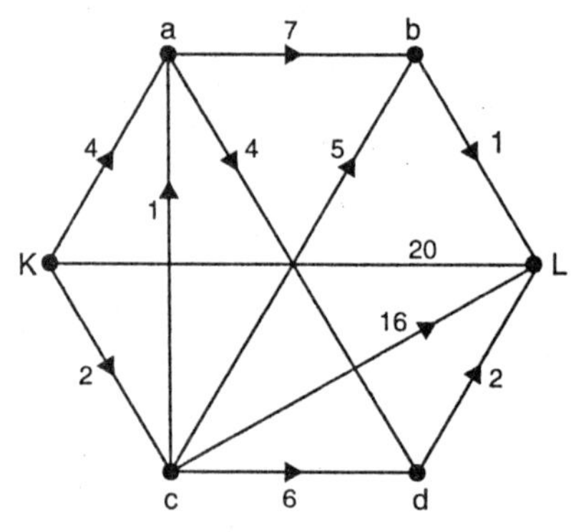

Fig. 104

Sol. Step I. Include the vertex K in S and determine all the direct paths from K to all other vertices without going through any other vertex.

S	Distance to all other vertices					
	K	*a*	*b*	*c*	*d*	L
K	0	4(K)	∞	2(K)	∞	20(K)

Step II. Include the vertex in S which is nearest to K and determine shortest paths to all vertices through this vertex and update the values. The nearest vertex is *c*.

S	Distance to all other vertices					
	K	*a*	*b*	*c*	*d*	L
K, *c*	0	3(K, *c*)	7(K, *c*)	2(K)	8(K, *c*)	18(K, *c*)

Step III. The vertex which is 2nd nearest to K is 9, included in S.

S	Distance to all other vertices					
	K	*a*	*b*	*c*	*d*	L
K, *c, a*	0	3(K, *c*)	7(K, *c*)	2(K)	7(K, *c, a*)	18(K, *c*)

Step IV. The vertex which is 3rd nearest to K is *b*, is included in S.

S	Distance to all other vertices					
	K	*a*	*b*	*c*	*d*	L
K, *c, a, b*	0	3(K, *c*)	7(K, *c*)	2(K)	7(K, *c, a*)	8(K, *c, b*)

Step V. The vertex which is next nearest to K is *d*, is included in S.

S	Distance to all other vertices					
	K	*a*	*b*	*c*	*d*	L
K, *c, a, b, d*	0	3(K, *c*)	7(K, *c*)	2(K)	7(K, *c, a*)	8(K, *c, b*).

Since, $n - 1$ vertices included in S. Hence we have found the shortest distance from K to all other vertices. Thus, the shortest distance between K and L is 8 and the shortest path is K, *c, b*, L.

TRAVELLING SALESMAN PROBLEM

Suppose a salesman wants to visit a certain number of cities alloted to him. He knows the distance of journey between every pair of cities. His problem is to select such a route that starts from his home city, passes through each city exactly once and return to his home city in the shortest possible distance. This problem is closely related to finding a Hamiltonian circuit of minimum length. If we represent the cities by vertices and road connecting two cities by edges we get a weighted graph where, with every edge ei a number wi (weight) is associated.

A physical interpretation of the above abstract is: consider a graph G as a map of n cities where $w(i, j)$ is the distance between cities i and j. A salesman wants to have a tour of the n cities which starts and ends at the same city and includes visiting each of the remaining $n - 1$ cities once and only once.

In the graph if we have n vertices (cities) then there are $(n - 1)!$ edges (routes), and the total number of Hamiltonian circuits in a complete graph of n vertices will be $(n - 1)!/2$.

Nearest Neighbour Method. This procedure gives reasonably good results for the travelling salesman problem. The method is as follows.

Step I. Select any arbitrary vertex and find the vertex that is nearest to this starting vertex to form an initial path of one edge.

Step II. Let v denote the latest vertex that was added to the path. Now, among the rest of the vertices that are not in the path, select the closest one to v and add to the path, the edge-connecting v and this vertex. Repeat this step until all the vertices of graph G are included in the path.

Step III. Join starting vertex and the last vertex added by an edge and form the circuit.

Example 49. *Use the nearest-neighbour method to solve the following travelling salesman problem, for the graph shown in Fig. 105, starting at vertex* v_1.

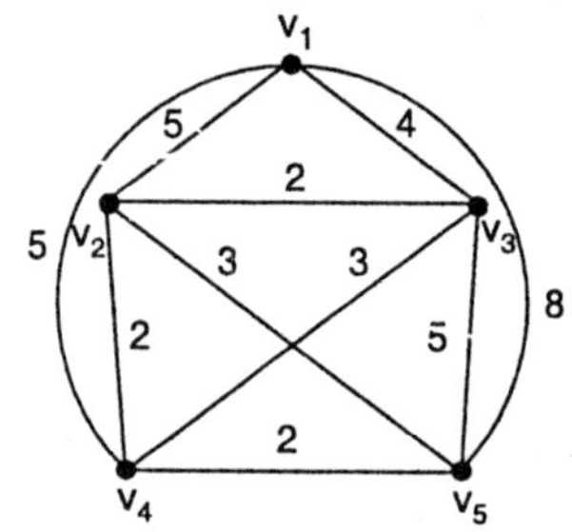

Fig. 105

Sol. We have to start from vertex v_1. By using nearest neighbour method, vertex by vertex construction of the tour or Hamiltonian circuit is shown in the Fig. 106.

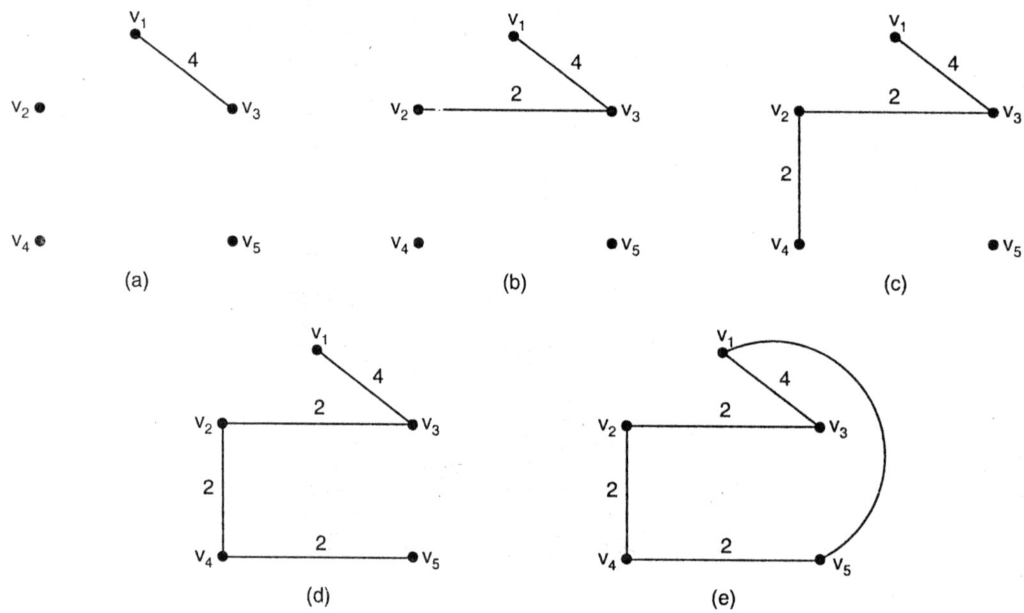

Fig. 106

The total distance of this route is 18.

EXERCISES

1. Prove that a circuit in a bipartite graph must have an even number of edges.
2. Prove that a cycle in a bipartite graph must have at least four edges.
3. Show that each graph is planar by redrawing it.

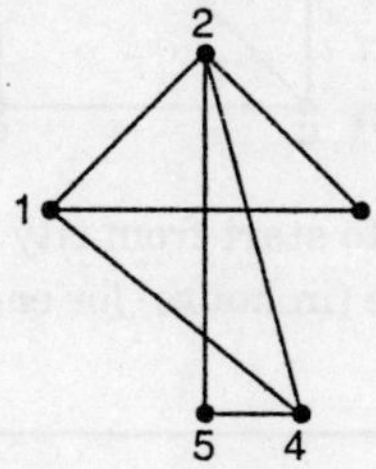

(i)

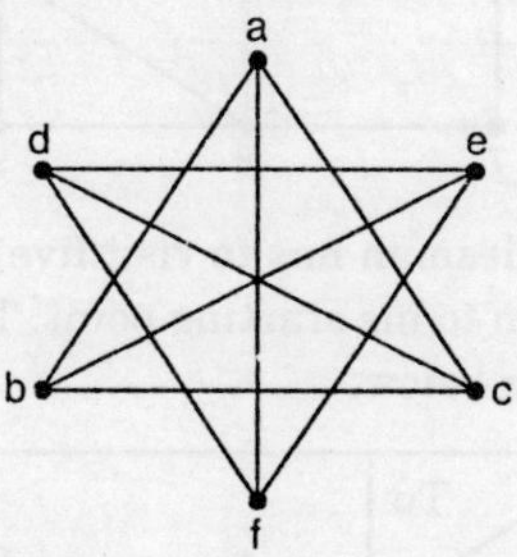

(ii)

4. Show that the following graph is no planar by finding a subgraph homeomorphic to $k_{3,3}$ or k_5.

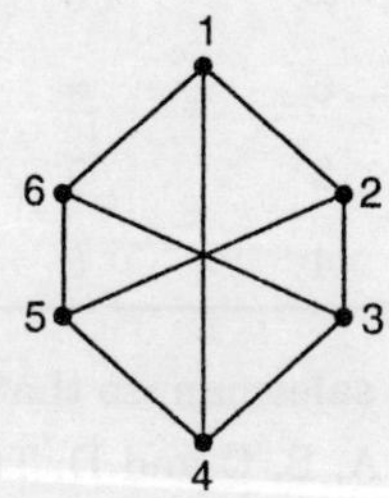

5. Determine the chromatic number of each graph

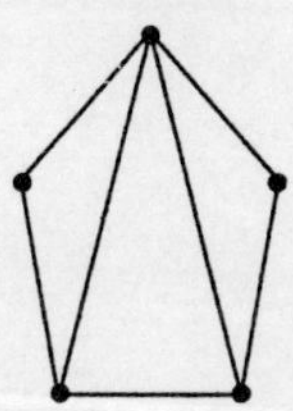
(i)

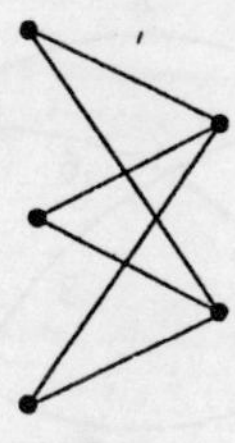
(ii)

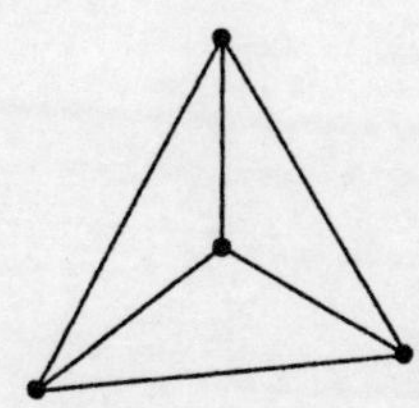
(iii)

6. Draw the following graphs.
   (i) K_7 (ii) $K_{1,8}$ (iii) $K_{4,4}$
7. For which values of m and n is $K_{m,n}$ regular ?
8. Show that K_n has a Hamiltonian circuit whenever $n \geq 3$.
9. Find the shortest path between a and z in the graph shown in figure.

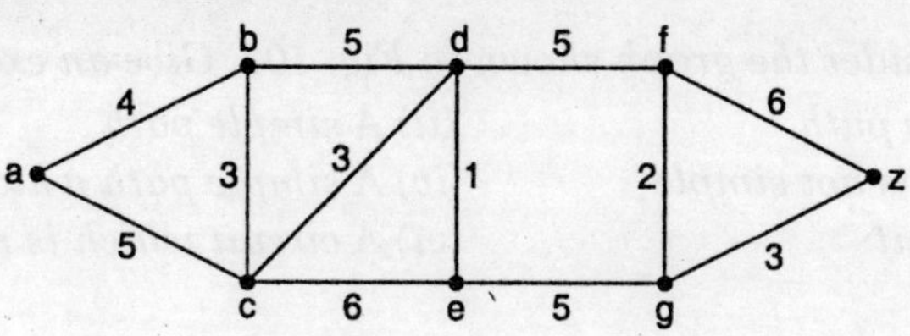

10. Determine whether each of the following graph has an Euler circuit and Hamiltonian circuit. Construct such a circuit when one exists.

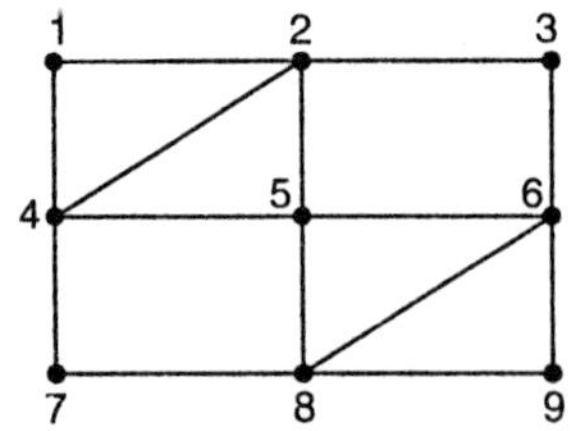

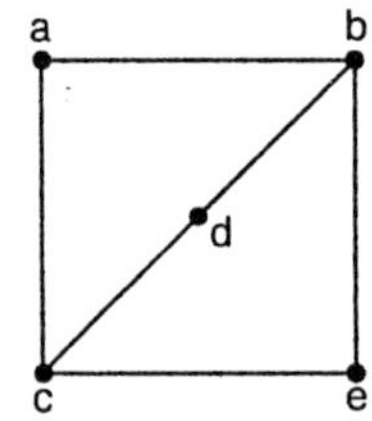

11. A travelling salesman has to visit five cities. He wishes to start from city A, visit each city once and then return to his starting point. The travelling time (in hours) for each city from a particular city is given below:

From \ To	A	B	C	D	E
A	∞	4	7	3	4
B	4	∞	6	3	4
C	7	6	∞	7	5
D	3	3	7	∞	7
E	4	4	5	7	∞

What is the sequence of visit of the salesman, so that total travel time is minimum.

12. A salesman has to visit four cities A, B, C and D. The distance between the four cities are as shown in figure.

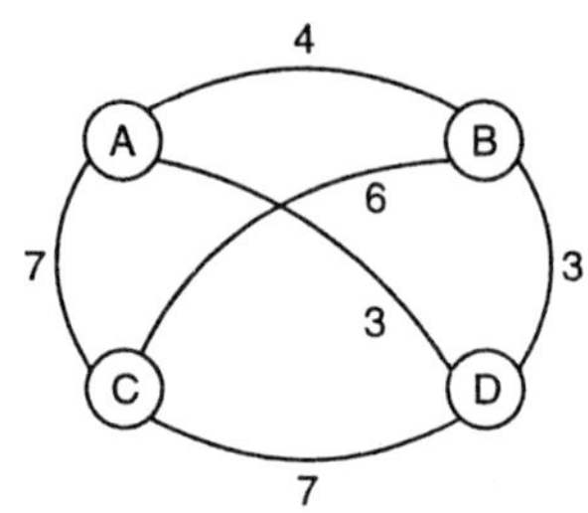

If the salesman starts from city A and has to come back to city A, which route should he select so that total distance travelled by him is minimized ?

SOLVED PROBLEMS

Problem 1. *Consider the graph shown in Fig. 107. Give an example of the following :*

(*i*) *An elementary path* (*ii*) *A simple path*

(*iii*) *A path which is not simple* (*iv*) *A simple path which is not elementary*

(*v*) *A simple circuit* (*vi*) *A circuit which is not simple.*

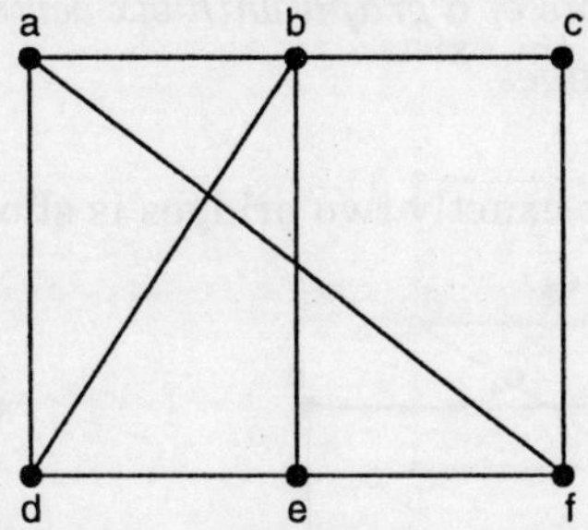

Fig. 107

Sol. There are many solutions to the above problems, but we will take only one for each.

(*i*) An elementary path is *a, b, c, f, e, d.*

(*ii*) A simple path is *a, f, e, b, d.*

(*iii*) A path which is not simple is *a, b, e, e b, d.*

(*iv*) A simple path which is not elementary is *a, b, e, f, c, b, d.*

(*v*) A simple circuit is *a, b, c, f, a.*

(*vi*) A circuit which is not simple is *a, b, e, f, c, b, d, a.*

Problem 2. *Give an example of a graph with six vertices that has exactly two cut points.*

Sol. The graph with exactly two cut points is shown in Fig. 108.

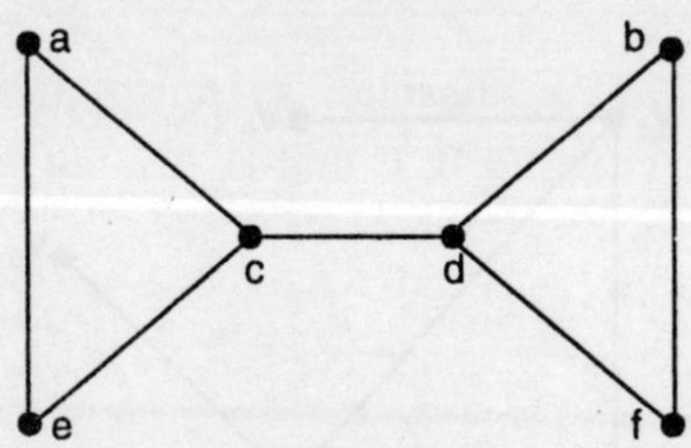

Fig. 108

The two cut points in this graph are *c* and *d*.

Problem 3. *Give an example of a graph with six vertices that has no cut points.*

Sol. The graph with no cut points is shown in Fig. 109.

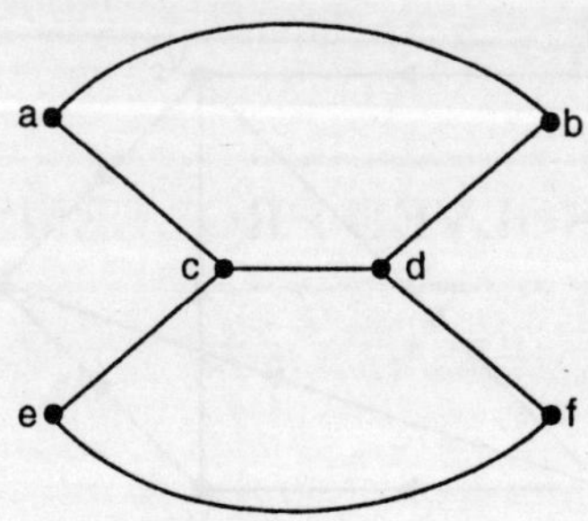

Fig. 109

Problem 4. *Give an example of a graph with six vertices for the following :*

(*a*) *that has exactly two bridges.*

(*b*) *that has no bridges.*

Sol. (*a*) The graph that has exactly two bridges is shown in Fig. 110.

The two bridges are e_1 and e_2.

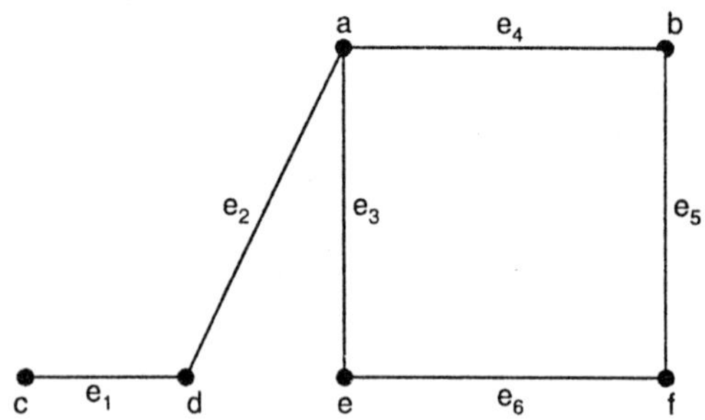

Fig. 110

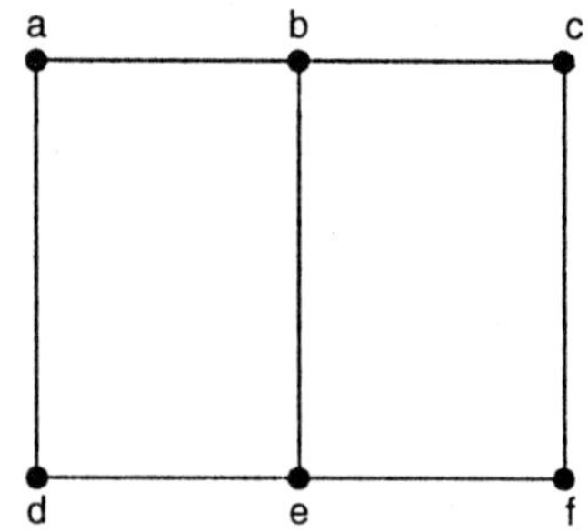

Fig. 111

(*b*) The graph with no bridges is shown in Fig. 111.

Problem 5. *Draw a graph whose every edge is a bridge.*

Sol. The graph shown in Fig. 112 is a graph whose every edge is a bridge because if any edge is removed from the graph *h*, we got two components or a disconnected graph.

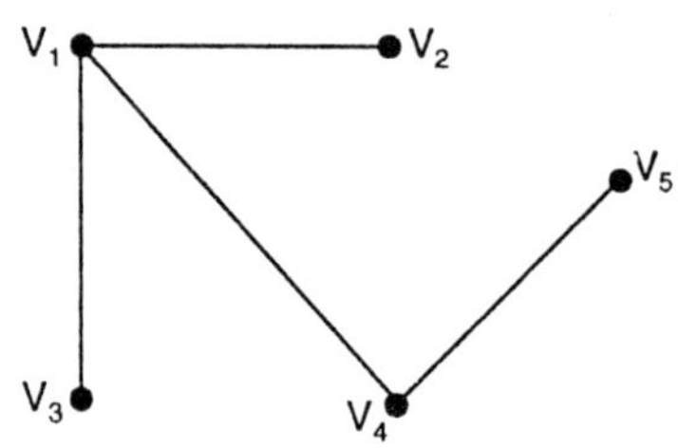

Fig. 112

Problem 6. *Consider the directed graph as shown in Fig. 113. Show the four different subgraphs of this graph having at least four vertices.*

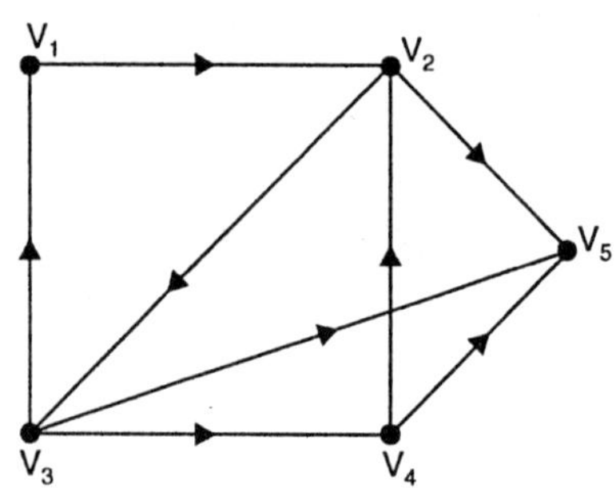

Fig. 113

Sol. The subgraphs of the directed graph are shown in Figs. 114, 115, 116 and 117.

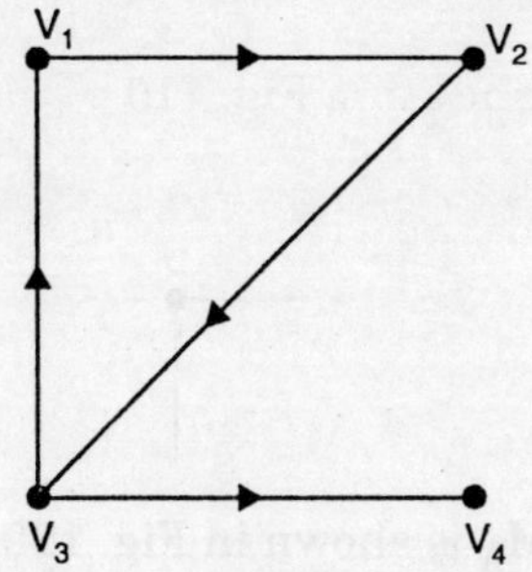

Fig. 114

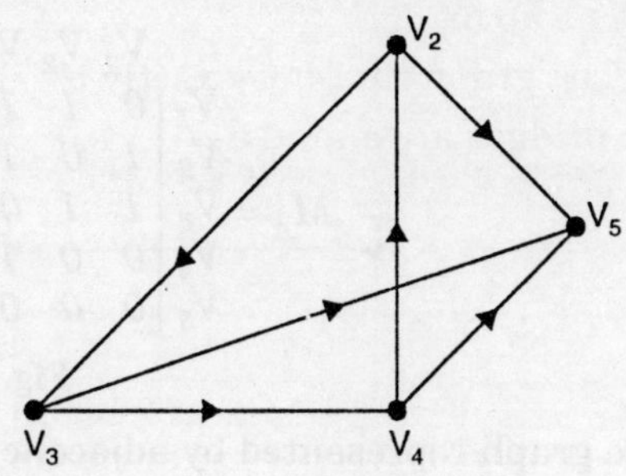

Fig. 115

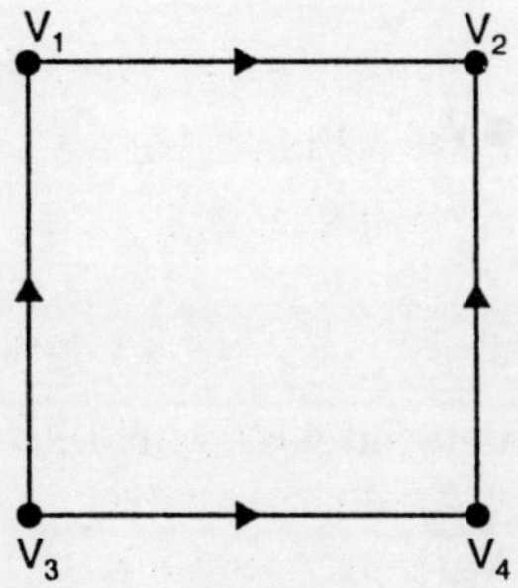

Fig. 116

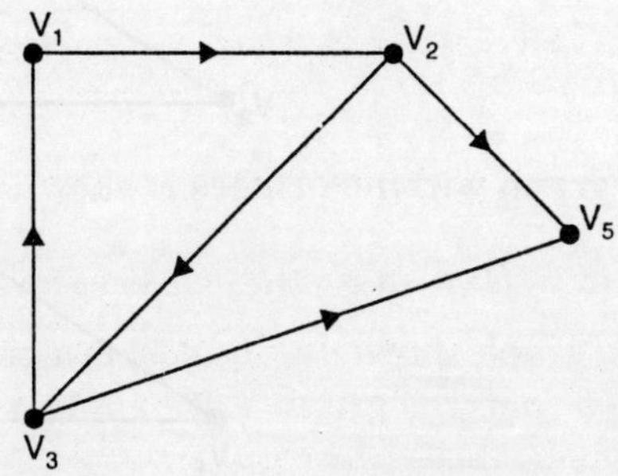

Fig. 117

Problem 7. *Consider the multigraph shown in Fig. 118. Show two different subgraphs of this multigraph which are itself multigraphs.*

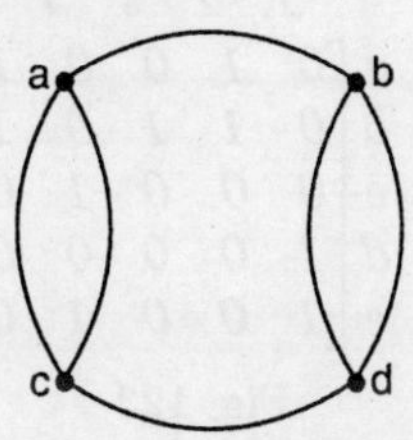

Fig. 118

Sol. The two different subgraphs of this multigraph which are itself multigraphs are shown in Figs. 119 and 120.

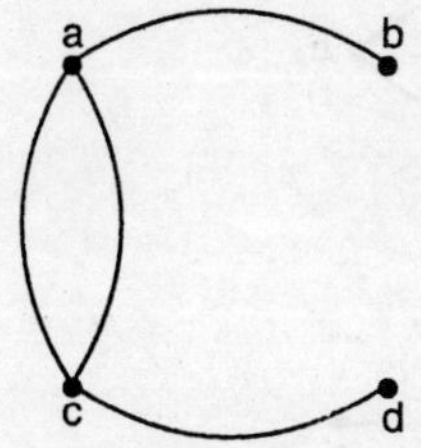

Fig. 119

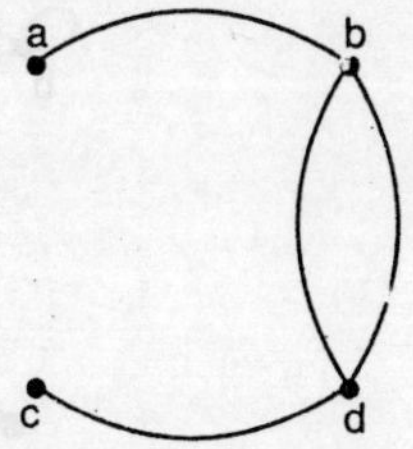

Fig. 120

Problem 8. *Draw the undirected graph represented by adjacency matrix M_A shown in Fig. 121.*

$$M_A = \begin{array}{c} \\ V_1 \\ V_2 \\ V_3 \\ V_4 \\ V_5 \end{array} \begin{array}{c} \begin{array}{ccccc} V_1 & V_2 & V_3 & V_4 & V_5 \end{array} \\ \begin{bmatrix} 0 & 1 & 1 & 0 & 0 \\ 1 & 0 & 1 & 0 & 0 \\ 1 & 1 & 0 & 1 & 0 \\ 0 & 0 & 1 & 0 & 1 \\ 0 & 0 & 0 & 1 & 1 \end{bmatrix} \end{array}$$

Fig. 121

Sol. The graph represented by adjacency matrix M_A is shown in Fig. 122.

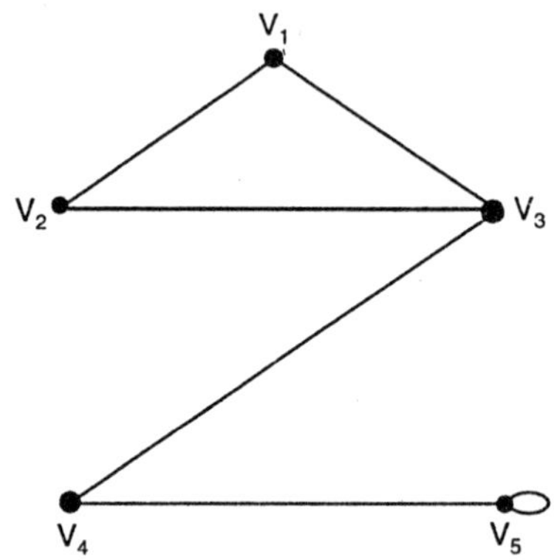

Fig. 122

Problem 9. *Draw the undirected graph represented by incidence matrix M_I shown in Fig. 123.*

$$M_I = \begin{array}{c} \\ a \\ b \\ c \\ d \\ e \end{array} \begin{array}{c} \begin{array}{cccccc} e_1 & e_2 & e_3 & e_4 & e_5 & e_6 \end{array} \\ \begin{bmatrix} 0 & 1 & 0 & 0 & 1 & 1 \\ 0 & 1 & 1 & 0 & 1 & 0 \\ 0 & 0 & 0 & 1 & 0 & 1 \\ 1 & 0 & 0 & 0 & 0 & 0 \\ 1 & 0 & 0 & 1 & 0 & 0 \end{bmatrix} \end{array}$$

Fig. 123

Sol. The graph represented by incidence matrix M_I is shown in Fig. 124.

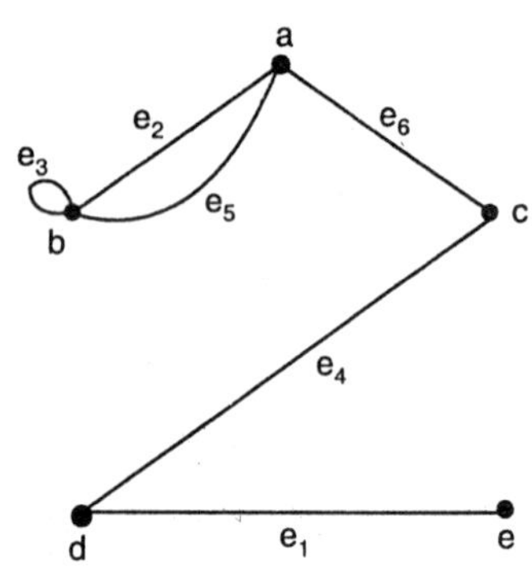

Fig. 124

Problem 10. *Draw the multigraph G whose adjacency matrix M_A is shown in Fig. 125.*

$$M_A = \begin{bmatrix} 1 & 3 & 0 & 0 \\ 3 & 0 & 1 & 1 \\ 0 & 1 & 2 & 2 \\ 0 & 1 & 2 & 0 \end{bmatrix}$$

Fig. 125

Sol. The multigraph corresponding to the adjacency matrix M_A is shown in Fig. 126.

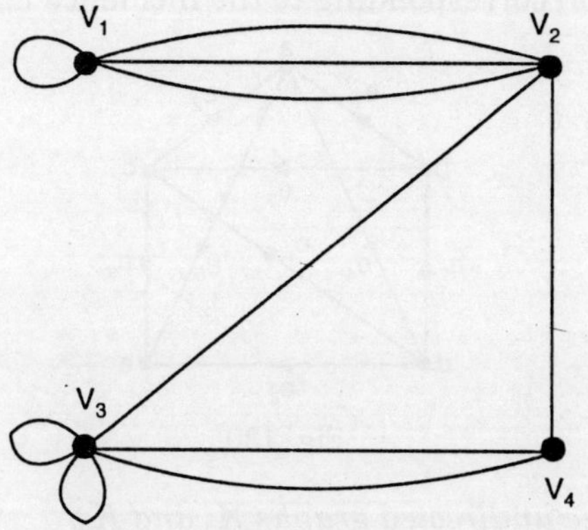

Fig. 126

Problem 11. *Draw the directed graph G whose adjacency matrix M_A is shown in Fig. 127.*

$$M_A = \begin{array}{c} \\ a \\ b \\ c \\ d \\ e \\ f \end{array} \begin{array}{c} \begin{array}{cccccc} a & b & c & d & e & f \end{array} \\ \begin{bmatrix} 0 & 0 & 1 & 0 & 0 & 0 \\ 0 & 0 & 0 & 0 & 1 & 0 \\ 0 & 0 & 0 & 1 & 0 & 0 \\ 0 & 0 & 0 & 1 & 1 & 0 \\ 0 & 0 & 0 & 0 & 1 & 1 \\ 0 & 1 & 0 & 0 & 0 & 1 \end{bmatrix} \end{array}$$

Fig. 127

Sol. The directed graph corresponding to the adjacency matrix M_A is shown in Fig. 128.

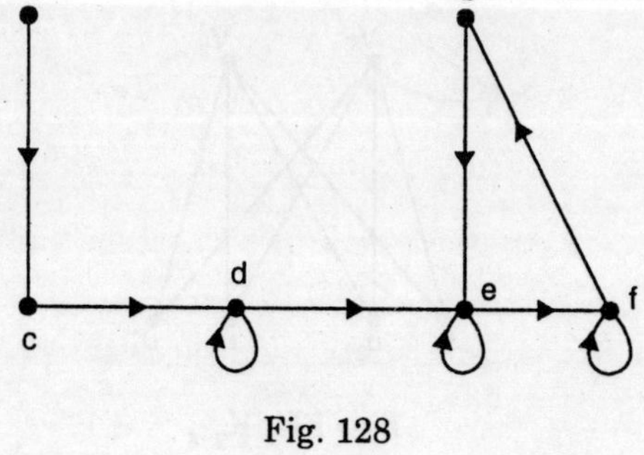

Fig. 128

Problem 12. *Draw the directed graph G whose incidence matrix M_I is shown in Fig. 129.*

$$M_I = \begin{array}{c} \\ a \\ b \\ c \\ d \\ e \end{array} \begin{array}{c} \begin{array}{ccccccccc} e_1 & e_2 & e_3 & e_4 & e_5 & e_6 & e_7 & e_8 & e_9 \end{array} \\ \begin{bmatrix} -1 & -1 & 0 & 1 & 1 & 0 & 0 & 0 & 0 \\ 1 & 0 & 1 & 0 & 0 & 0 & -1 & 0 & 0 \\ 0 & +1 & -1 & 0 & 0 & -1 & 0 & 0 & 1 \\ 0 & 0 & 0 & -1 & 0 & 1 & 1 & 1 & 0 \\ 0 & 0 & 0 & 0 & -1 & 0 & 0 & -1 & -1 \end{bmatrix} \end{array}$$

Fig. 129

Sol. The directed graph corresponding to the incidence matrix M_I is shown in Fig. 130.

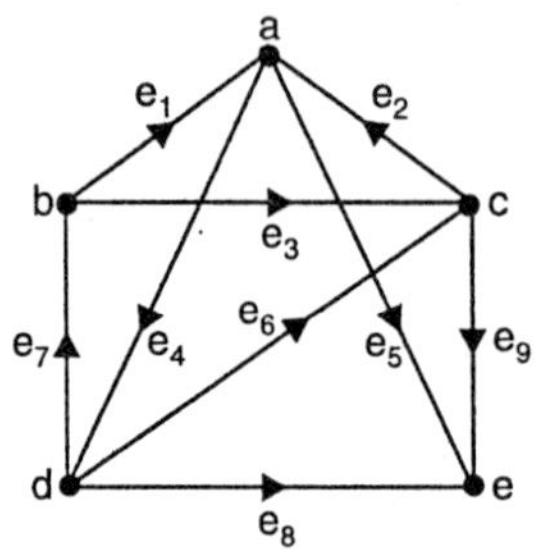

Fig. 130

Problem 13. *Draw the undirected graphs K_3 and K_5.*

Sol. Undirected graphs K_3 and K_5 are shown in Figs. 131 and 132.

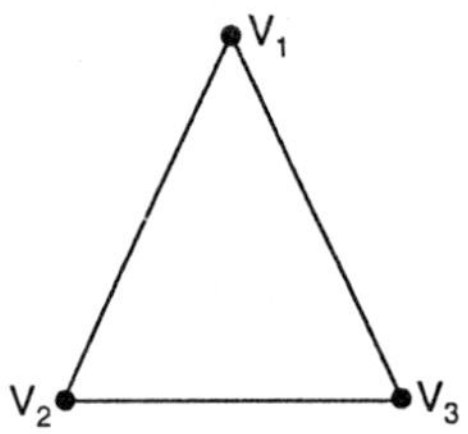

Fig. 131. K_3

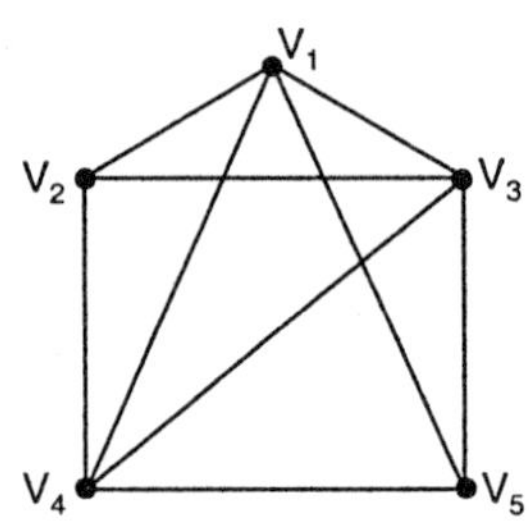

Fig. 132. K_5

Problem 14. *Draw the complete bipartite graphs $K_{2,3}$, $K_{2,4}$ and $K_{2,5}$.*

Sol. The complete bipartite graphs $K_{2,3}$, $K_{2,4}$ and $K_{2,5}$ are shown in Figs. 133, 134 and 135 respectively.

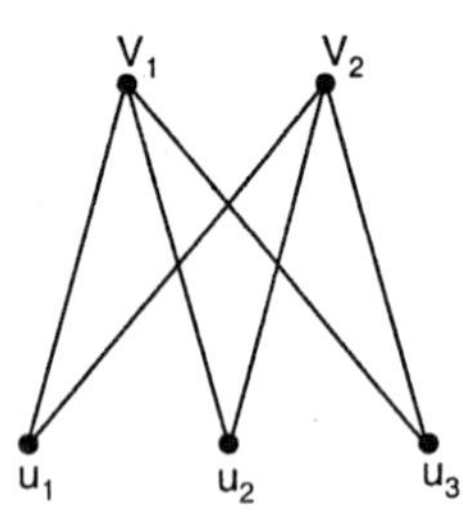

Fig. 133. $K_{2,3}$

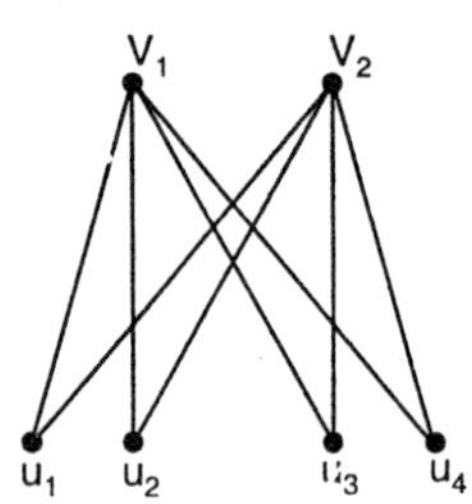

Fig. 134. $K_{2,4}$

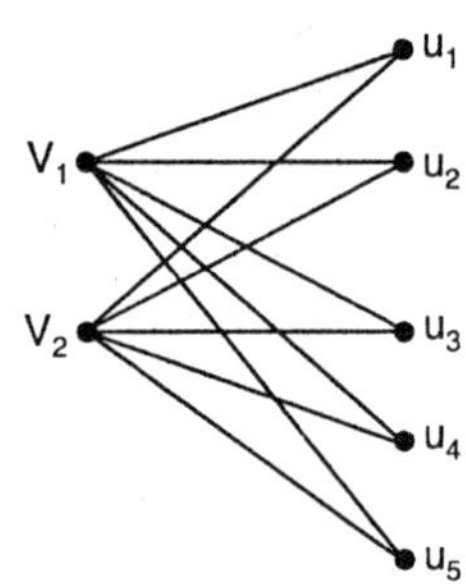

Fig. 135. $K_{2,5}$

Problem 15. *Give an example of a graph that has an Euler circuit which is also a Hamiltonian circuit.*

Sol. The graph having an Euler circuit which is also a Hamiltonian circuit is shown in Fig. 136.

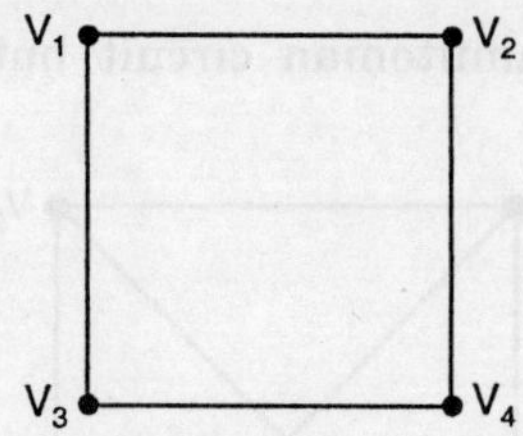

Fig. 136

In this graph V_1, V_2, V_3, V_4, V_1 is both an Euler circuit as well as Hamiltonian circuit.

Problem 16. *Give an example of a graph that has an Euler circuit and a Hamiltonian circuit, which are distinct.*

Sol. The graph having an Euler circuit and a Hamiltonian circuit which are distinct is shown in Fig. 137.

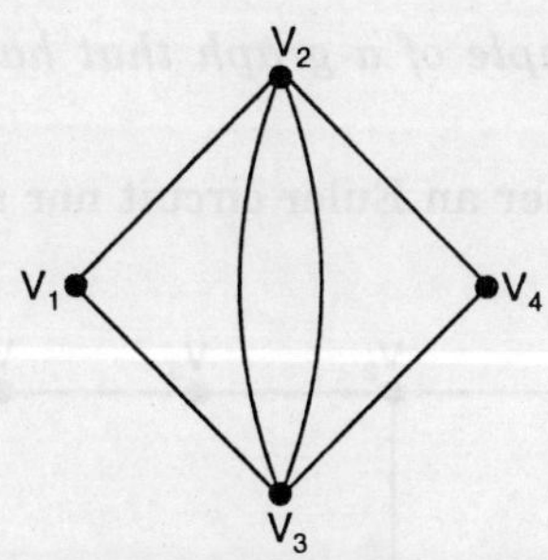

Fig. 137

The Euler circuit is $V_1, V_5, V_2, V_5, V_4, V_2, V_1$.
The Hamiltonian circuit is V_1, V_2, V_4, V_5, V_1.

Problem 17. *Give an example of a graph which has an Euler circuit but not a Hamiltonian circuit.*

Sol. The graph having an Euler circuit but not a Hamiltonian circuit is shown in Fig. 138.

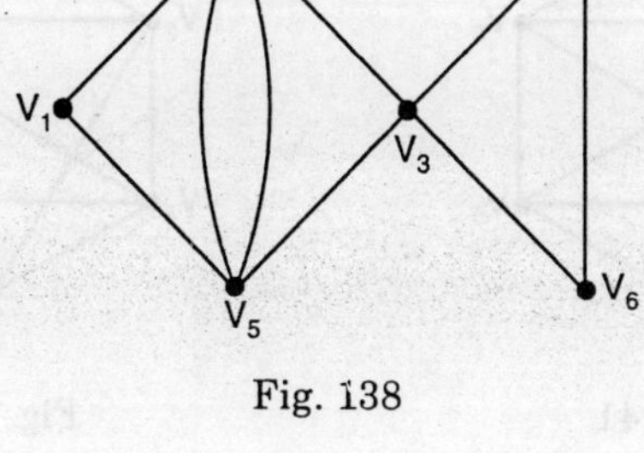

Fig. 138

The Euler circuit is $V_1, V_5, V_2, V_5, V_3, V_4, V_6, V_3, V_2, V_1$.
There is no Hamiltonian circuit.

Problem 18. *Give an example of a graph which has a Hamiltonian circuit but not an Euler circuit.*

Sol. The graph having a Hamiltonian circuit but not an Euler circuit is shown in Fig. 139.

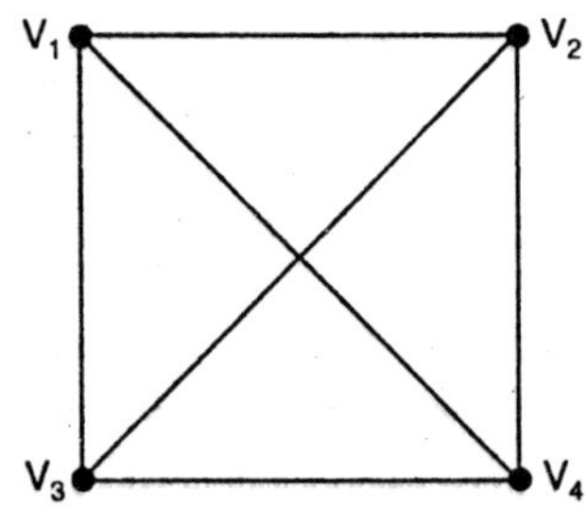

Fig. 139

The Hamiltonian circuit is V_1, V_2, V_4, V_3, V_1.
There is no Euler circuit.

Problem 19. *Give an example of a graph that has neither an Euler circuit nor a Hamiltonian circuit.*

Sol. The graph having neither an Euler circuit nor a Hamiltonian circuit is shown in Fig. 140.

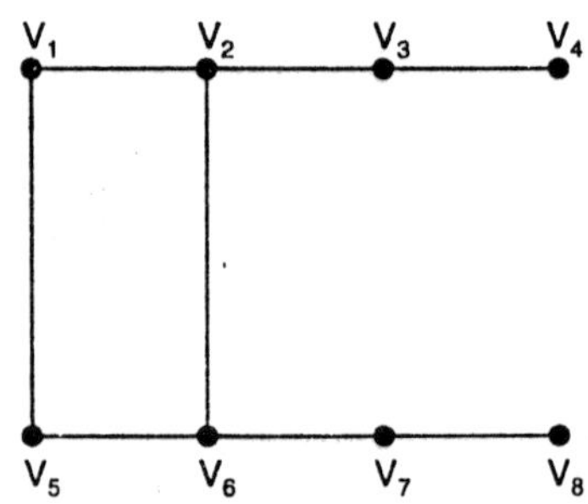

Fig. 140

Problem 20. *Draw a planar representation of graphs shown in Figs. 141 and 142.*

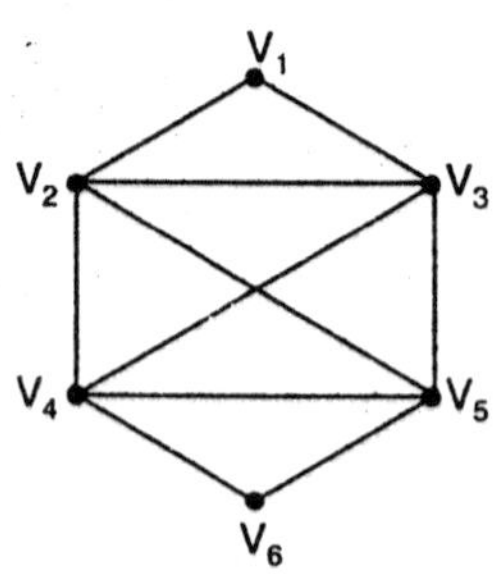

Fig. 141

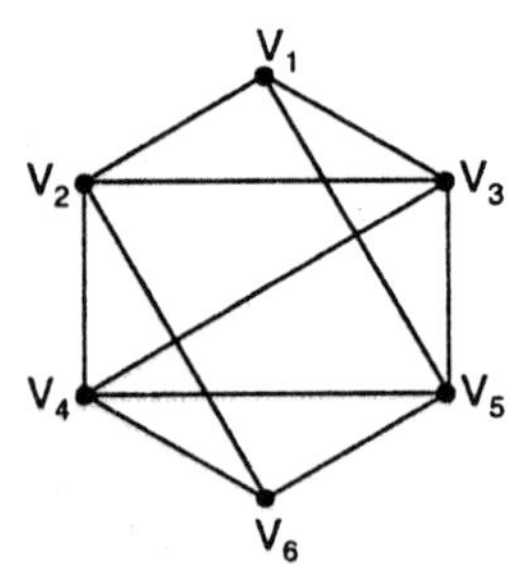

Fig. 142

Sol. The planar representation of graph shown in Fig. 143 is shown in Fig. 144.

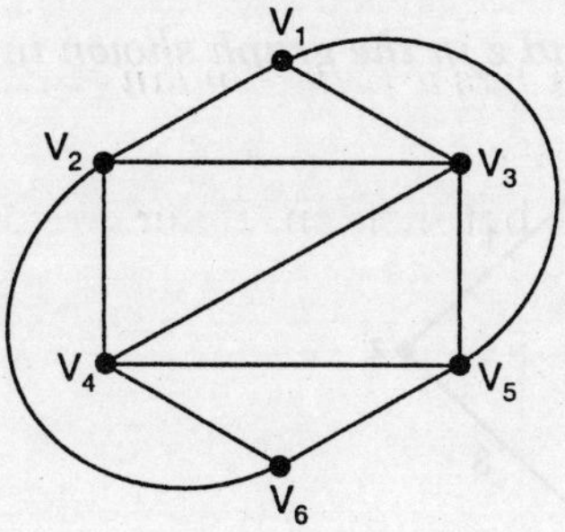

Fig. 143

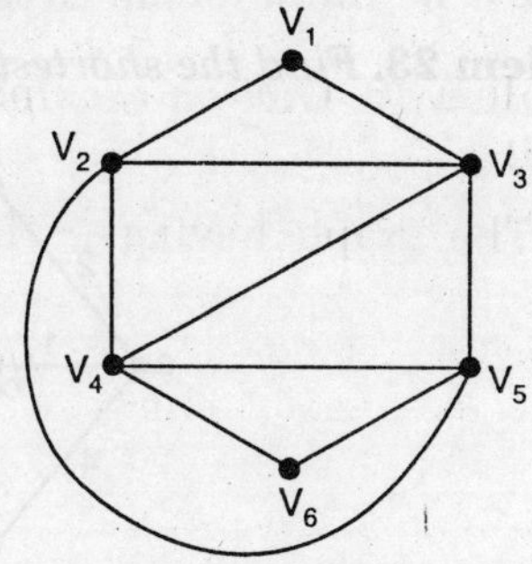

Fig. 144

The planar representation of graph shown in Fig. 142 is shown in Fig. 143.

Problem 21. *Show that* $V - E + R = 2$ *for the connected planar graphs shown in Figs. 145. and 146.*

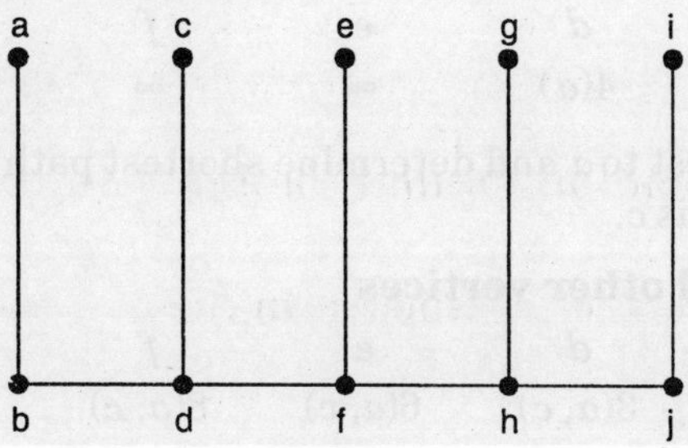

Fig. 145

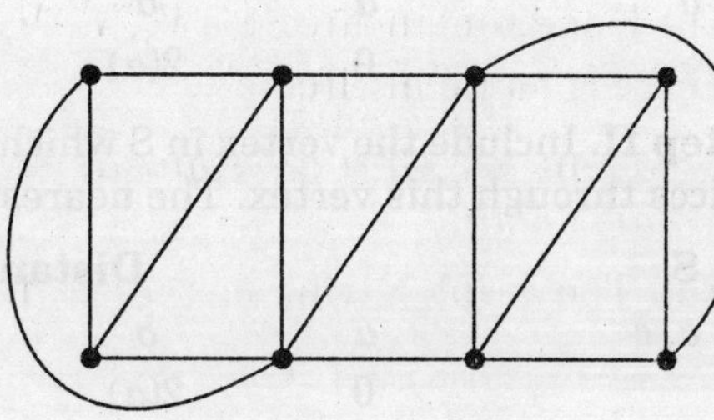

Fig. 146

Sol. (*i*) The graph shown in Fig. 143 contains vertices V = 10, edges E = 9 and regions R = 1. Putting the values, we have 10 – 9 + 1 = 2. Hence proved.

(*ii*) The graph shown in Fig. 144 contains vertices V = 8, edges E = 15 and regions R = 9. Putting the values, we have 8 – 15 + 9 = 2. Hence proved.

Problem 22. *Show that* $e \geq 3V - 6$ *for the connected planar graphs shown in Figs. 147 and 148.*

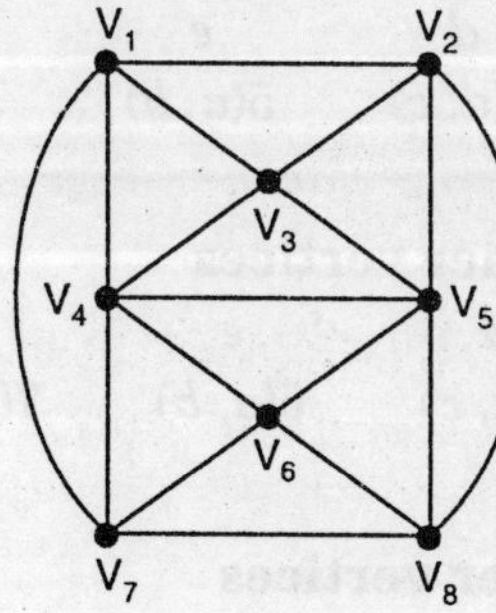

Fig. 147

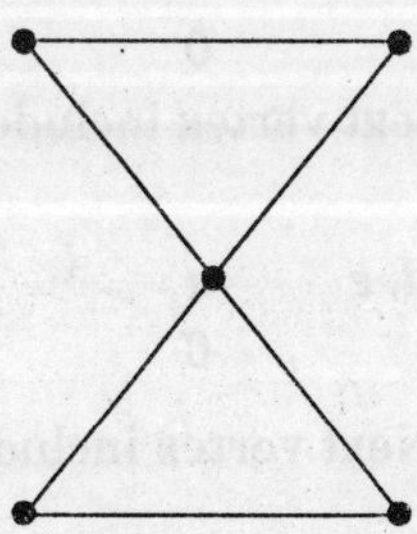

Fig. 148

Sol. (*i*) The graph shown in Fig. 147 contains vertices V = 8 and edges $e = 17$. Putting the values we have $e = 3 \times 8 - 6 = 18 > 17$. Hence proved.

(*ii*) The graph shown in Fig. 148 contains vertices V = 5 and edges $e = 6$. Putting the values, we have $3 \times 5 - 6 = 11 > 6$. Hence proved.

Problem 23. *Find the shortest path between a and z in the graph shown in Fig. 149.*

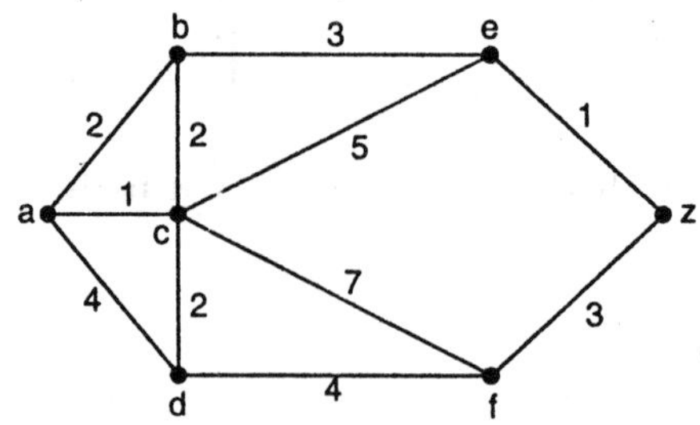

Fig. 149

Sol. Step I. Include the vertex a in S and determine all the direct paths from a to all other vertices without going through any other vertices.

S	Distance to all other vertices						
a	a	b	c	d	e	f	z
	0	2(a)	1(a)	4(a)	∞	∞	∞

Step II. Include the vertex in S which is nearest to a and determine shortest path to all the vertices through this vertex. The nearest vertex is c.

S	Distance to all other vertices						
a, c	a	b	c	d	e	f	z
	0	2(a)	1(a)	3(a, c)	6(a, c)	8(a, c)	∞

Step III. Include the vertex in S which is 2nd nearest to S and determine shortest path to all the vertices through this vertex. The 2nd nearest vertex is b.

S	Distance to all other vertices						
a, c, b	a	b	c	d	e	f	z
	0	2(a)	1(a)	3(a, c)	5(a, b)	8(a, c)	∞

Step IV. Next vertex included in S is d.

S	Distance to all other vertices						
a, c, b, d	a	b	c	d	e	f	z
	0	2(a)	1(a)	3(a, c)	5(a, b)	7(a, c)	∞

Step V. Next vertex included in S is e.

S	Distance to all other vertices						
a, c, b, d, e	a	b	c	d	e	f	z
	0	2(a)	1(a)	3(a, c)	5(a, b)	7(a, c)	6(a, b, e)

Step VI. Next vertex included in S is f.

S	Distance to all over vertices						
a, c, b, d, e, f	a	b	c	d	e	f	z
	0	2(a)	1(a)	3(a, c)	5(a, b)	7(a, c)	6(a, b, e).

This ends our procedure as $n - 1$ vertices are included in S. Thus, the shortest distance between a and z is 6 and the shortest path is a, b, e, z.

Chapter 11

Trees

GENERAL TREES

A graph which has no cycle is called an acyclic graph. A tree is an acyclic graph or graph having no cycles.

A tree or general tree is defined as a non-empty finite set of elements called vertices or nodes having the property that each node can have minimum degree 1 and maximum degree n. It can be partitioned into $n + 1$ disjoint subsets such that the first subset contains the root of the tree and the remaining n subsets contains the elements of the n subtree. (Fig. 1)

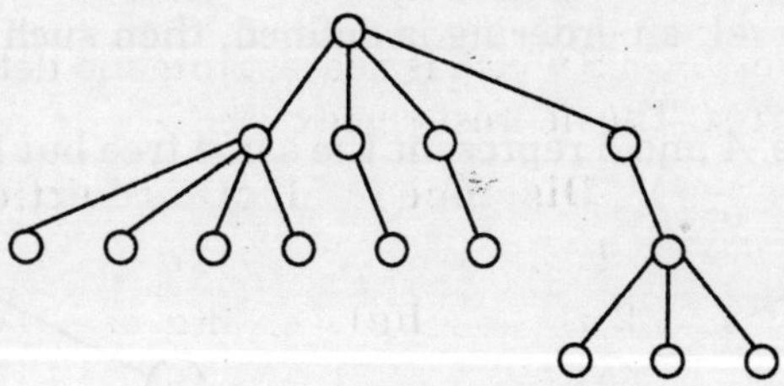

Fig. 1. General Tree

DIRECTED TREES

A directed tree is an acyclic directed graph. It has one node with indegree 1, while al other nodes have indegree 1 as shown in Figs. 2 and 3.

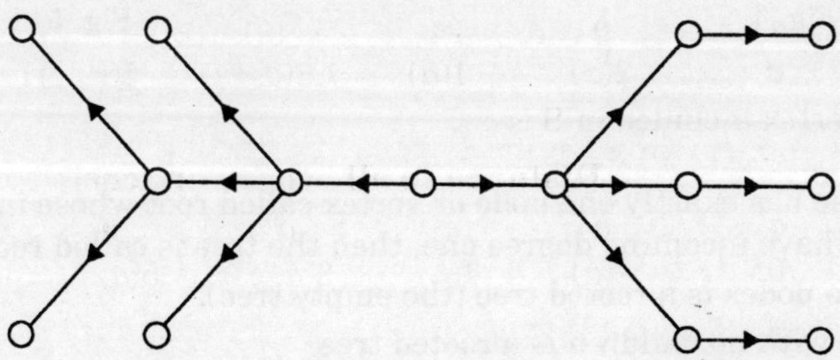

Directed trees

Fig. 2

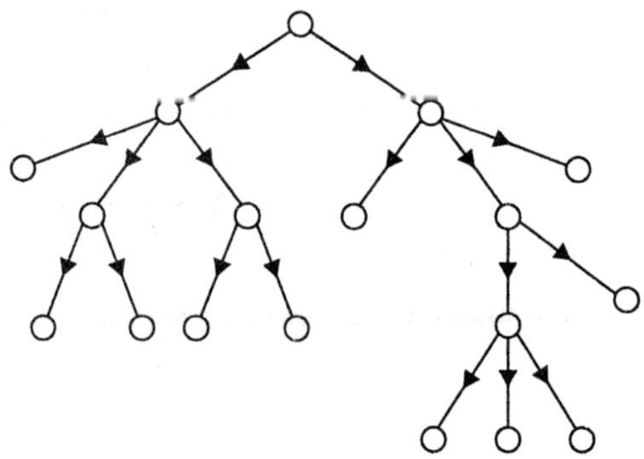

Fig. 3

The node which has outdegree 0 is called an external node or a terminal node or a leaf. The nodes which has outdegree greater than or equal to one are called internal nodes or branch nodes.

ORDERED TREES

If in a tree at each level, an ordering is defined, then such a tree is called an ordered tree.

e.g., the trees shown in Figs. 4 and 5 represent the same tree but have different orders.

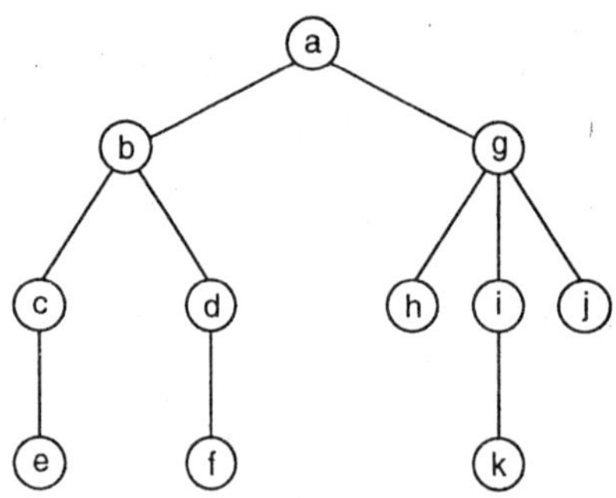

Fig. 4

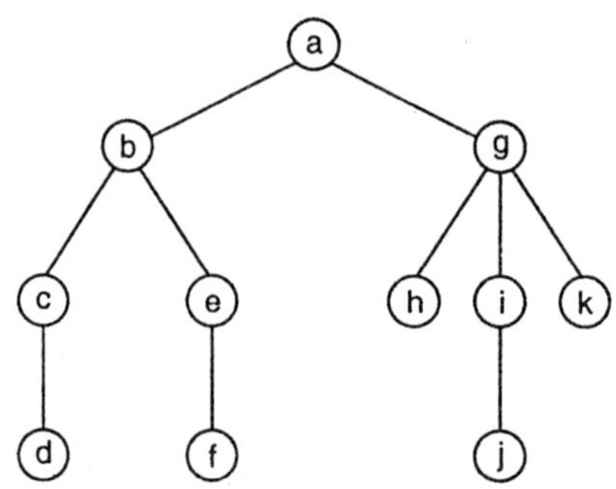

Fig. 5

ROOTED TREES

If a directed tree has exactly one node or vertex called root whose incoming degree is 0 and all other vertices have incoming degree one, then the tree is called rooted tree.

* A tree with no nodes is a rooted tree (the empty tree).

* A single node with no children is a rooted tree.

PATH LENGTH OF A VERTEX

The path length of a vertex in a rooted tree is defined to be the number of edges in the path from the root to the vertex.

Example 1. *Find the path lengths of the nodes b, f, l, q in Fig. 6.*

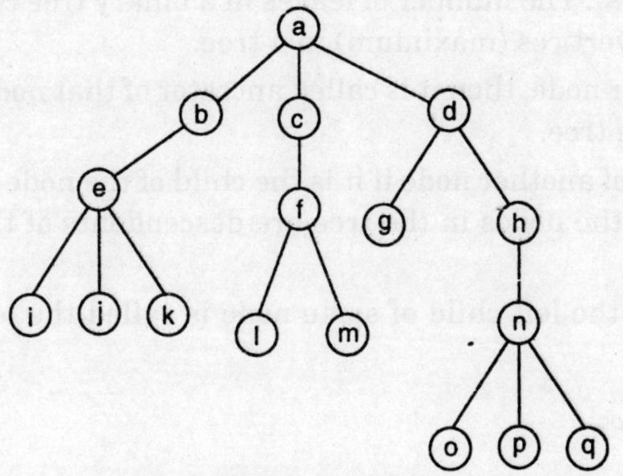

Fig. 6

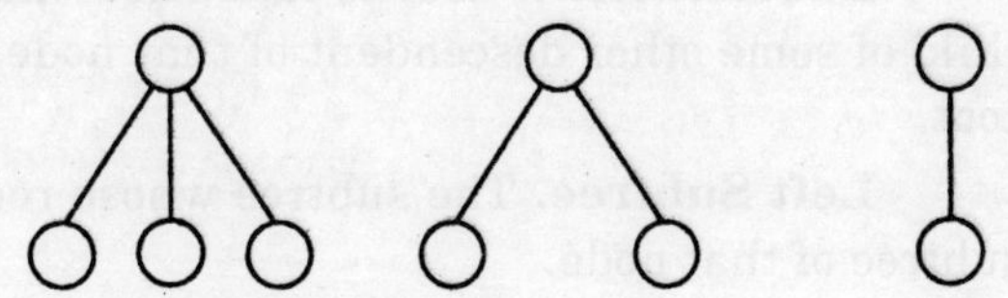

Fig. 7

Sol. The path length of node *b* is one.

The path length of node *f* is two.

The path length of node *l* is three.

The path length of node *q* is four.

Theorem I. *Prove that there is one and only one path between every pair of vertices in a tree T.*

Proof. We know that T is a connected graph, in which there must exist at least one path between every pair of vertices. Now assume that there exists two different paths from some node *a* to some node *b* of T. The union of these two paths will contain a cycle and therefore T cannot be a tree. Hence, there is only one path between every pair of vertices in a tree.

FOREST

If the root and the corresponding edges connecting the nodes are deleted from a tree, we obtain a set of disjoint trees. This set of disjoint trees is called a forest. (Fig. 7)

BINARY TREE

If the outdegree of every node is less than or equal to 2, in a directed tree then the tree is called a binary tree. A tree consisting of no nodes (empty tree) is also a binary tree. A binary tree is shown in Figs. 8 and 9.

BASIC TERMINOLOGY

Root. A binary tree has a unique node called the root of the tree.

Left Child. The node to the left of the root is called its left child.

Right Child. The node to the right of the root is called its right child.

Parent. A node having left child or right child or both is called parent of the nodes.

Siblings. Two nodes having the same parent are called siblings.

Leaf. A node with no children is called a leaf. The number of leaves in a binary tree can vary from one (minimum) to half the number of vertices (maximum) in a tree.

Ancestor. If a node is the parent of another node, then it is called ancestor of that node. The root is an ancestor of every other node in the tree.

Descendent. A node is called descendent of another node if it is the child of the node or child of some other descendent of that node. All the nodes in the tree are descendents of the root.

Left Subtree. The subtree whose root is the left child of some node is called the left subtree of that node.

Example 2. *For the tree as shown in Fig. 8.*

(*i*) *Which node is the root ?*

(*ii*) *Which nodes are leaves ?*

(*iii*) *Name the parent node of each node.*

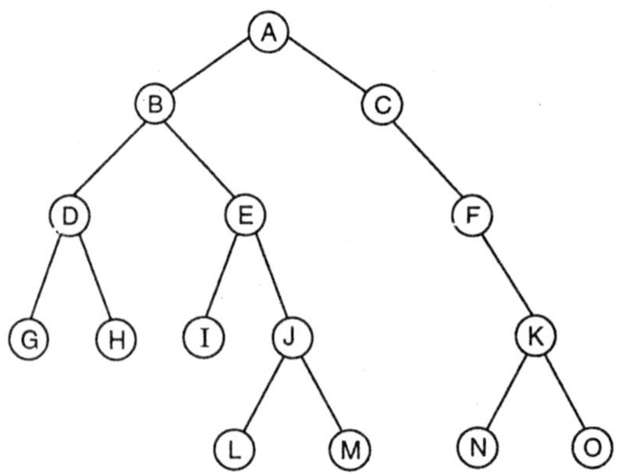

Fig. 8

Sol. (*i*) The node A is the root node.

(*ii*) The nodes G, H, I, L, M, N, O are leaves.

(*iii*)

Nodes	Parent
B, C	A
D, E	B
F	C
G, H	D
I, J	E
K	F
L, M	J
N, O	K

Right Subtree. The subtree whose root is the right child of some node is called the right subtree of that node.

Level of a Node. The level of a node is its distance from the root. The level of root is defined as zero. The level of all other nodes is one more than its parent node. The maximum number of nodes at any level N is 2^N.

Depth or Height of a Tree. The depth or height of a tree is defined as the maximum number of nodes in a branch of tree. This is one more than the maximum level of the tree *i.e.*, the depth of root is one. The maximum number of nodes in a binary tree of depth d is $2^d - 1$, where $d \geq 1$.

External Nodes. The nodes which has no children are called external nodes or terminal nodes.

Internal Nodes. The nodes which has one or more than one children are called internal nodes or non-terminal nodes.

Example 3. *For the tree as shown in Fig. 9.*

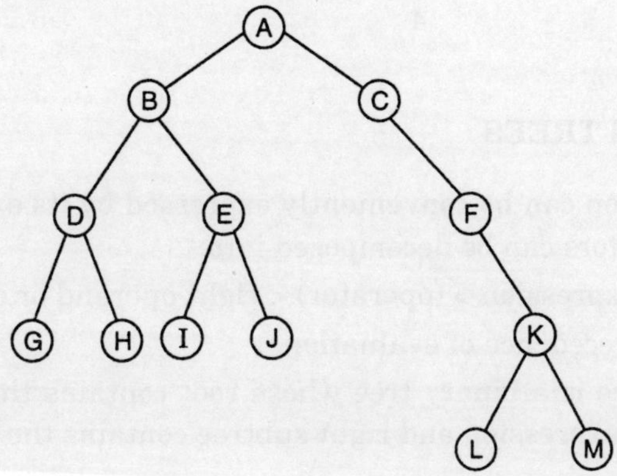

Fig. 9

(*i*) *List the children of each node.* (*ii*) *List the siblings.*

(*iii*) *Find the depth of each node.* (*iv*) *Find the level of each node.*

Sol. (*i*) The children of each node is as follows :

Node	**Children**
A	B, C
B	D, E
C	F
D	G, H
E	I, J
F	K
K	L, M

(*ii*) The siblings are as follows :

Siblings

B and C

D and E

G and H

I and J

L and M are all siblings.

(*iii*)

Node	Depth or Height
A	1
B, C	2
D, E, F	3
G, H, I, J, K	4
L, M	5

(*iv*)

Node	Level
A	0
B, C	1
D, E, F	2
G, H, I, J, K	3
L, M	4

BINARY EXPRESSION TREES

Algebraic expression can be conveniently expressed by its expression tree. An expression having binary operators can be decomposed into

< left operand or expression > (operator) < right operand or expression >

depending upon precedence of evaluation.

The expression tree is a binary tree whose root contains the operator and whose left subtree contains the left expression and right subtree contains the right expression.

Example 4. *Construct the binary expression tree for the expression* $(a + b) * (d / c)$.

Sol. The binary expression tree for the expression $(a + b) * (d/c)$ is shown in Fig. 10.

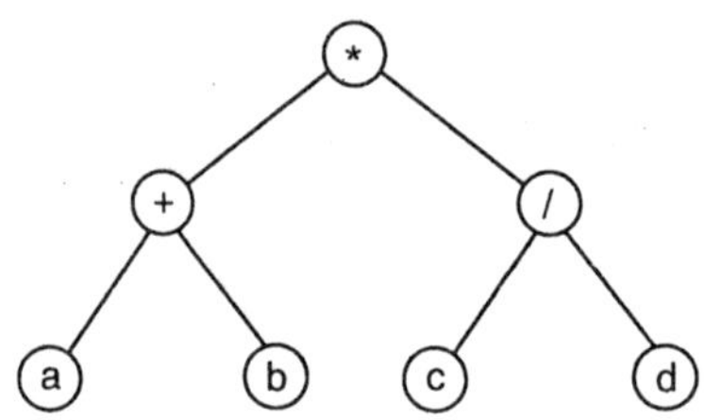

Fig. 10

COMPLETE BINARY TREE

Complete binary tree is a binary tree if all its levels, except possibly the last, have the maximum number of possible nodes as for left as possible. The depth of complete binary tree having n nodes is $\log_2 n + 1$.

For example : The tree shown in Fig. 11 is a complete binary tree.

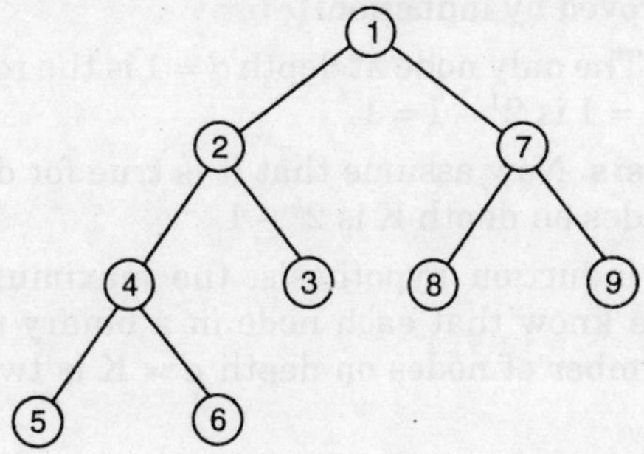

Fig. 11

FULL BINARY TREE

Full binary tree is a binary tree in which all the leaves are on the same level and every non-leaf node has two children.

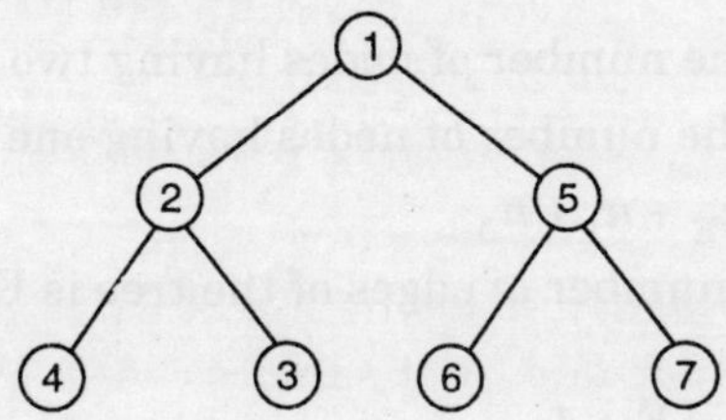

Fig. 12

For example : The tree shown in Fig. 12 is a full binary tree.

Theorem II. *Prove that the maximum number of nodes on level n of a binary tree is 2^n, where $n \geq 0$.*

Proof. This can be proved by induction.

Basis of Induction. The only node at level $n = 0$ is the root node. Thus, the maximum number of nodes on level $n = 0$ is $2^0 = 1$.

Induction Hypothesis. Now assume that it is true for level j, where $n \geq j \geq 0$. Therefore, the maximum no. of nodes on level j is 2^j.

Induction Step. By induction hypothesis, the maximum number of nodes on level $j - 1$ is 2^{j-1}. Since, we know that each node in binary tree has maximum degree 2. Therefore, the maximum number of nodes on level j is twice the maximum number of level $j - 1$.

Hence, at level j, the maximum number of nodes is

$$= 2 \cdot 2^{j-1}$$

$$= 2^j. \text{ Hence proved.}$$

Theorem III. *Prove that the maximum number of nodes in a binary tree of depth d is $2^d - 1$, where $d \geq 1$.*

Proof. This can be proved by induction.

Basis of Induction. The only node at depth $d = 1$ is the root node. Thus, the maximum number of nodes on depth $d = 1$ is $2^1 - 1 = 1$.

Induction Hypothesis. Now assume that it is true for depth K, $d > k \geq 1$. Therefore, the maximum number of nodes on depth K is $2^k - 1$.

Induction Step. By induction hypothesis, the maximum number of nodes on depth $K - 1$ is $2^{k-1} - 1$. Since, we know that each node in a binary tree has maximum degree 2, therefore, the maximum number of nodes on depth $d = K$ is twice the maximum number of nodes on depth $K - 1$.

So, at depth $d = K$, the maximum number of nodes is $= (2 \cdot 2^{K-1}) - 1$

$$= 2^{K-1+1} - 1 = 2^K - 1. \text{ Hence proved.}$$

Theorem IV. *Prove that in a binary tree, if n_E is the number of external nodes or leaves and n_I is the no. of internal nodes, then $n_E = n_I + 1$.*

Proof. Let n be the total number of nodes in the tree. Then, we may have three types of nodes in the tree.

n_E = the number of nodes having zero degree.

n_I = the number of nodes having two degree.

n_0 = the number of nodes having one degree.

So, we have $$n = n_E + n_I + n_0 \quad \text{...(1)}$$

Let us assume that the number of edges of the tree is E. So, with these E edges we can connect E + 1 nodes.

Hence $$n = E + 1 \quad \text{...(2)}$$

Since, all edges are either from a node of degree one or from a node of degree two,

therefore, $$E = n_0 + 2n_I \quad \text{...(3)}$$

Put this value in the eq. (2), we have

$$n = n_0 + 2n_I + 1 \quad \text{...(4)}$$

Subtract eq. (4) from eq. (1), we get

$$n_E = n_I + 1$$

Hence proved.

TRAVERSING BINARY TREES

Traversing means to visit all the nodes of the tree. There are three standard methods to traverse the binary trees. These are as follows :

1. Preorder traversal
2. Postorder traversal
3. Inorder traversal.

1. Preorder traversal. The preorder traversal of a binary tree is a recursive process. The preorder traversal of a tree is

(*i*) Visit the root of the tree.

(*ii*) Traverse the left subtree in preorder.

(*iii*) Traverse the right subtree in preorder.

2. Postorder traversal. The postorder traversal of a binary tree is a recursive process. The postorder traversal of a tree is

(*i*) Traverse the left subtree in postorder.

(*ii*) Traverse the right subtree in postorder.

(*iii*) Visit the root of the tree.

3. Inorder traversal. The inorder traversal of a binary tree is a recursive process. The inorder traversal of a tree is

(*i*) Traverse in inorder the left subtree.

(*ii*) Visit the root of the tree.

(*iii*) Traverse in inorder the right subtree.

Example 5. *Determine the preorder, postorder and inorder traversal of the binary tree as shown in Fig. 13.*

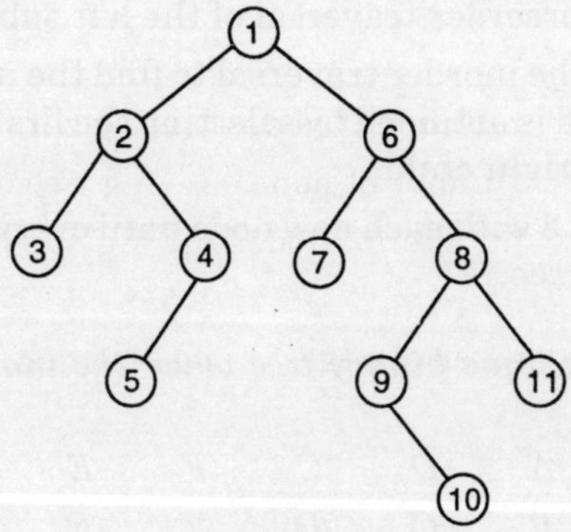

Fig. 13

Sol. The preorder, postorder and inorder traversal of the tree is as follows :

Preorder	:	1	2	3	4	5	6	7	8	9	10	11
Postorder	:	3	5	4	2	7	10	9	11	8	6	1
Inorder	:	3	2	5	4	1	7	6	9	10	8	11.

Example 6. *Give the preorder, inorder and postorder traversals of the tree shown in Fig. 14.*

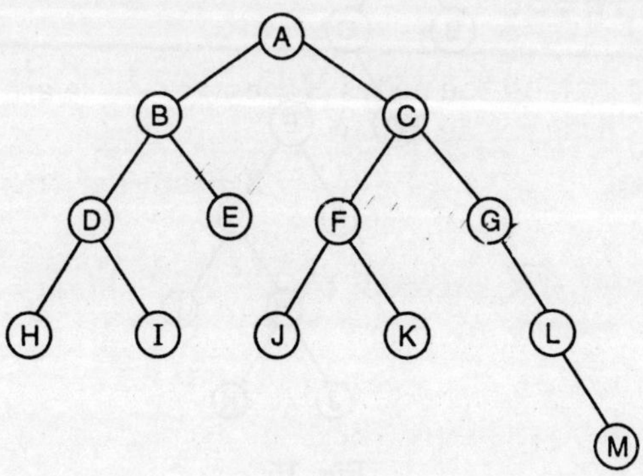

Fig. 14

Sol.

Preorder	:	A,	B,	D,	H,	I,	E,	C,	F,	J,	K,	G,	L, M
Inorder	:	H,	D,	I,	B,	E,	A,	J,	F,	K,	C,	G,	L, M
Postorder	:	H,	I,	D,	E,	B,	J,	K,	F,	M,	L,	G,	C, A.

ALGORITHMS

(*a*) Algorithm to Draw a Unique Binary Tree When Inorder and Preorder Traversal of the Tree is Given

1. We know that the root of the binary tree is the first node in its preorder. Draw the root of the tree.

2. To find the left child of the root node, first use the inorder traversal to find the nodes in the left subtree of the binary tree. (All the nodes that are left to the root node in the inorder traversal are the nodes of the left subtree). After that the left child of the root is obtained by selecting the first node in the preorder traversal of the left subtree. Draw the left child.

3. In the same way, use the inorder traversal to find the nodes in the right subtree of the binary tree. Then the right child is obtained by selecting the first node in the preorder traversal of the right subtree. Draw the right child.

4. Repeat the steps 2 and 3 with each new node until every node is not visited in preorder. Finally, we obtain the unique tree.

Example 7. *Draw the unique binary tree when the inorder and preorder traversal is given as follows :*

Inorder	:	*B*	*A*	*D*	*C*	*F*	*E*	*J*	*H*	*K*	*G*	*I*
Preorder	:	*A*	*B*	*C*	*D*	*E*	*F*	*G*	*H*	*J*	*K*	*I.*

Sol. We know that the root of the binary tree is the first node in preorder traversal. Now check A, in the inorder traversal, all the nodes that are left of A, are nodes of left subtree and all the nodes that are right of A, are nodes of right subtree. Read the next node in preorder and check its position against the root node, if it is left of root node, then draw it as left child, otherwise draw it as right child. Repeat the above process for each new node until all the nodes of preorder traversal are read and finally we obtain the binary tree as shown in Fig. 15.

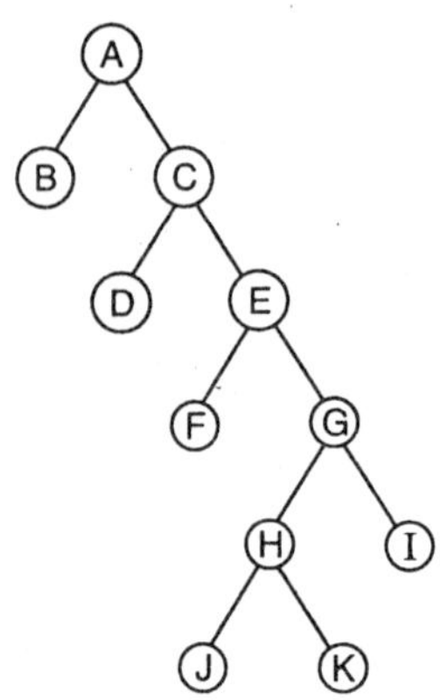

Fig. 15

Example 8. *Draw the unique binary tree when the following is given :*

Inorder	:	*d*	*b*	*h*	*e*	*a*	*f*	*c*	*i*	*j*	*g*
Preorder	:	*a*	*b*	*d*	*e*	*h*	*c*	*f*	*g*	*i*	*j*

Sol. The first node in preorder is *a* and hence *a* is the root node. Unique binary tree is shown in Fig. 16.

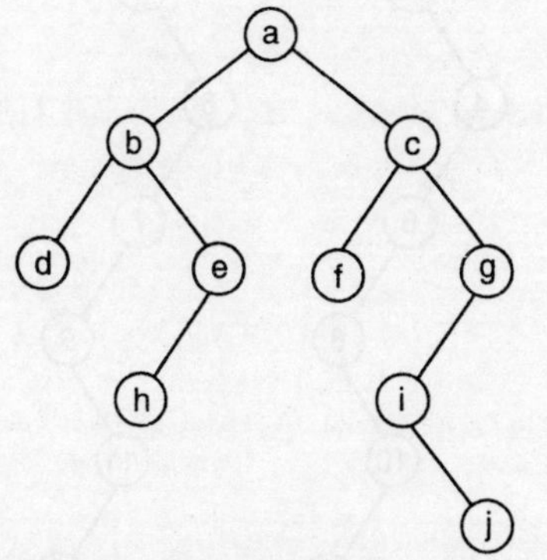

Fig. 16

(*b*) Algorithm to Draw a Unique Binary Tree When Inorder and Postorder Traversal of the Tree is Given

1. We know that the root of the binary tree is the last node in its postorder. Draw the root of the tree.

2. To find the right child of the root node, first use the inorder traversal to find the nodes in the right subtree of the binary tree. (All the nodes that are right to the root node in the inorder traversal are the nodes of the right subtree). After that the right child of the root is obtained by selecting the last node in the postorder traversal of the right subtree. Draw the right child.

3. In the same way, use the inorder traversal to find the nodes in the left subtree of the binary tree. Then the left child is obtained by selecting the last node in the postorder traversal of the left subtree. Draw the left child.

4. Repeat the steps 2 and 3 with each new node until every node is not visited in postorder. After visiting the last node, we obtain the unique tree.

Example 9. *Draw the unique binary tree for the given Inorder and Postorder traversal*

Inorder	:	*4*	*6*	*10*	*12*	*8*	*2*	*1*	*5*	*7*	*11*	*13*	*9*	*3*
Postorder	:	*12*	*10*	*8*	*6*	*4*	*2*	*13*	*11*	*9*	*7*	*5*	*3*	*1*

Sol. We know that the root node is the last node in postorder traversal. Hence one is the root node.

Now check the inorder traversal, we know that root is at the centre, hence all the nodes that are left to the root node in inorder traversal are the nodes of left subtree and all that are right to the root node are the nodes of the right subtree.

Now, visit the next node from back in postorder traversal and check its position in inorder traversal, if it is on left of root then draw it as left child and if it is on right, then draw it as right child.

Repeat the above process for each new node and we obtain the binary tree as shown in Fig. 17.

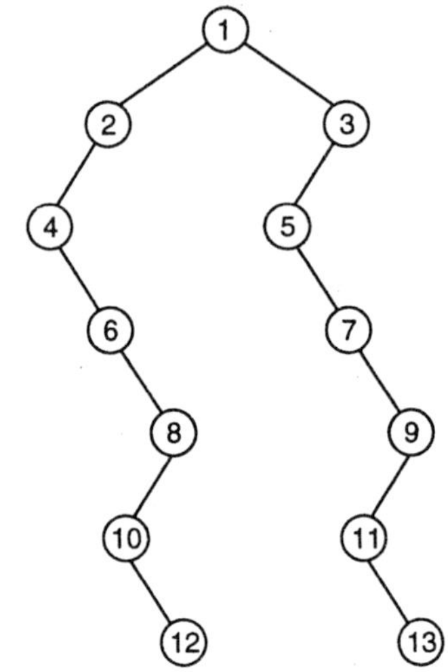

Fig. 17

Example 10. *Draw the binary tree when Inorder and Postorder traversal is given :*

Inorder : *m k n j o l u s v q t p r*

Postorder : *m n k o u v s t q r p l j.*

Sol. We know that the last node in Postorder is the root node, hence *j* is the root. Now applying the algorithm as above, we obtain the tree shown in Fig. 18.

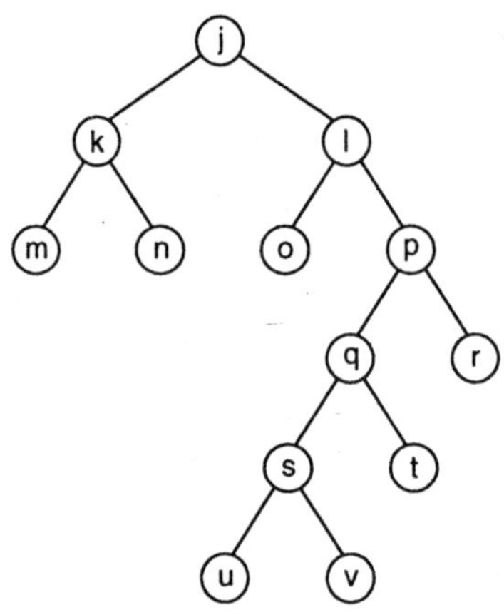

Fig. 18

(*c*) **Algorithm to Convert General Tree into the Binary Tree**

1. Starting from the root node, the root of the tree is also the root of the binary tree.

2. The first child C_1 (from left) of the root node in the tree is the left child C_l of the root node in binary tree and the sibling of the C_1 is the right child of C_l and so on.

3. Repeat the step 2 for each new node.

Example 11. *Convert the following tree as shown in Fig. 19 into binary tree.*

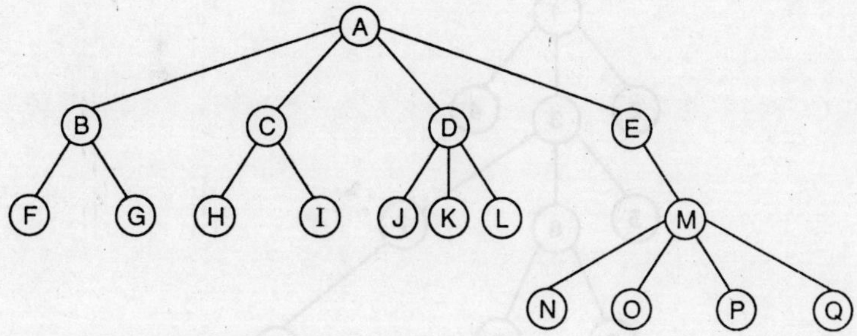

Fig. 19

Sol. The root of the tree is the root of the binary tree. Hence A is the root of the binary tree. Now B becomes the left child of A in binary tree, C becomes the right child of B, D becomes right child of C, and E becomes right child of D in the binary tree and similarly applying the algorithm we obtain the binary tree as shown in Fig. 20.

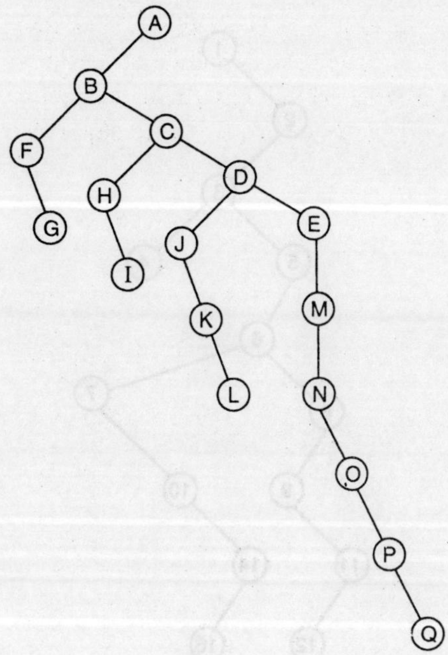

Fig. 20

Example 12. *Convert the general tree as shown in Fig. 21 into binary tree.*

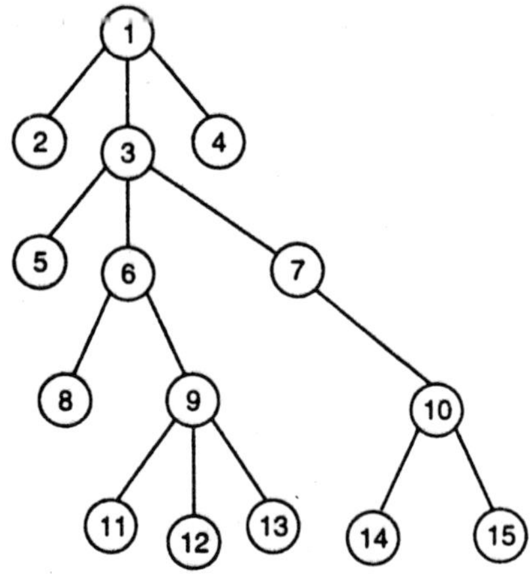

Fig. 21

Sol. The root node 1 in general tree is the root node of the binary tree. Now applying the above algorithm we obtain the binary tree as shown in Fig. 22.

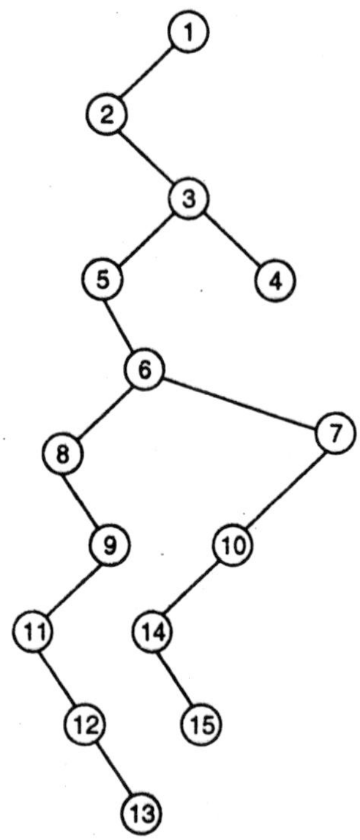

Fig. 22

EXERCISES

1. Given the tree as shown in figure.

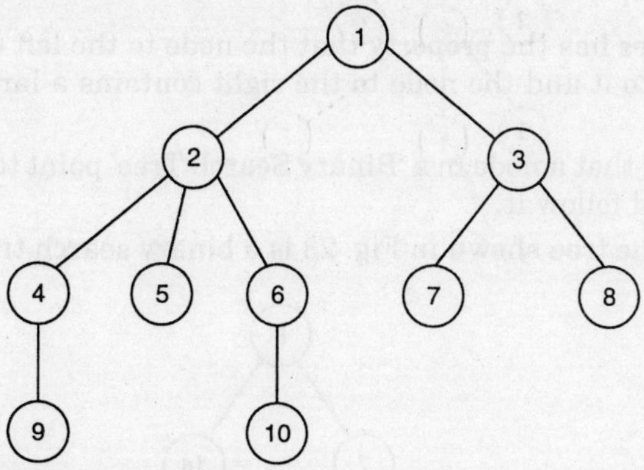

(*i*) Determine the parents of 5 and 3.

(*ii*) Determine the children of 2 and 6.

(*iii*) Determine the descendents of 4 and 3.

(*iv*) Determine the siblings of 4 and 3.

(*v*) Determine the leaves.

(*vi*) Determine internal nodes.

(*vii*) Draw the subtree rooted at 3.

2. Determine all the trees with six vertices.

3. Construct a binary tree whose inorder and preorder traversal is given below :

Inorder :	*d*	*g*	*b*	*e*	*i*	*h*	*j*	*a*	*c*	*f*
Preorder :	*a*	*b*	*d*	*g*	*e*	*h*	*i*	*j*	*c*	*f*

4. Represent the following expressions in a binary tree.

(*i*) $(x + 4) * [(3 * y + z)/(4 + p)]$

(*ii*) $[(a - b) \uparrow 2]/(a + b)$

5. A tree has $2n$ vertices of degree 1, $3n$ vertices of degree 3, and n vertices of degree 4. Determine the number of vertices and edges in the tree.

6. If T be a tree consisting of e edges and n vertices, write a formula relating e and n.

7. A tree has two vertices of degree 2, one vertices of degree 3, and three vertices of degree 4. Draw the tree.

8. Construct the ordered rooted tree whose preorder traversal is *a, b, f, c, g, h, i, d, e, j, k, l,* where *a* has four children, *c* has three children, *j* has two children, *b* and *c* have one child each, and all other vertices are leaves.

9. Give a binary tree representation of the tree which is as shown in figure.

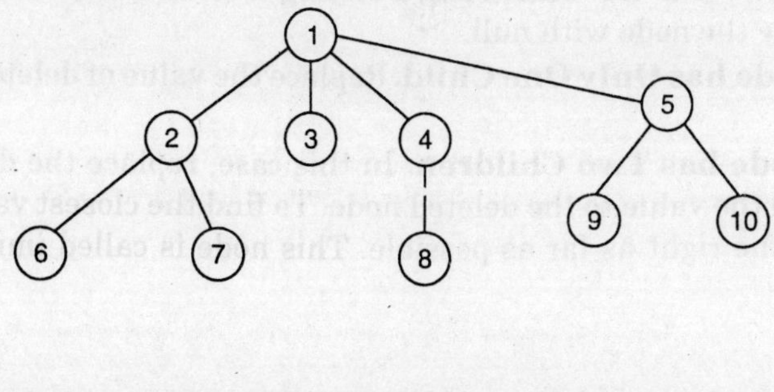

10. Construct a complete binary tree of height 4.
11. How many edges are there in a forest of t trees containing a total of n vertices.

BINARY SEARCH TREES

Binary search trees has the property that the node to the left contains a smaller value than the node pointing to it and the node to the right contains a larger value than the node pointing to it.

It is not necessary that a node in a 'Binary Search Tree' point to the nodes whose values immediately precede and follow it.

For example : The tree shown in Fig. 23 is a binary search tree.

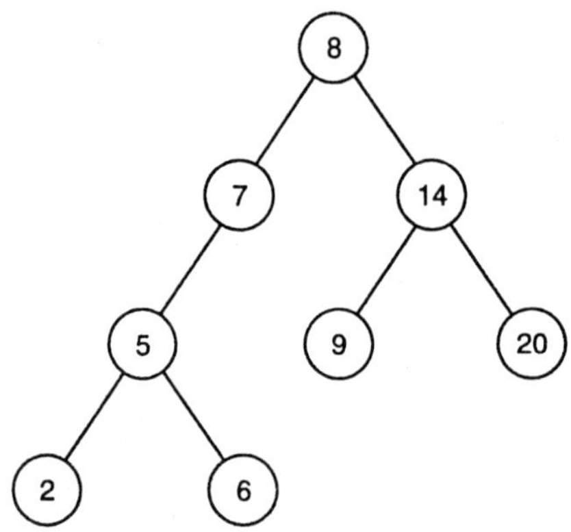

Fig. 23

Inserting into a Binary Search Tree. Consider a binary search tree T. Suppose we have given an ITEM of information to insert in T. The ITEM is inserted as a leaf in the tree. The following steps explains a procedure to insert an ITEM in the binary search tree T.

1. Compare the ITEM with the root node.
2. If ITEM > ROOT NODE, proceed to the right child and it becomes root node for the right subtree.
3. If ITEM < ROOT NODE, proceed to the left child.
4. Repeat the above steps until we meet a node which has no left and right subtree.
5. Now if the ITEM is greater than node, then the ITEM is inserted as the right child and if the ITEM is less than node, then the ITEM is inserted as the left child.

Deleting in a Binary Search Tree. Consider a binary search tree T. Suppose we want to delete a given ITEM from binary search tree. To delete an ITEM from a binary search tree, we have three cases, depending upon the number of children of the deleted node.

1. Deleted Node has no Children. Deleting a node which has no children is very simple, as just replace the node with null.

2. Deleted Node has Only One Child. Replace the value of deleted node with the only child.

3. Deleted Node has Two Children. In this case, replace the deleted node with the node that is closest in the value to the deleted node. To find the closest value, we move once to the left and then to the right as far as possible. This node is called immediate predecessor.

Now replace the value of deleted node with immediate predecessor and then delete the replaced node by using case 1 or 2.

Example 13. *Show the binary search tree after inserting 3, 1, 4, 6, 9, 2, 5, 7 into an initially empty binary search tree.*

Sol. The insertion of the above nodes in the empty binary search tree is shown in Fig. 24.

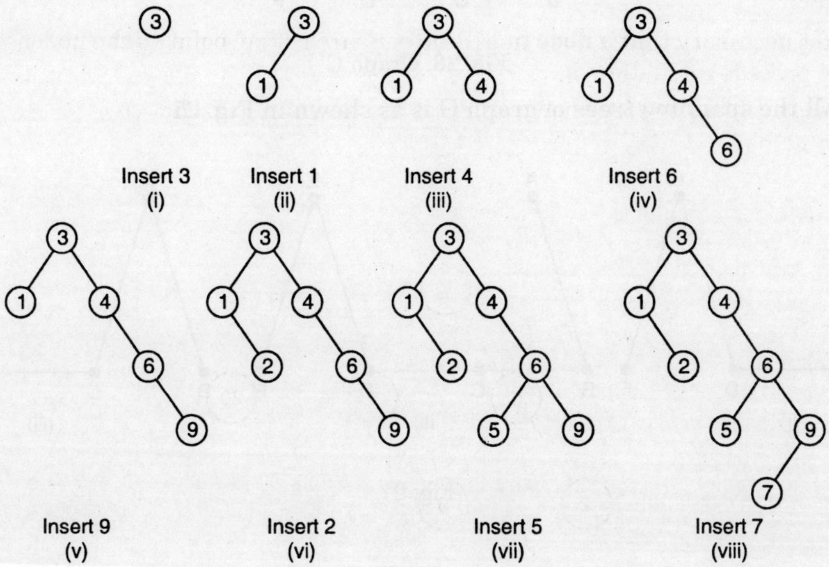

Fig. 24

Example 14. *Show the binary tree shown in Fig. 24 (viii) after deleting the root node.*

Sol. To delete the root node, first replace the root node with the closest element of the root. For this, first move one step left and then to the right as far as possible to the node. Then delete the replaced node. The tree after deletion is shown in Fig. 25.

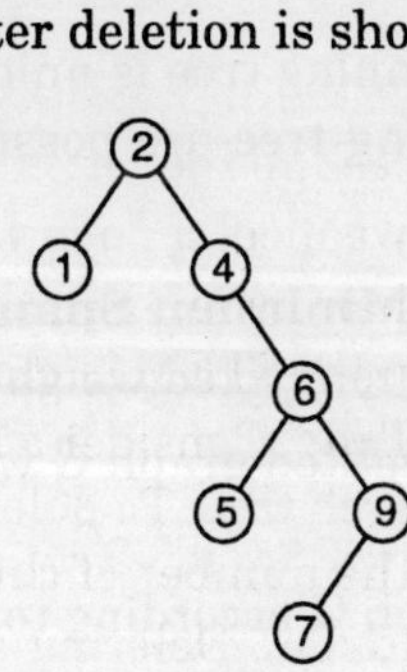

Fig. 25

SPANNING TREE

Consider a connected graph G = (V, E). A spanning tree T is defined as a subgraph of G if T is a tree and T includes all the vertices of G.

Example 15. *Draw all the spanning trees of the graph G shown in Fig. 26.*

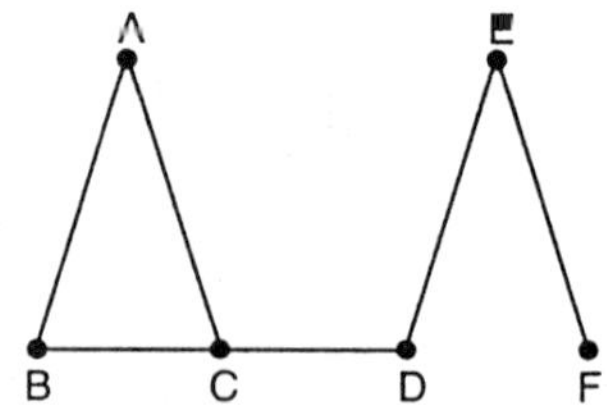

Fig. 26. Graph G

Sol. All the spanning trees of graph G is as shown in Fig. 27.

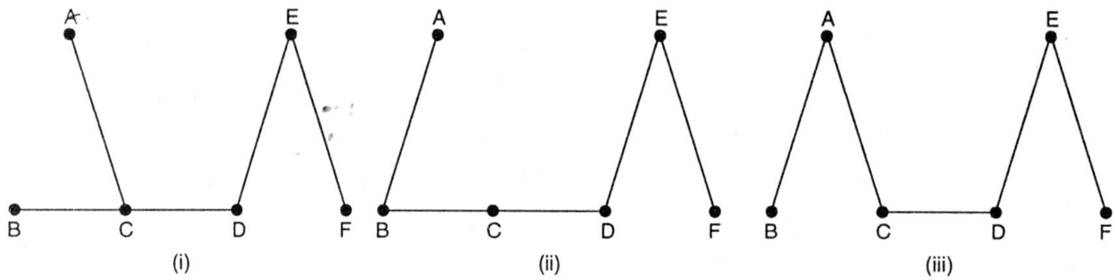

Fig. 27

MINIMUM SPANNING TREE

Consider a connected weighted graph G = (V, E). A minimal spanning tree T of the graph G is a tree whose total weight is smallest among all the spanning trees of the graph G. The total weight of the spanning tree is the sum of the weights of the edges of the spanning trees.

The minimum weight of the spanning tree is unique but the spanning tree may not be unique because more than one spanning tree are possible when more than one edges exist having the same weight.

Kruskal's Algorithm to Find Minimum Spanning Tree. This algorithm finds the minimum spanning tree T of the given connected weighted graph G.

1. Input the given connected weighted graph G with n vertices whose minimum spanning tree T, we want to find.

2. Order all the edges of the graph G according to increasing weights.

3. Initialise T with all vertices but do not include any edge.

4. Add each of the graph G in T which does not form a cycle until $n - 1$ edges are added.

Example 16. *Determine the minimum spanning tree of the weighted graph shown in Fig. 28.*

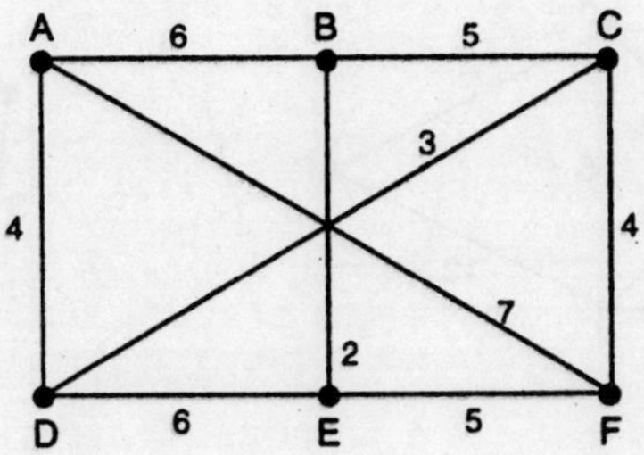

Fig. 28

Sol. Using Kruskal's algorithm, arrange all the edges of the weighted graph in increasing order and initialise spanning tree T with all the six vertices of G. Now start adding the edges of G in T which do not form a cycle and having minimum weights until five edges are not added as there are six vertices. (Fig. 29).

Edges	Weights	Added or Not
(B, E)	2	Added
(C, D)	3	Added
(A, D)	4	Added
(C, F)	4	Added
(B, C)	5	Added
(E, F)	5	Not added
(A, B)	6	Not added
(D, E)	6	Not added
(A, F)	7	Not added.

Minimum Spanning Tree

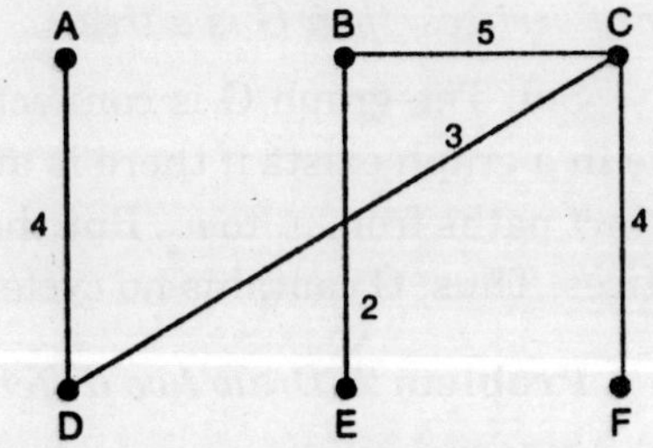

Fig. 29

EXERCISES

1. Construct all possible spanning trees of the graphs as shown in figure.

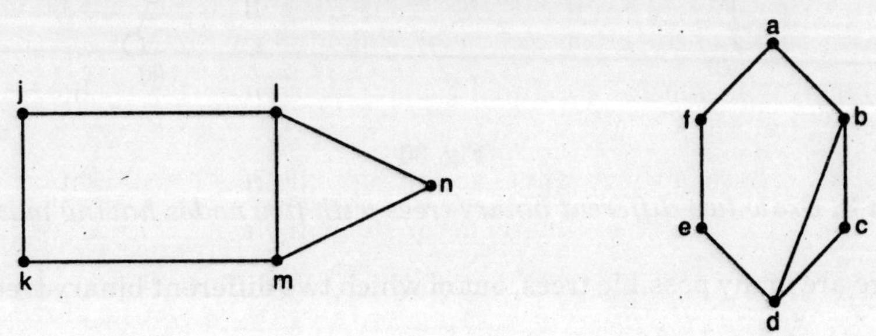

(*i*) (*ii*)

2. Determine the minimum spanning tree of the graphs as shown in figure.

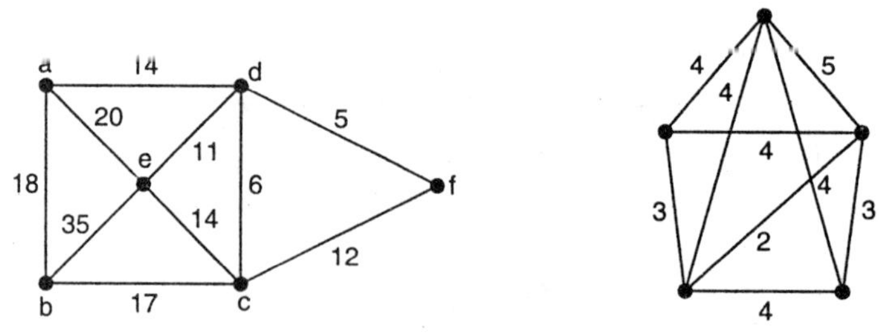

(i) (ii)

3. Construct a binary search tree for the words in the sentence, "The principal and the teachers meet during holidays".
4. Construct a binary search tree for the words mango, orange, banana, papaya, pear, peach and apple using alphabetical order.

SOLVED PROBLEMS

Problem 1. *Show that if in a graph G there exists one and only one path between every pair of vertices, then G is a tree.*

Sol. The graph G is connected since there is a path between every pair of vertices. A cycle in a graph exists if there is at least one pair of vertices (v_1, v_2) such that there exist two distinct paths from v_1 to v_2. But the graph G has one and only one path between every pair of vertices. Thus, G contains no cycle. Hence, G is a tree.

Problem 2. *Draw two different binary trees with five nodes having only one leaf.*

Sol. The two trees out of many possible trees with five nodes having only one leaf is shown in Fig. 30.

(i) (ii)

Fig. 30

Problem 3. *Draw two different binary trees with five nodes having maximum number of leaves.*

Sol. There are many possible trees, out of which two different binary trees are shown in Fig. 31.

Fig. 31

Problem 4. *How will you differentiate between a general tree and a binary tree ?*

Sol.

General Tree	*Binary Tree*
1. There is no such tree having zero nodes or an empty general tree.	1. There may be an empty binary tree.
2. If some node has a child, then there is no such distinction.	2. If some nodes has a child, then it is distinguished as a left child or a right child.
3. The trees shown in Fig. 32. are same, when we consider them as general trees.	3. The trees shown in Fig. 32 are distinct, when we consider them as binary trees, because in (*i*), 4 is right child of 2 while in (*ii*), 4 is left child of 2.
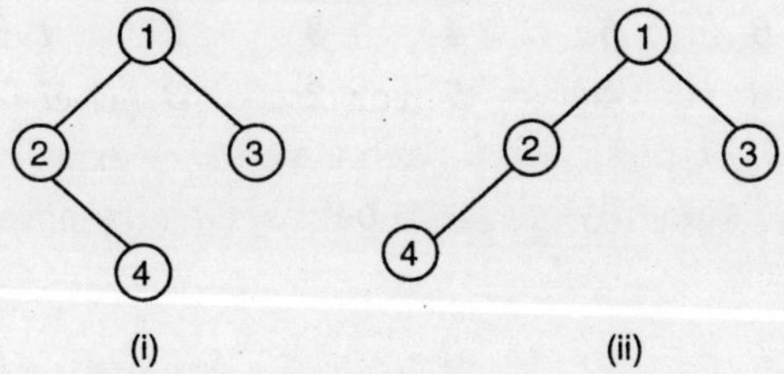 Fig. 32	

Problem 5. *Determine the value of expression tree shown in Fig. 33.*

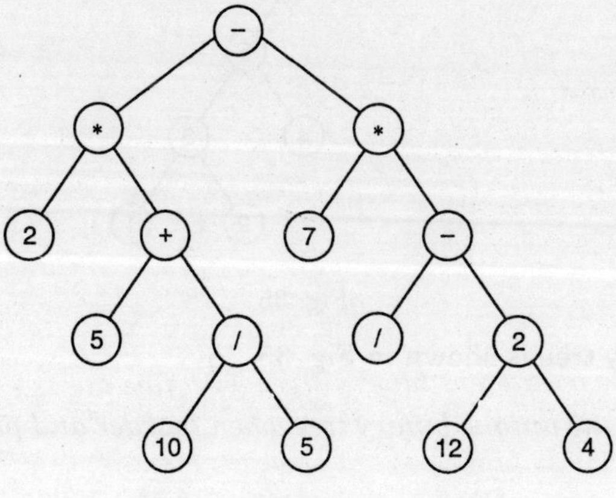

Fig. 33

Sol. The value of expression tree is 2.

Problem 6. *Determine the value of expression tree shown in Fig. 34.*

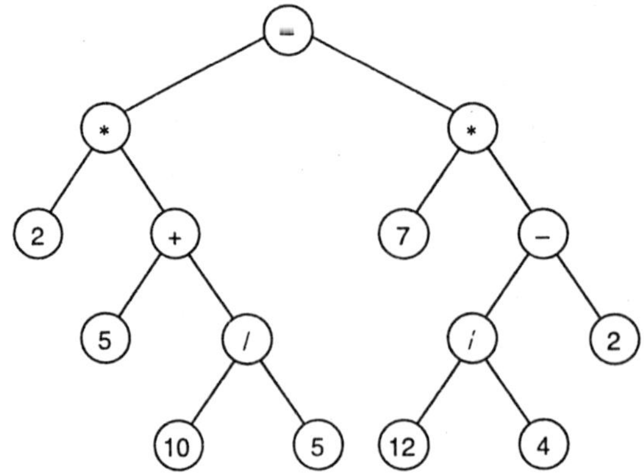

Fig. 34

Sol. The value of expression tree is 7.

Problem 7. *Draw the unique binary tree when inorder and preorder traversal of tree is given as follows :*

Preorder	:	*10*	*9*	*8*	*7*	*6*	*5*	*4*	*3*	*2*	*1*
Inorder	:	*10*	*8*	*9*	*7*	*6*	*4*	*5*	*2*	*3*	*1*

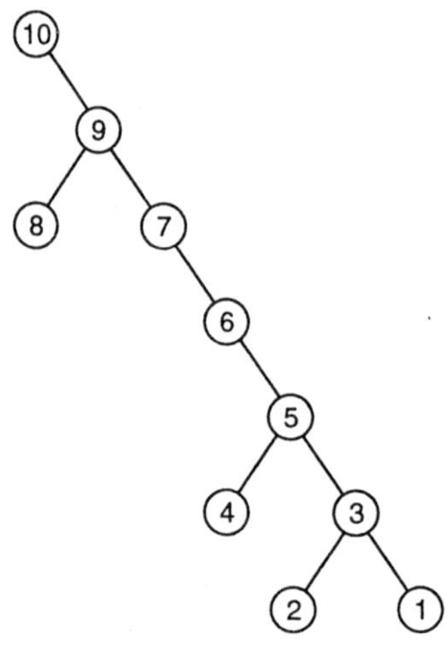

Fig. 35

Sol. Unique binary tree is shown in Fig. 35.

Problem 8. *Draw the unique binary tree when inorder and preorder traversal of tree is given as follows :*

Inorder	:	*3*	*	*a*	+	*b*	*	*6*	*	c	+	*d*
Preorder	:	*	+	*	*3*	*a*	*b*	+	*	*6*	c	*d*

Sol. Unique binary tree is shown in Fig. 36.

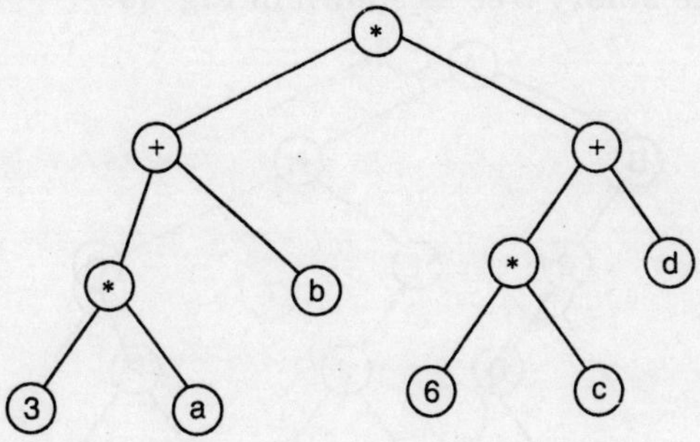

Fig. 36

Problem 9. *Draw the binary expression tree, when inorder and postorder traversal of the tree is given as follows :*

Postorder :	*k*	*l*	– *m*	+	*n*	*p*	*q*	*r*	*t* – /	+	– *
Inorder :	*k*	–	*l* + *m*	*	*n*	–	*p*	+	*q* / *r*	–	*

Sol. Binary expression tree is shown in Fig. 37.

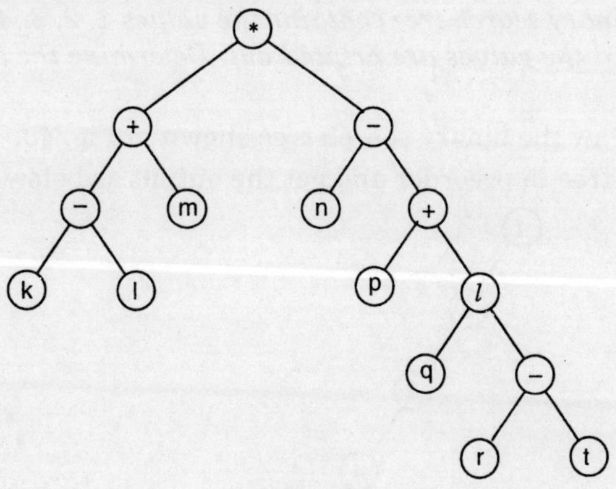

Fig. 37

Problem 10. *Convert the forest shown in Fig. 38 into binary tree.*

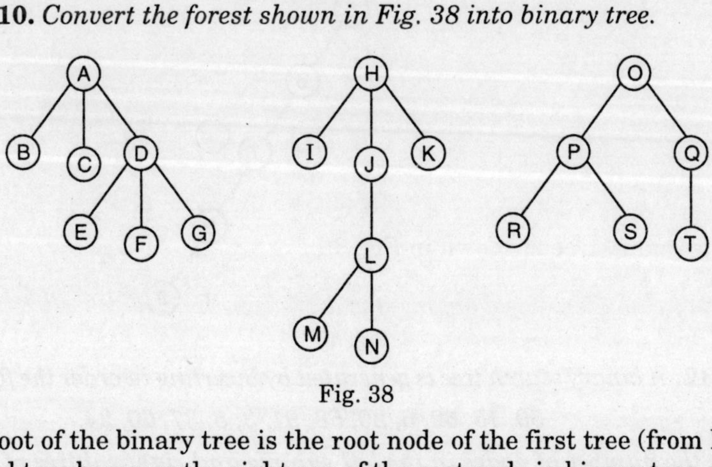

Fig. 38

Sol. The root of the binary tree is the root node of the first tree (from left) and the root node of the second tree becomes the right son of the root node in binary tree and the root node of the third tree becomes the right son of the right son in binary tree. Repeat this procedure for each level and we obtain the binary tree as shown in Fig. 39.

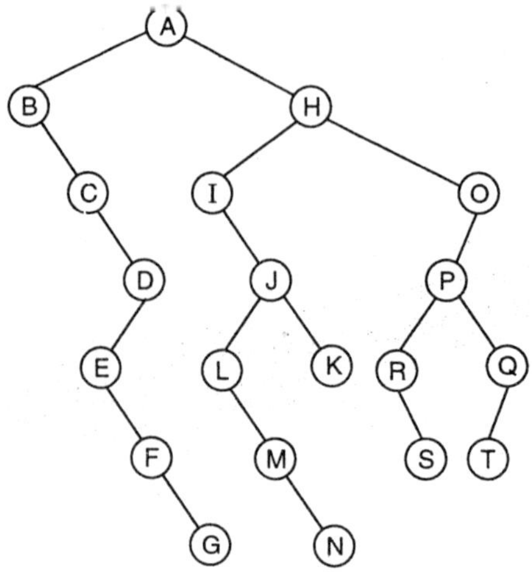

Fig. 39

Problem 11. *A binary search tree contains the values 1, 2, 3, 4, 5, 6, 7, 8, 9. The tree is traversed in preorder and the values are printed out. Determine the sequence of the print out values.*

Sol. First of all draw the binary search tree shown in Fig. 40.

Now traverse the tree in preorder and get the output as below 1, 2, 3, 4, 5, 6, 7, 8, 9.

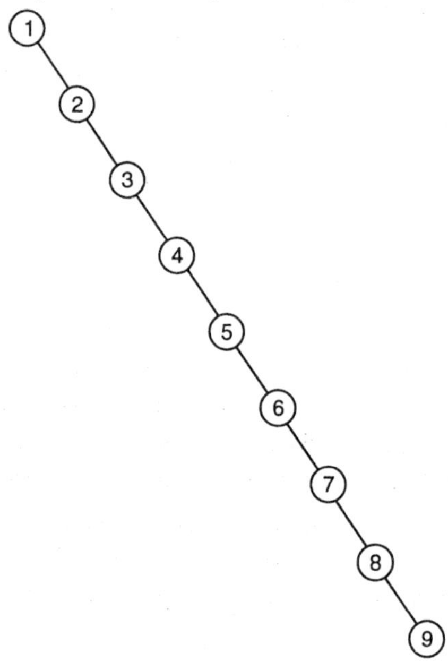

Fig. 40

Problem 12. *A binary search tree is generated by inserting in order the following integers :*

50, 15, 62, 5, 20, 58, 91, 3, 8, 37, 60, 24.

Determine the number of nodes in the left subtree and right subtree of the root.

Sol. First of all draw the binary search tree as shown in Fig. 41.

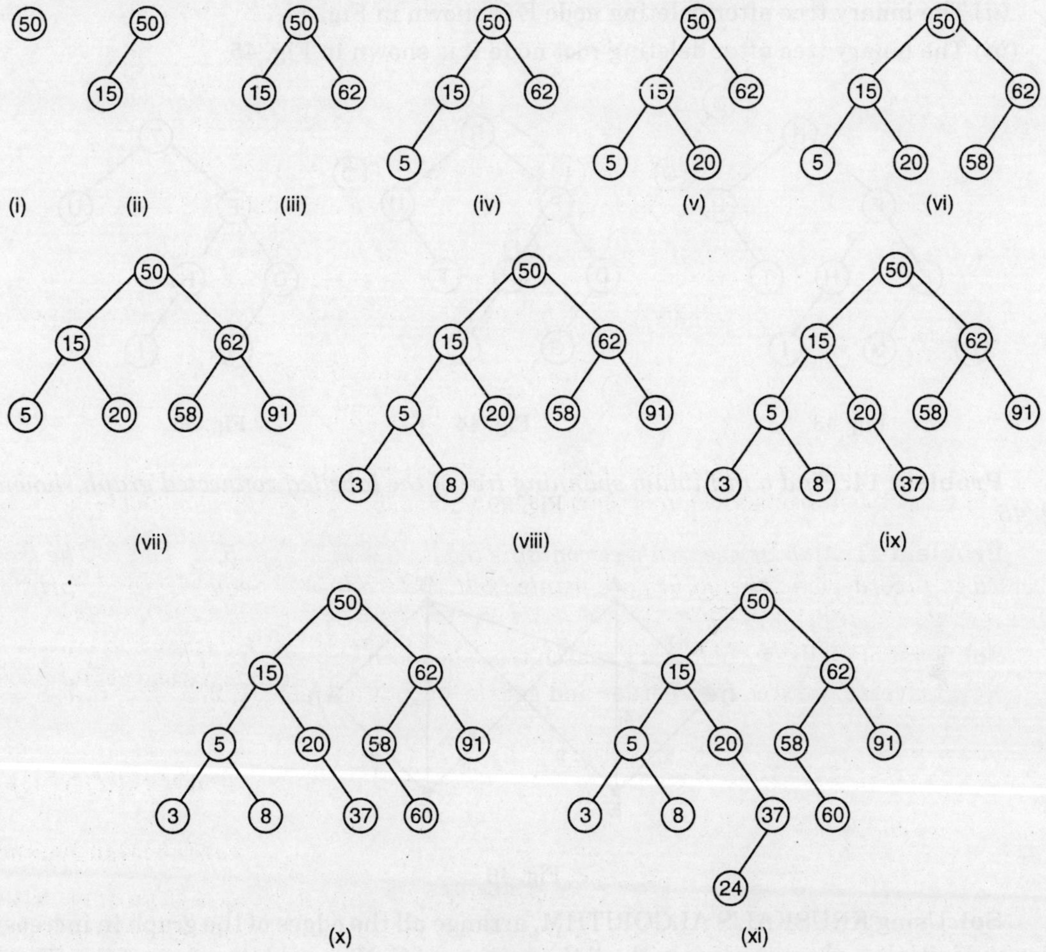

Fig. 41

Thus, the number of nodes in left subtree of the root is 7 and right subtree of the root is 4.

Problem 13. *Consider the binary tree as shown in Fig. 42. Draw the binary tree for each of the following operations, if applied to the binary tree.*

(i) Delete the node V *(ii) Delete the node E*

(iii) Delete the root node R.

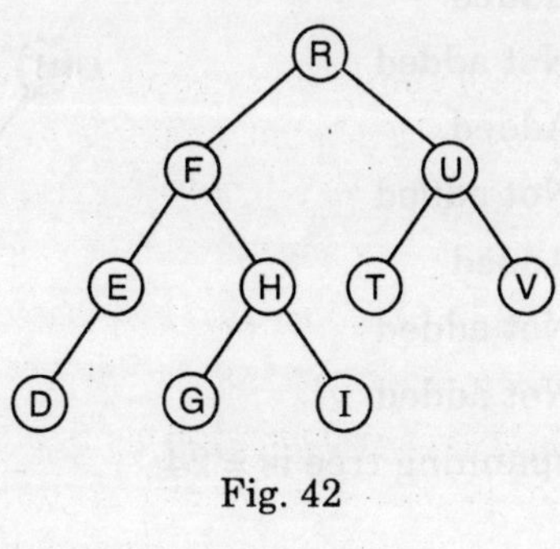

Fig. 42

Sol. (*i*) The binary tree after deleting node V is shown in Fig. 43.

(*ii*) The binary tree after deleting node E is shown in Fig. 44.

(*iii*) The binary tree after deleting root node R is shown in Fig. 45.

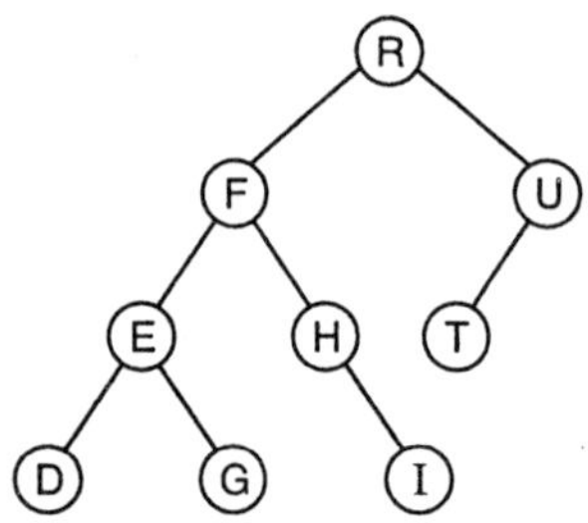

Fig. 43

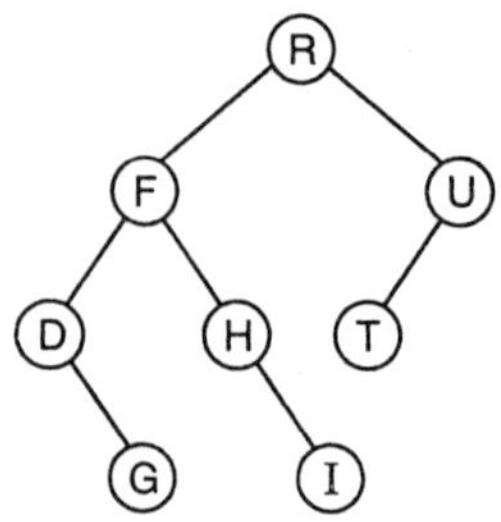

Fig. 44

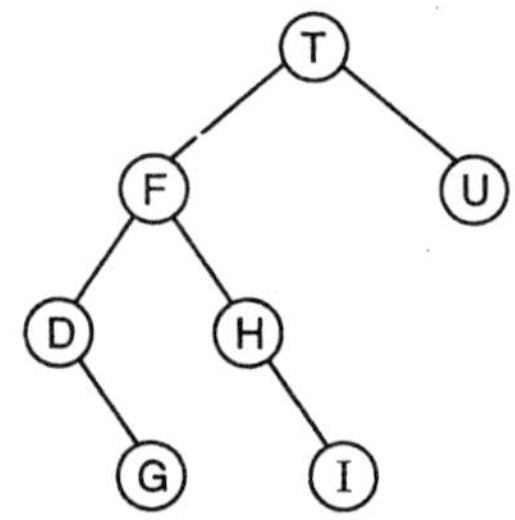

Fig. 45

Problem 14. *Find a minimum spanning tree of the labelled connected graph shown in Fig. 46.*

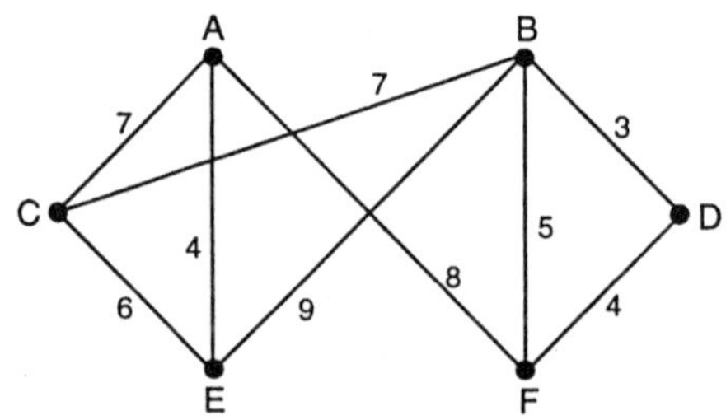

Fig. 46

Sol. Using KRUSKAL'S ALGORITHM, arrange all the edges of the graph in increasing order and initialize spanning tree with all the vertices of G. Now, add the edges of G in T which do not form a cycle and have minimum weight until $n - 1$ edges are not added, where n is the number of vertices. The spanning tree is shown in Fig. 47.

Edges	**Weights**	**Added or Not**
(B, D)	3	Added
(A, E)	4	Added
(D, F)	4	Added
(B, F)	5	Not added
(C, E)	6	Added
(A, C)	7	Not added
(B, C)	7	Added
(A, F)	8	Not added
(E, B)	9	Not added

Minimum Spanning Tree

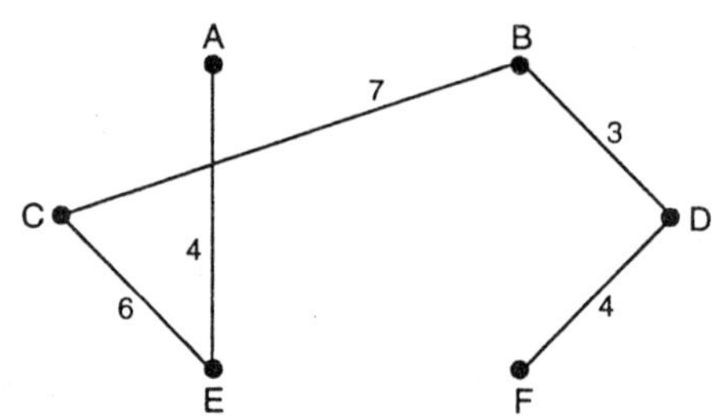

Fig. 47

The minimum weight of spanning tree is = 24.

Problem 15. *Find all the spanning trees of graph G and find which is the minimal spanning tree of G shown in Fig. 48.*

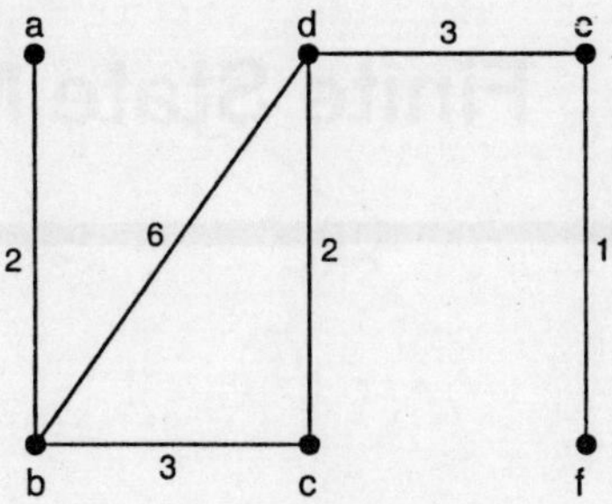

Fig. 48

Sol. There are total three spanning trees of the graph G which are as shown in Fig. 49.

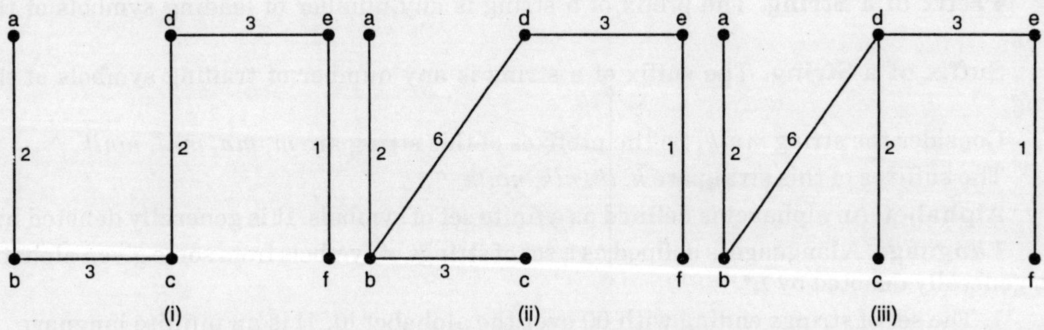

Fig. 49

To find the minimal spanning tree, use the KRUSKAL'S ALGORITHM. The minimal spanning tree is shown in Fig. 50.

Edges	**Weights**	**Added or Not**
(E, F)	1	Added
(A, B)	2	Added
(C, D)	2	Added
(B, C)	3	Added
(D, E)	3	Added
(B, D)	6	Not added.

Minimal Spanning Tree

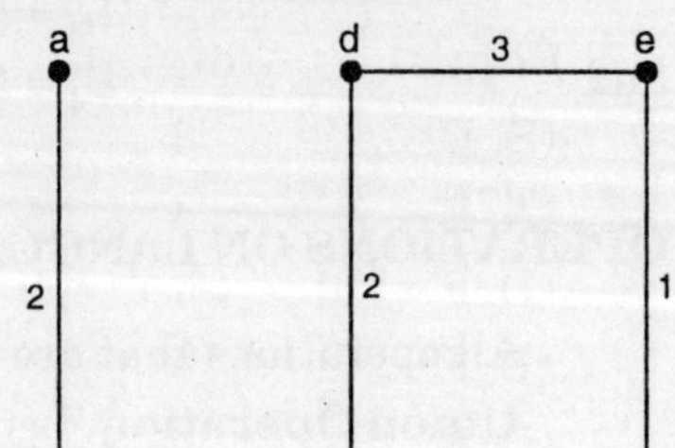

Fig. 50

The first one is the minimal spanning having the minimum weight = 11.

Chapter 12

Finite State Machines

STRING

A string is a finite sequence of symbols *i.e.,* letters and digits in any order. The length of a string w, is the number of symbols composing the string and is denoted by $| w |$.

e.g., 1. Consider the symbols k, l, m, n, then $mnlk, kmn, lmn$ are all strings.

2. The length of these strings are 4, 3 and 3 respectively.

Prefix of a String. The prefix of a string is any number of leading symbols of that string.

Suffix of a String. The suffix of a string is any number of trailing symbols of that string.

e.g., Consider the string $mnlk$. So the prefixes of this string are $m, mn, mnl, mnlk, \wedge$.

The suffixes of this string are $k, lk, nlk, mnlk, \wedge$.

Alphabet. An alphabet is defined as a finite set of symbols. It is generally denoted by Σ.

Language. A language is defined as a set of strings of symbols from some given alphabet. It is generally denoted by Σ^*.

e.g., 1. The set of strings ending with 00 over the alphabet {0, 1} is an infinite language.

The members of this language are 00, 100, 000, 0000, 1000, 1100 ...... etc.

2. Consider the alphabet $\Sigma = \{0, 1\}$. The following sets are all languages over the alphabet Σ.

$$L_1 = \{0, 00, 01, 10, 100, 101\}$$
$$L_2 = \{010, 11011\}$$
$$L_3 = \{\ \}$$

and
$$L_4 = \{0^n\ 10^n \mid n \geq 1\}$$

OPERATIONS ON LANGUAGES

All operations that are applicable on sets can be applied to languages also.

Union Operation. Let us assume that L_1 and L_2 be two languages. Then $L_1 \cup L_2$ is also a language that contains the sentences in either L_1 or L_2.

Intersection Operation. Again if we have L_1 and L_2 be two languages. Then $L_1 \cap L_2$ is a language that contains all the sentences that are in both L_1 and L_2.

Concatenation Operation. Let L_1 and L_2 be two languages over the alphabet Σ. Then the concatenation of L_1 and L_2, denoted by L_1L_2, is the language defined as

$$L_1L_2 = \{w_1 w_2 \mid w_1 \in L, \text{ and } w_2 \in L_2\}$$

It means that L_1L_2 denotes the set of all strings which come from the concatenation of a word L_1 with a string from L_2.

e.g., Consider the languages

$$L_1 = \{a, b\} \text{ and } L_2 = \{ab, bb\}$$

Then $L_1L_2 = \{aab, abb, bab, bbb\}$

and $L_2L_1 = \{aba, abb, bba, bbb\}$.

Kleen Closure Operation. It is a unary operation and is denoted by L^*. It is defined as the infinite union of $L^1, L^2, L^3, L^4, \ldots\ldots, L^\infty$ and is denoted as

$$L^* = L^0 \cup L^1 \cup L^2 \cup \ldots\ldots \cup L^\infty$$

$$= \bigcup_{i=0}^{\infty} L^i$$

where $L^0 = \{\wedge\}$; $L^1 = L$; $L^2 = LL$; $L^3 = LLL$

and $L^{n+1} = L^nL$ for $n > 1$.

REGULAR EXPRESSIONS AND REGULAR LANGUAGES

Regular Expressions. The class of regular expressions over Σ is defined recursively as follows :

(*a*) The symbol $\wedge$ and ϕ are regular expressions over Σ.

(*b*) Every letter $a \in \Sigma$ is a regular expression over Σ.

(*c*) If r_1 and r_2 are regular expressions over Σ, then so are $(r_1 + r_2)$, $(r_1 \cdot r_2)$ and $(r_1)^*$.

where + indicates alternation, · denotes concatenation and * denotes closure.

(*d*) The regular expressions are only those that are obtained using rules (*a*) to (*c*).

e.g.,
1. The 11 is a regular expression over the alphabet $\Sigma = \{0, 1\}$ representing the set $\{1, 1\}$.
2. The $(0 + 1)^*$ is a regular expression over the alphabet $\Sigma = \{0, 1\}$ representing all strings of 0's and 1's.
3. The $(0 + 1)^*$ 00 is a regular expression denoting all strings of 0's and 1's ending with 00.
4. The $(1 + 10)^*$ is a regular expression denoting all string of 0's and 1's that begins with 1 and not having two consecutive zeroes.
5. The $0^*\ 1^*$ is a regular expression denoting all strings of 0's and 1's having any number of zeroes followed by any number of 1's.

Regular Language. Consider a language L over the alphabet Σ. Then L is called a regular over Σ if there exists a regular expression r over Σ such that $L = L(r)$.

e.g.,
1. If $r = 01^*$, then $L(r)$ consists of all strings of 0's and 1's that begin with 0 and followed by any no. if 1's.

2. If $r = 0 + 1^*$, then L(r) consists of strings containing a single 0 or any no. of 1's.
3. If $r = (0 + 1)^*10$, then L(r) consists of all strings of 0's and 1's that ends with 10.

FINITE STATE AUTOMATA

A finite state automata M is denoted by a 5-tuple $(Q, \Sigma, \delta, q_0, F)$, where

1. Q is a finite set of states.
2. Σ is a finite input alphabet.
3. q_0 in Q is the initial state.
4. F is a subset of Q and is the set of final states.
5. δ is transition function or next-state function mapping $Q \times \Sigma$ to Q.

To make a finite state automata, either we have to denote it by indicating its five parts or we can make a state diagram of it.

STATE DIAGRAM OF A FINITE STATE AUTOMATA

A finite state automata can be defined by means of its state diagram. The state diagram is a labeled directed graph, where

— The vertices of the graph are the states in Q.
— The arrow from the state q_i to q_j with label a represents a transition or next-state function, which defines $\delta(q_i, a) = q_j$.
— The initial state q_0 is represented by an arrow.
— The final state or accepting state is represented by a double circle.

Example 1. *Make a finite state automata M that accepts the language of all strings of 0's and 1's that ends with 00.*

Sol. The automata M is defined by representing its five parts, which is as follows.

1. $Q = \{q_0, q_1, q_2\}$
2. $\Sigma = \{0, 1\}$, input symbols
3. q_0, is the initial state
4. $F = \{q_2\}$, is the final state
5. δ, the transition function is represented by the table as shown in figure.

	Inputs	
States	*0*	*1*
q_0	q_1	q_0
q_1	q_2	q_0
q_2	q_2	q_0

Example 2. *Draw the state diagram of the automata M, defined in the previous example.*

Sol.

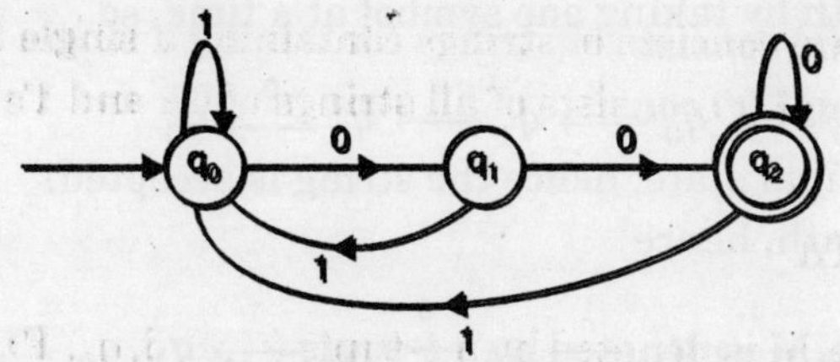

This defines the automaton M, which accepts all the strings of 0's and 1's that ends with double 0.

In this the final state q_2, is represented by a double circle, and the initial state q_0 is represented by special arrow.

Note. From now onwards, we will represent the finite state automata with its state diagram unless and until not specified.

Example 3. ***Construct a finite state automata M, that accepts the language L(r) of all strings of 0's and 1's which contains 000 as substring.***

Sol. The finite state automata M, accepting the language L(*r*) is defined as follows :

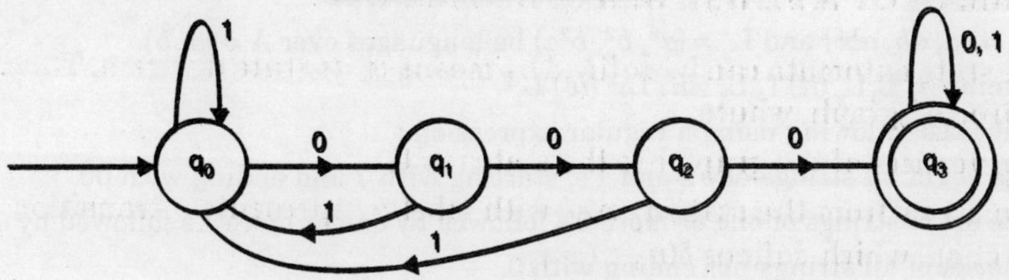

RELATIONSHIP BETWEEN REGULAR LANGUAGES AND FINITE AUTOMATA

Every automaton M defines a language over Σ, where Σ is an input alphabet.

If M is the finite Automaton, then L(M) is the language accepted by automaton, which is the collection of all strings constructed from Σ, which are accepted by M.

By acceptance of strings, we mean that M recognises the string *w* iff it halts in the final state.

A language L over an alphabet Σ is called regular iff there exists a finite state automaton M such that L = L (M).

Example 4. ***Determine whether or not the automata M as shown in figure accepts the following strings or not :***

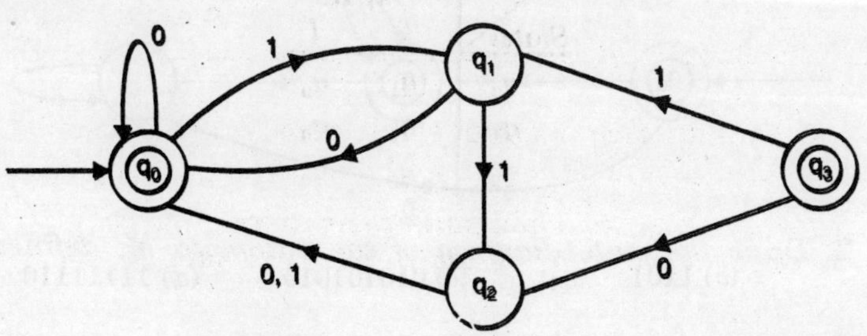

(a) 0100 ***(b) 1101*** (c) ε ***(d) 010101010.***

Sol. (*a*) Trace the path by taking one symbol at a time, so

$$\text{Path} = q_0 \xrightarrow{0} q_0 \xrightarrow{1} q_1 \xrightarrow{0} q_0 \xrightarrow{0} q_0$$

Since it halts in the final state, hence the string is accepted.

(*b*) Again trace the path, hence

$$\text{Path} = q_0 \xrightarrow{1} q_1 \xrightarrow{1} q_2 \xrightarrow{0} q_0 \xrightarrow{1} q_1$$

Since it does not halt in the final state at the end, hence the string is not accepted.

(*c*) This time since the initial state is the final state hence without inputing anything we were in the final state hence it is accepted.

(*d*) This time the path is

$$\text{Path} : q_0 \xrightarrow{0} q_0 \xrightarrow{1} q_1 \xrightarrow{0} q_0 \xrightarrow{1} q_1 \xrightarrow{0} q_0 \xrightarrow{1} q_1 \xrightarrow{0} q_0 \xrightarrow{1} q_1 \xrightarrow{0} q_0$$

Since q_0 is the final state at the end, hence it is accepted.

EXERCISES

1. Let $L_1 = \{a, ab, ab^2\}$ and $L_2 = \{a^2, b^2, b^2a\}$ be languages over $A = \{a, b\}$.
   Determine (*i*) L_1L_2 (*ii*) L_2L_1 (*iii*) L_2^2 (*iv*) L_1^2.
2. Describe the following using a regular expression :
   (*i*) The set of all strings of 0's and 1's, starting with 1 and ending with 00.
   (*ii*) The set of strings of one or more 0's followed by one or more 1's followed by one or more 2's .
   (*iii*) The set of all strings not ending with 0.
3. Describe in English the strings accepted by the following regular expressions :
   (*i*) 0 * 10 * (*ii*) (0 + 1)*00 (0 + 1)*11(0 + 1)* (*iii*) 1*0.
4. Describe the following sets by regular expressions :
   (*i*) {10, 11} (*ii*) {^, 0, 00, 000, ...} (*iii*) {^, 0, 1, 01, 10, 11, 00, ...}.
5. Make a finite state automata M that accepts the given set of strings over {0, 1} :
   (*i*) Even no. of 0's (*ii*) odd no. of 1's
   (*iii*) odd no. of 0's and 1's (*iv*) odd number of 0's even number of 1's.
6. Describe the elements of the set A* for the following values of A :
   (*i*) {10} (*ii*) {0, 10} (*iii*) {0 0 0}.
7. Determine whether each of the following strings is recognised by the deterministic finite-state automaton as shown in figure :

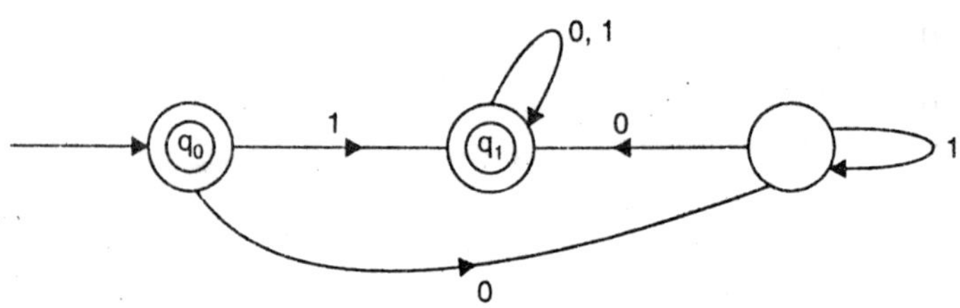

(*a*) 010 (*b*) 1101 (*c*) 010101010 (*d*) 11111110.

FINITE STATE MACHINES—(FSM)

A finite state machine also called FSM, is similar to a finite state automaton except that the finite state machine prints an output, which is different from the input alphabet.

Thus the FSM is defined as follows :

A finite state machine M consists of six tuple

$(Q, \Sigma, q_0, O, \delta, f)$, where

1. Q, is finite set of states.
2. Σ, is finite input alphabet
3. q_0, is the initial state.
4. O, is finite set of output symbols.
5. δ, is a truvrition function mapping $Q \times \Sigma \to Q$.
6. f, is an output function mapping $Q \times \Sigma \to O$.

Example. *The following defines a finite state machine by representing its 6 parts :*

1. $Q = \{q_0, q_1, q_2, q_3\}$
2. $\Sigma = \{0, 1\}$
3. q_0 is the initial state.
4. $O = \{0, 1\}$.
5. The transition function $\delta : Q \times \Sigma \to Q$ is defined as

$$\delta(q_0, 0) = q_0 \qquad \delta(q_2, 0) = q_1$$
$$\delta(q_0, 1) = q_1 \qquad \delta(q_2, 1) = q_2$$
$$\delta(q_1, 0) = q_3 \qquad \delta(q_3, 0) = q_2$$
$$\delta(q_1, 1) = q_0 \qquad \delta(q_3, 1) = q_1$$

6. The output function $f : Q \times \Sigma \to O$ is defined as

$$f(q_0, 0) = 1 \qquad f(q_2, 0) = 0$$
$$f(q_0, 1) = 0 \qquad f(q_2, 1) = 1$$
$$f(q_1, 0) = 1 \qquad f(q_3, 0) = 0$$
$$f(q_1, 1) = 1 \qquad f(q_3, 1) = 0$$

REPRESENTATION OF THE FINITE STATE MACHINE

There are mainly two ways to represent the finite state machine.

1. State Table
2. State Diagram.

1. State table of a finite state machine. A finite state machine can be represented by a table called the state table of a machine. The state table of the finite state machine, M, combines the transition function δ and the output function f into a single table and represent the function

$F : Q \times \Sigma \to Q \times O$ defined as

$$F(q_i, a_j) = (\delta(q_i, a_j), f(q_i, a_j))$$

Example 5. *Construct the state table for the finite state machine given in the previous example.*

Sol.

State	δ		f	
	Input		Input	
	0	1	0	1
q_0	q_1	q_0	1	0
q_1	q_3	q_0	1	1
q_2	q_1	q_2	0	1
q_3	q_2	q_1	0	0

2. State diagram of a finite state machine. The finite state machine can be represented by a labeled directed graph called the state diagram of the machine M.

The state diagram of an FSM M = {Q, Σ, q_0, O, δ, *f*) is a labeled directed graph, where the vertices of the graph are states of M and if

$$\delta(q_i, a_j) = q_k \text{ and } f(q_i, a_j) = O_t$$

then there is an arc from q_i to q_k which is labeled with the pair a_j/O_t.

The initial state of the FSM is represented by an arrow as in FSA.

Example 6. *Construct the state diagram for the finite state machine whose state table is shown in previous example.*

Sol.

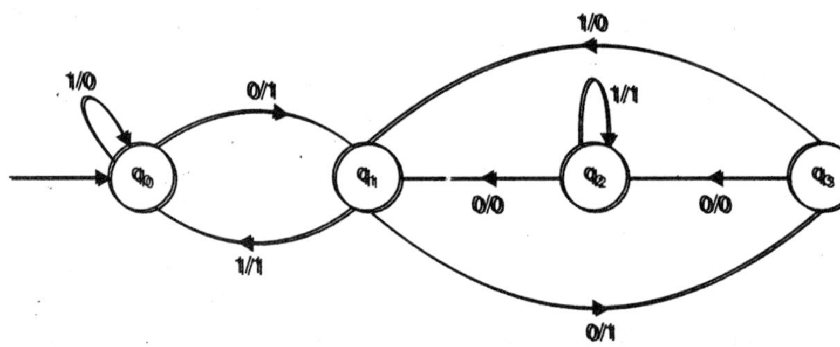

Example 7. *Give the state table for the finite state machine whose state diagram is shown in figure.*

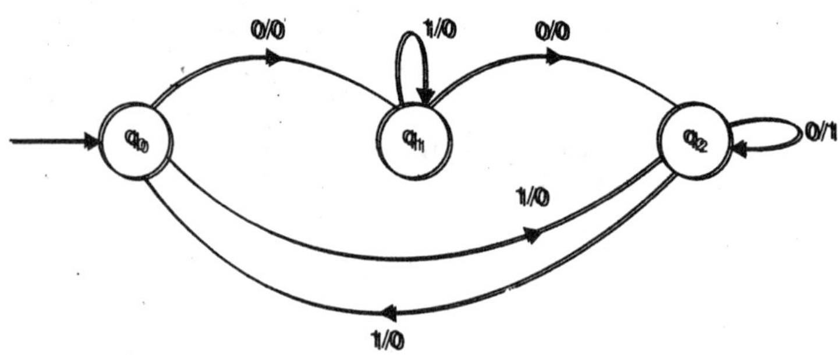

Sol. The state table for the FSM is shown in figure.

State	δ		f	
	Input		Input	
	0	*1*	*0*	*1*
q_0	q_1	q_0	0	0
q_1	q_0	q_2	0	1
q_2	q_1	q_1	1	0

Example 8. *Construct the state diagram for the finite state machine whose state table is shown in figure.*

State	δ		f	
	Input		Input	
	0	*1*	*0*	*1*
q_0	q_1	q_0	1	1
q_1	q_2	q_0	0	0
q_2	q_0	q_3	1	0
q_3	q_1	q_2	0	1

Sol. The state diagram of the finite state machine is given in the figure.

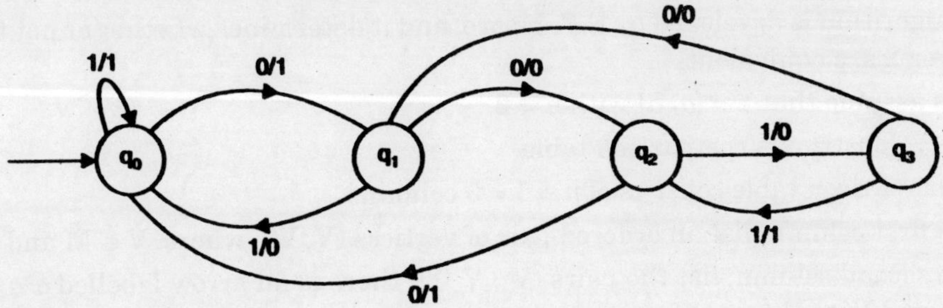

Example 9. *Given the finite state machine as shown in figure. Determine the output for each of the following input strings :*

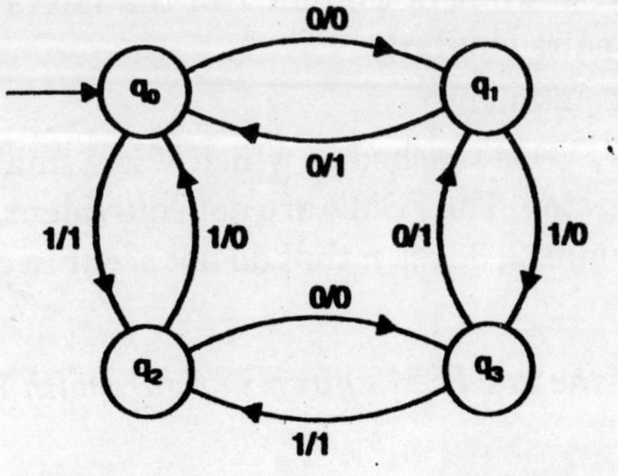

(a) 1010 *(b) 11011011* *(c) 1010001001.*

Sol. (*a*) Beginning at q_0, follow the path which are labeled by the input symbols and write the output symbols. Thus.

$$q_0 \xrightarrow{1/1} q_2 \xrightarrow{0/0} q_3 \xrightarrow{1/1} q_2 \xrightarrow{0/0} q_3$$

So the output for this string is 1010.

(*b*) Follow the same procedure as in (*a*). Thus

$$q_0 \xrightarrow{1/1} q_2 \xrightarrow{1/0} q_0 \xrightarrow{0/0} q_1 \xrightarrow{1/0} q_3 \xrightarrow{1/1} q_2 \xrightarrow{0/0} q_3 \xrightarrow{1/1} q_2 \xrightarrow{1/0} q_0$$

So the output for the string 11011011 is 10001010.

(*c*) The output for the string 1010001001 is as follows.

$$q_0 \xrightarrow{1/1} q_2 \xrightarrow{0/0} q_3 \xrightarrow{1/1} q_2 \xrightarrow{0/0} q_3 \xrightarrow{0/1} q_1 \xrightarrow{0/1} q_0 \xrightarrow{1/1} q_2 \xrightarrow{0/0} q_3 \xrightarrow{0/1} q_1 \xrightarrow{1/0} q_3$$

The output is 1010111010.

EQUIVALENT MACHINES

Two finite state machines are said to be equivalent if, starting from their initial states, both will produce the same output sequence when they have the identical input sequence.

FSM EQUIVALENCE ALGORITHM

This algorithm is developed by E.F. Moore, and it determines whether or not two FSM's M and M′ over Σ are equivalent.

Let us assume that $\Sigma = \{a, b\}$ with $n = 2$.

Step 1. Construct a comparison table

The comparison table consists of $n + 1 = 3$ columns.

In the first column, list an ordered pair of vertices (V, V′), where $V \in M$ and $V' \in M'$.

In the second column, list the pairs (V_a, V_a') if there is an arrow labelled $a \in \Sigma$ leading from V to V_a and V′ to V_a'.

In column three, list the pairs (V_b, V_b') if there is an arrow labelled $b \in \Sigma$ leading from V to V_b and V′ to V_b'. Repeat this exercise for all other columns in the table. If (V_a, V_a') does not occur previously in column 1, place it in column 1 of the next row and repeat the process). Treat the (V_b, V_b') and other pairs similarly.

Step 2. Establish the equivalence

If in the table a pair (V, V′) is reached in which V is a final vertex of M and V′ is a non final vertex of A′ or vice-versa, stop. The FSM's are not equivalent. Otherwise, terminate when there are no new pairs in columns 2, 3, ..., *n* that do not occur in column 1. In this case, M and M′ are equivalent.

Example 10. *Consider the two FSM's given in figure with Σ = {0, 1}. Determine whether they are equivalent or not.*

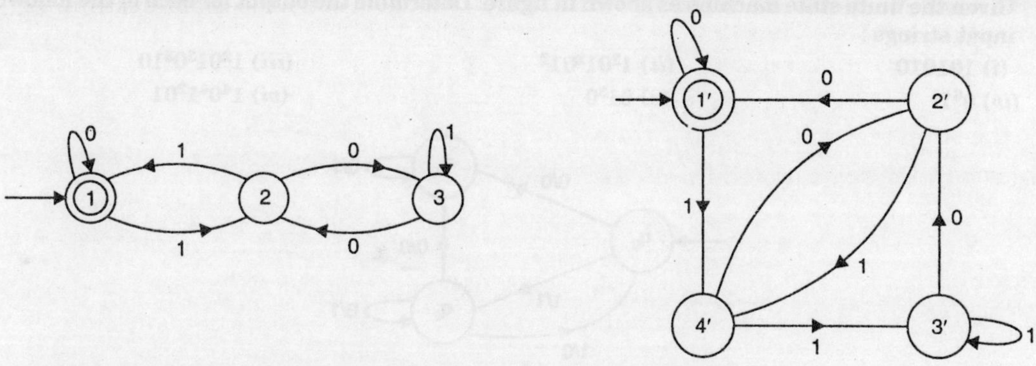

Sol. Construct the comparison table as shown in figure.

(V, V')	(V_0, V_0')	V_1, V_1'
(1, 1′)	(1, 1′)	(2, 4′)
(2, 4′)	(3, 2′)	(1, 3′)

In the table, in second column, we see that 1 is a final vertex in FSM M but 3′ is not a final vertex which indicates the non equivalence.

EXERCISES

1. Construct the state diagram of the finite state machine M, whose transition table is shown in the figure.

$\Sigma = \{a, b\}$, $Q = \{q_0, q_1, q_2\}$, $O = \{x, y, z\}$

	δ		*f*	
State	*Input*		*Input*	
	0	*1*	*0*	*1*
q_0	q_1	q_2	x	y
q_1	q_2	q_1	x	z
q_2	q_0	q_1	z	y

2. Construct the state table for the finite state machines whose state diagram is shown in figure.

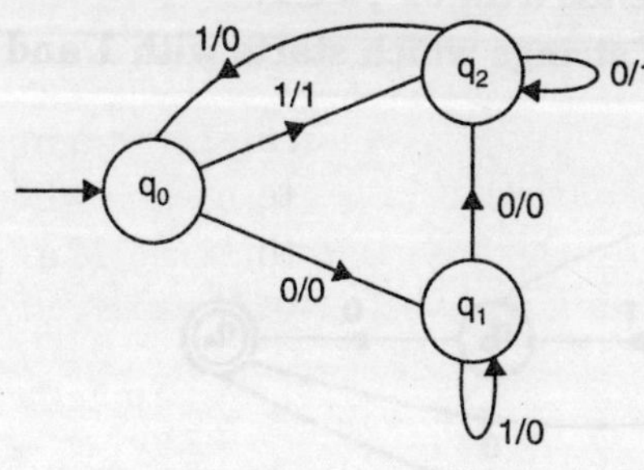

(*i*)

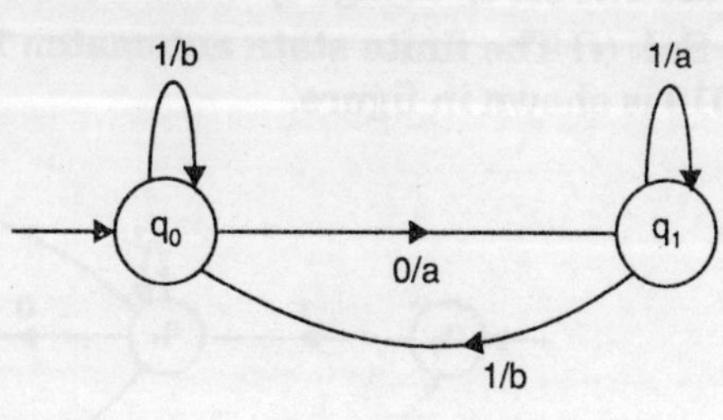

(*ii*)

3. Given the finite state machine as shown in figure. Determine the output for each of the following input strings :

(i) 101010 (ii) $1^2 01^2 01^2$ (iii) $1^2 01^3 0^3 10$

(iv) $0^6 1^5$ (v) $01^3 0$ (vi) $1^4 0^4 1^2 01$

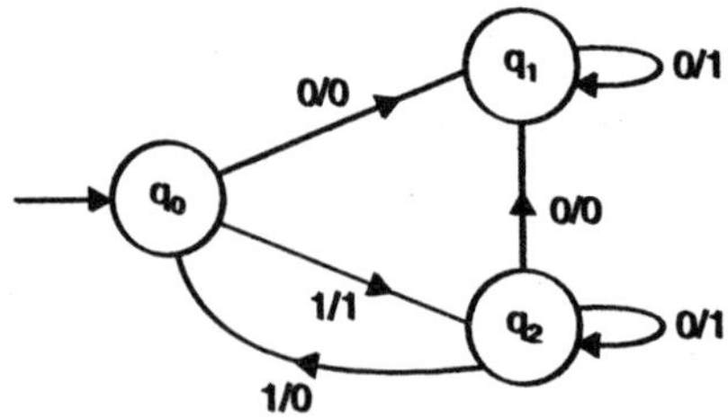

4. Construct a finite state machine that gives an output of 1 if the number of input symbols read so far is divisible by 3 and an output of 0 otherwise.

5. Construct a finite state machine that changes every other bit, starting with the second bit, of an input string and leaves the other bits unchanged.

SOLVED PROBLEMS

Problem 1. *Consider the languages $L_1 = \{a, a^2, a^3, a^4\}$ and $L_2 = \{b, b^2, ab^2\}$ over the alphabet $\Sigma = \{a, b\}$.*

Determine the languages (i) L_1L_2 (ii) L_1L_1 (iii) L_2L_1 (iv) L_2L_2.

Sol. (i) The strings in L_1 are concatenated with L_2 and the language become

$$L_1L_2 = \{ab, ab^2, aab^2, a^2b, a^2b^2, a^3b^2, a^3b, a^4b^2, a^4b, a^5b^2\}$$

(ii) The strings in L_1 are concatenated with L_2 and the language become

$$L_1L_1 = \{a^2, a^3, a^4, a^5, a^6, a^7, a^8\}$$

(iii) The strings in L_2 are concatenated with L_1 and the language become

$$L_2L_1 = \{ba, ba^2, ba^3, ba^4, b^2a, b^2a^2, b^2a^3, b^2a^4, ab^2a, ab^2a^2, ab^2a^3, ab^2a^4\}$$

(iv) The strings in L_2 are concatenated with L_2 and the language become

$$L_2L_2 = \{b^2, b^3, bab^2, b^4, b^2ab^2, ab^3, ab^4, ab^2ab^2\}.$$

Problem 2. *Construct the finite state automaton, which recognizes the following languages :*

(i) The set of strings of 0's and 1's each of which starts with a1 and ends with 010.

(ii) The set of strings of 0's and 1's each of which ends with 01^k, $k \geq 1$.

Sol. (i) The finite state automaton for the set of strings which starts with 1 and ends with 010 is shown in figure.

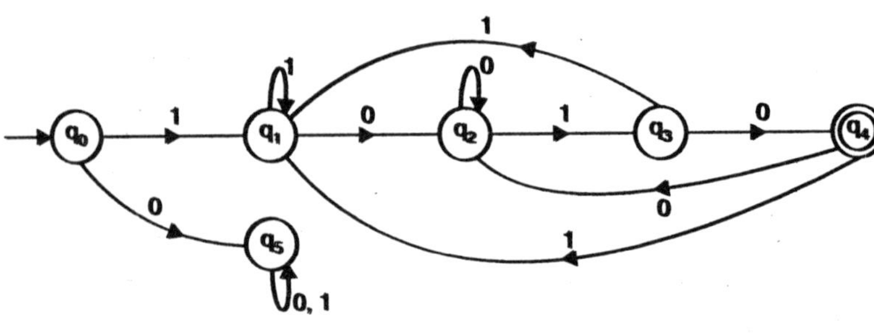

(*ii*) The finite state automaton for the set of strings each of which ends with 01^k, $k \geq 1$ is shown in figure.

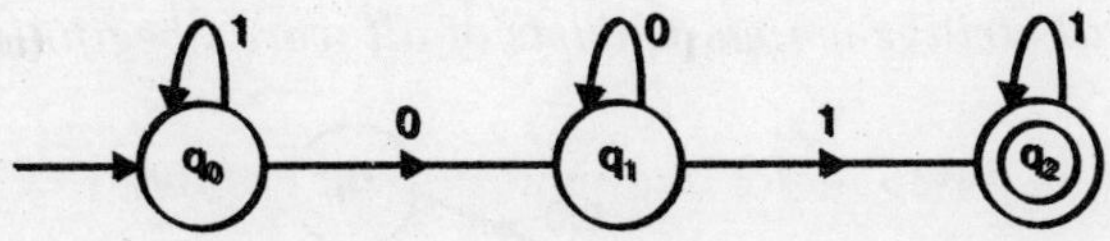

Problem 3. ***Construct the finite state automaton, which recognizes the following languages :***

(i) The set of strings of 0's and 1's none of which contains the substring 010.

(ii) The set of strings of 0's and 1's in each of which there is an even number of 0's and 1's.

Sol. (*i*) The finite state automaton for the set of strings none of which contains the substring 010 is shown in figure.

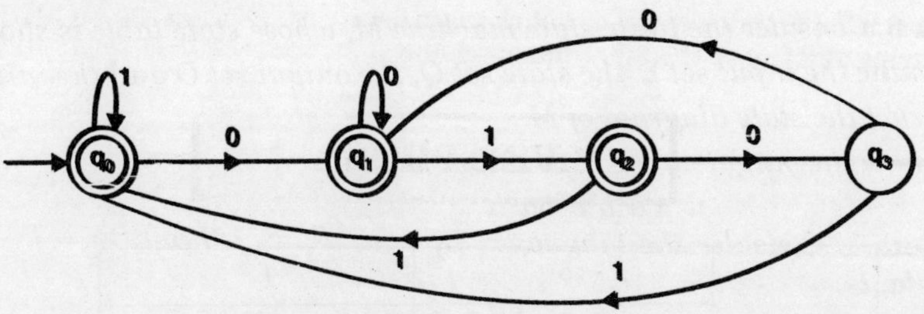

Note. The trick here is to make an automaton which accepts the language of strings which ends with 010 and make all these states final which are non final in case of 010 and non final which are final.

(*ii*) The automaton for the set of strings in each of which there is an even number of 0's and 1's is shown in figure.

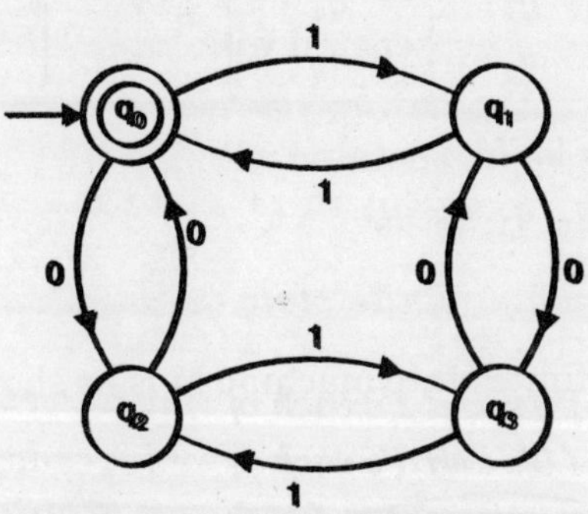

Problem 4. ***Consider the language $L = \{a^2, b^2, ab\}$. Determine***

(i) L^0 ***(ii)*** L^2 ***(iii)*** L^3.

Sol. (*i*) The language $L^0 = \{\wedge\}$

(*ii*) The language $L^2 = \{a^4, a^2b^2, a^3b, b^2a^2, b^4, b^2ab, aba^2, ab^3, abab\}$ the language.

(*iii*) The language $L^3 = \{a^6, a^4b^2, a^5b, a^2b^2a^2, a^2b^4, a^2b^2ab, a^3ba^2, a^3b^3, a^3bab, a^2a^4, b^2a^2b^2, b^2a^3b, b^4a^2, b^6, b^4ab, b^2aba^2, b^2ab^3, b^2abab, aba^4, aba^2b^2, aba^3b, ab^3a^2, ab^5, ab^3ab, ababa^2, abab^3, ababab\}$.

Problem 5. *Describe the following sets by regular expression R over $\Sigma = \{a, b\}$*

(i) L_1 = the set of all strings which consists of all words with exactly two a's.

(ii) L_2 = the set of all strings which consists of all words beginning with 'a' followed by zero or more b's.

(iii) $L_3 = \{\wedge, a, aa, aaa, aaaa, ...\}$

(iv) $L_4 = \{ab, ba\}$

(v) L_5 = The set of all words without a double a.

Sol. (*i*) The regular expression for L_1 is $R = b^*ab^*ab^*$

(*ii*) The regular expression for L_2 is $R = ab^*$

(*iii*) The regular expression for L_3 is $R = a^*$

(*iv*) The regular expression for L_4 is $R = ab + ba$

(*v*) The regular expression for L_5 is $R = b^*(abb^*)^*(\wedge + a)$.

Problem 6. *Consider the finite state machine M, whose state table is shown in figure.*

(a) Determine the input set Σ, the state set Q, the output set O and the initial state of M.

(b) Construct the state diagram of M.

(c) Determine the output string if the input string

$$w = a^2b^2a^2ba^3b$$

States	δ		f	
	Input		Input	
	a	b	a	b
q_0	q_2	q_1	y	z
q_1	q_2	q_3	x	y
q_2	q_2	q_1	y	z
q_3	q_3	q_0	z	x

Sol. (*a*) The input set $\Sigma = \{a, b\}$

The finite state set $Q = \{q_0, q_1, q_2, q_3\}$

The output set $O = \{x, y, z\}$

The initial state is q_0.

(*b*) The state diagram of finite state machine M is as shown in figure.

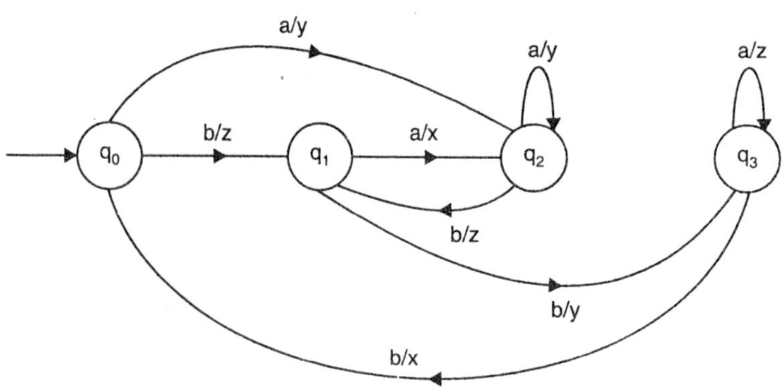

(*c*) The output string corresponding to the input string $w = a^2b^2\, a^2ba^3b$ is determined by following the path in the state diagram.

$$\text{Path} = q_0 \xrightarrow{a/y} q_2 \xrightarrow{a/y} q_2 \xrightarrow{b/z} q_1 \xrightarrow{b/y} q_3 \xrightarrow{a/z} q_3 \xrightarrow{a/z} q_3 \xrightarrow{b/y}$$

$$q_1 \xrightarrow{a/x} q_2 \xrightarrow{a/y} q_2 \xrightarrow{a/y} q_2 \xrightarrow{b/z} q_1$$

Thus the string is $y^2zyz^2yxy^2z$.

Problem 7. *Consider the finite state machine M, whose state table is shown in figure.*

State	δ		*f*	
	Input		*Input*	
	0	*1*	*0*	*1*
q_0	q_1	q_3	*a*	*b*
q_1	q_1	q_2	*a*	*a*
q_2	q_3	q_4.	*b*	*b*
q_3	q_1	q_0	*b*	*b*
q_4	q_3	q_4	*b*	*b*

(*a*) *Determine the input set Σ, the state set Q, the output set O and the initial state of M.*

(*b*) *Construct the state diagram of M.*

(*c*) *Determine the output string if the input string*

$$w = 110111000101.$$

Sol. (*a*) The input set Σ = {0, 1}

The state set Q = $\{q_0, q_1, q_2, q_3, q_4\}$

The output set O = {*a*, *b*}

The initial state is q_0.

(*b*) The state diagram of the finite state machine M is shown in figure.

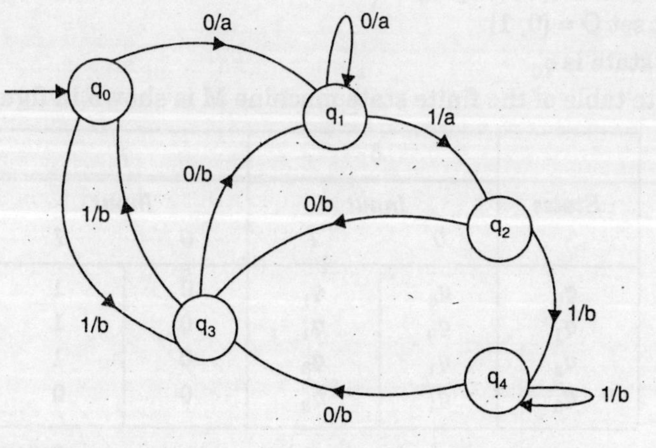

(*c*) The output string corresponding to the input string of $w = 1^2\ 0\ 1^3\ 0^3\ 1\ 0\ 1$ is determined by following the path in the state diagram.

$$\text{Path} = q_0 \xrightarrow{1/b} q_3 \xrightarrow{1/b} q_0 \xrightarrow{0/a} q_1 \xrightarrow{1/a} q_2 \xrightarrow{1/b} q_4 \xrightarrow{1/b} q_4 \xrightarrow{0/b}$$
$$q_3 \xrightarrow{0/b} q_1 \xrightarrow{0/a} q_1 \xrightarrow{1/a} q_2 \xrightarrow{0/b} q_3 \xrightarrow{1/b} q_0$$

Thus the string is $b^2a^2b^4a^2b^2$.

Problem 8. *Describe the elements of the set A* for the following values of A.*

(i) {a b} *(ii) {a a a}* *(iii) {b, bab}.*

Sol. (*i*) The set of all strings consisting of zero or more consecutive pairs of *ab*.

(*ii*) The set of all strings consisting of all *a*'s such that the number of *a*'s is divisible by three. It also includes a null string.

(*iii*) The set of all strings that begin and end with an *b* and have at least two *b*'s between every pair of *a*'s.

Problem 9. *Consider the finite state machine M, whose state diagram is shown in figure.*

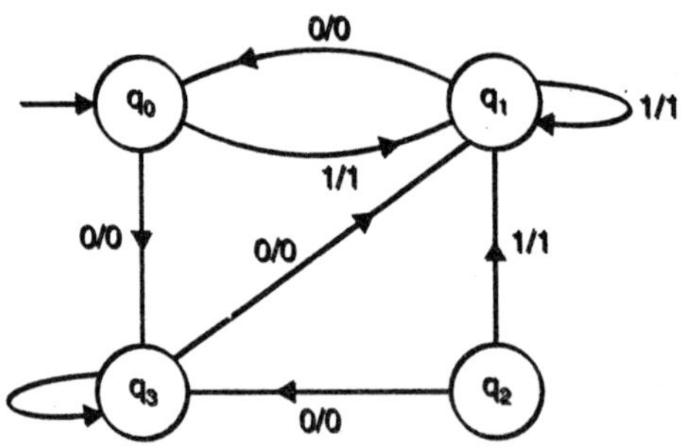

(*a*) *Determine the input set Σ, the state set Q, the output set O and the initial state of M.*

(*b*) *Construct the state table of M.*

(*c*) *Determine the output string if the input string*

$$w = 1^2 0^2 1^2 0^2 101.$$

Sol. (*a*) The input set $\Sigma = \{0, 1\}$

The state set $Q = \{q_0, q_1, q_2, q_3\}$

The output set $O = \{0, 1\}$

The initial state is q_0.

(*b*) The state table of the finite state machine M is shown in figure.

States	δ		*f*	
	Input		*Input*	
	0	*1*	*0*	*1*
q_0	q_3	q_1	0	1
q_1	q_0	q_1	0	1
q_2	q_1	q_3	0	1
q_3	q_1	q_3	0	0

(*c*) The output string corresponding to the input string $w = 1^2 0^2 1^2 0^2 101$ is determined by following the path in the state diagram.

$$\text{Path} = q_0 \xrightarrow{1/1} q_1 \xrightarrow{1/1} q_1 \xrightarrow{0/0} q_0 \xrightarrow{0/0} q_3 \xrightarrow{1/0} q_3 \xrightarrow{1/0} q_3 \xrightarrow{0/0}$$

$$q_1 \xrightarrow{0/0} q_3 \xrightarrow{1/0} q_3 \xrightarrow{0/0} q_1 \xrightarrow{1/1} q_1$$

Thus the string is $1^2 0^8 1$.

Problem 10. ***Consider the finite state machine M, whose state diagram is shown in figure.***

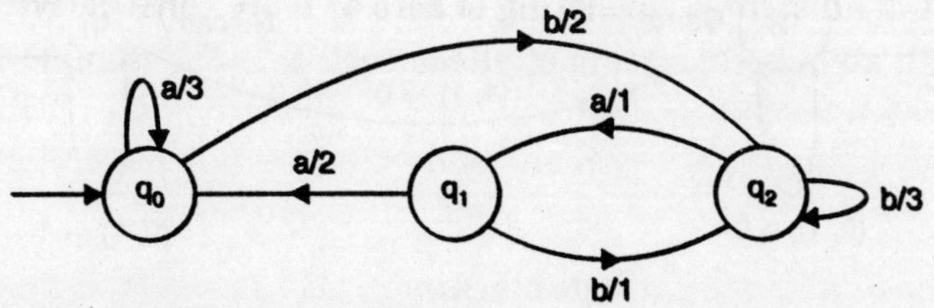

(a) Determine the input set Σ, the state set Q, the output set O and the initial state of M.

(b) Construct the state table of M.

(c) Determine the output string if the input string

$$w = aba^2b^2ababa^2.$$

Sol. (*a*) The input set $\Sigma = \{a, b\}$

The state set $Q = \{q_0, q_1, q_2\}$

The output set $O = \{1, 2, 3\}$

The initial state is q_0.

(*b*) The state table of the finite state machine M is shown in figure.

States	δ Input *a*	δ Input *b*	*f* Input *a*	*f* Input *b*
q_0	q_0	q_2	3	2
q_1	q_0	q_2	2	1
q_2	q_1	q_2	1	3

(*c*) The output string corresponding to the input string $w = aba^2b^2ababa^2$ is determined by following the path in the state diagram.

$$\text{Path} = q_0 \xrightarrow{a/3} q_0 \xrightarrow{b/2} q_2 \xrightarrow{a/1} q_1 \xrightarrow{a/2} q_0 \xrightarrow{b/2} q_2 \xrightarrow{b/3} q_2 \xrightarrow{a/1}$$

$$q_1 \xrightarrow{b/1} q_2 \xrightarrow{a/1} q_1 \xrightarrow{b/1} q_2 \xrightarrow{a/1} q_1 \xrightarrow{a/2} q_0$$

Thus the string is $3212^2 31^5 2$.

Problem 11. ***Construct a finite state machine M, that adds two binary numbers.***

Sol. A finite state machine to perform binary addition can be constructed using just two states. The start state q_0 is used to remember that the previous carry is 0. The second state, q_1 is used to remember that the previous carry is 1. Since the inputs to the machine are pairs of bits, there are four possible inputs. These four possible input pairs are (0, 0), (0, 1), (1, 0) and (1, 1).

The transitions and the outputs are constructed from the sum of the two bits represented by the input and the carry represented by the state. The transition diagram of the FSM, M is shown in figure.

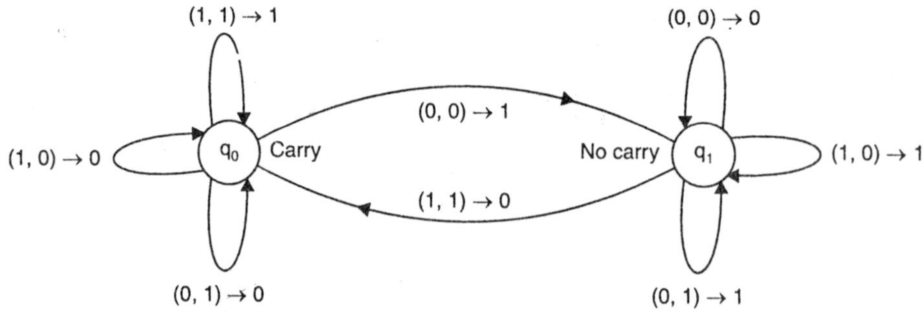

Index

D

E

F

G

H

I

P

R

S

T